T0327643

ANTENNAS FOR PORTABLE DEVICES

ANTENNAS FOR PORTABLE DEVICES

Zhi Ning Chen

Institute for Infocomm Research

Singapore

John Wiley & Sons, Ltd

Telephone (+44) 1243 779777

Email (for orders and customer service enquiries): cs-books@wiley.co.uk
Visit our Home Page on www.wiley.com

Other Wiley Editorial Offices

John Wiley & Sons Inc., 111 River Street, Hoboken, NJ 07030, USA

Jossey-Bass, 989 Market Street, San Francisco, CA 94103-1741, USA

Wiley-VCH Verlag GmbH, Boschstr. 12, D-69469 Weinheim, Germany

John Wiley & Sons Australia Ltd, 42 McDougall Street, Milton, Queensland 4064, Australia

John Wiley & Sons (Asia) Pte Ltd, 2 Clementi Loop #02-01, Jin Xing Distripark, Singapore 129809

John Wiley & Sons Canada Ltd, 6045 Freemont Blvd, Mississauga, ONT, L5R 4J3 Canada

Anniversary Logo Design: Richard J. Pacifico

A catalogue record for this book is available from the British Library

ISBN 978-0-470-03073-8

Typeset in 10/12pt Times by Integra Software Services Pvt. Ltd, Pondicherry, India
Printed and bound in Great Britain by CPI Antony Rowe, Chippenham and Eastbourne
This book is printed on acid-free paper responsibly manufactured from sustainable forestry in which at least two trees are planted for each one used for paper production.

Contents

Foreword

The tremendous success enjoyed by the cellular phone industry and advances in radio frequency integrated circuits have in recent years fostered the development of various wireless technologies, including RFID, mobile internet, body-centric communications, and UWB, which are operated at microwave frequencies. For aesthetic reasons, all these systems require small antennas that can be embedded into the mobile units. Furthermore, for minimally invasive microwave thermal therapies, small and thin antennas are much preferred.

Ten years ago, Dr Zhi Ning Chen the editor of this book was a research fellow at the City University of Hong Kong, working on the design of dielectric resonator antennas. Back then, we were already impressed by the creativity he showed in antenna research and by his leadership skills. His achievements in designing many innovative antennas for wireless applications have been outstanding. This edited book represents another significant achievement, bringing together contributions from key players in the topical areas of antenna designs for RFID tags, laptop computers, wearable devices, UWB systems, and microwave thermal therapies. Major issues and design considerations are discussed and explained in the various chapters.

I am sure that this book will be proven to be of considerable value to practising engineers, graduate students, and professors engaging in modern antenna research. I am delighted to extend my hearty congratulations to Dr Chen and all the authors of the chapters of the book.

Kwai-Man Luk
Head and Chair Professor
Department of Electronic Engineering
City University of Hong Kong
Hong Kong SAR, PR China

Acknowledgements

It is, as always, a pleasure to express my appreciation to those people who have helped and encouraged me in some way in the completion of this project. As the editor of this book, I would first of all like to express my heartfelt gratitude to my generous co-authors, my close friends. Without their excellent and professional contributions and collaboration, this book would not have been published on time. They have given generously their time and energy to share their experiences with us.

I would like to thank Sarah Hinton and Olivia Underhill from Wiley for encouraging me to propose this book right after finishing my first book, *Broadband Planar Antennas: Design and Applications*, published by Wiley in February 2006. Sarah was in charge of that work. I am grateful to Mark Hammond, also from Wiley, for his continuous support while the present work was under way. My grateful thanks are also due to our reviewers, content editor, copy-editor, typesetter as well as cover designers for their helpful and professional comments and work on this book.

As a researcher for the Institute for Infocomm Research, I would like to thank the senior management and my colleagues for their continuous and kind support and understanding. The Institute has provided me with generous facilities for research and development work since I joined in 1999. Most of our work on Chapters 3 and 7 was finished at the Institute.

As a supervisor, I would like to express my gratitude to my ex-students for their contribution to research on ultra-wideband and radio-frequency identification antennas. They are Ning Yang, Xuan Hui Wu, Dong Mei Shan, Terence See, Ailian Cai, Tao Wang, Yan Zhang, and Hui Feng Li.

Finally, I am immensely grateful to my wife, Lin Liu and our twin sons, Shi Feng and Shi Ya, for their understanding and support during the period when I was devoting all my weekends and holidays to preparing, writing, and editing this book. I hope its success, and my promise to spend more time with them in future, will compensate them for all they have lost.

Brian Collins would like to thank his colleagues at Antenova Ltd for their support and helpful suggestions as well as for access to their experimental results. He would also like to thank CST GmbH for providing the simulation results in Chapter 2 illustrating the interaction of fields with the human body.

Duixian Liu and *Brian Gaucher* would like to thank the IBM Yamato ThinkPad design team for their contributions of range and performance testing as well as the production level models used in testing. Peter Lee, Thomas Studwell and Thomas Hildner of IBM Raleigh had both the foresight and tenacity to understand how important wireless would be before it happened, and stuck by their convictions, providing the support to further this work. They also thank Frances O'Sullivan, Peter Hortensius, and Jeffrey Clark of IBM Raleigh, Arimasa Naitoh and Sohichi Yokota of IBM Yamato, Japan, and Ellen Yoffa, Modest Oprysko, and Mehmet Soyuer of the IBM Watson Research Center in Yorktown Heights for their leadership and vision on the ThinkPad antenna integration project. Much material was provided by Hitachi Cables of Japan, particularly Mr. Hisashi Tate. Without his patient and prompt support, the chapter would be incomplete. Mr. Shohei Fujio of the IBM Yamato lab in Japan was also kind enough to provide his plots and drawings. Mr. Hideyuki Usui and Mr. Kazuo Masuda of Lenovo Japan (formerly of the IBM Yamato lab) gave generously of their time for laptop wireless discussions as well as providing related information.

Xianming Qing wishes to thank his wife, Xiaoqing Yang, and sons (Qing Ke and Qing Yi) for their understanding and support during the preparation of this book. He would also like to thank Mr Terence See for his helpful comments, which resulted in welcome improvements to Chapter 3.

Koichi Ito and *Kazuyuki Saito* would like to thank Prof. Yutaka Aoyagi and Mr. Hirotoshi Horita, Tokyo Dental College, Japan, for their contributions to the use of antennas in clinical trials. They would also like to thank Dr Toshio Tsuyuguchi, School of Medicine, Chiba University, Japan, and Prof. Hideaki Takahashi, Brain Research Institute, Niigata University, Japan, for their valuable comments from the clinical side.

List of Contributors

Akram Alomainy	Queen Mary, University of London, United Kingdom
Zhi Ning Chen	Institute for Infocomm Research, Singapore
Brian Collins	Antenova Limited, United Kingdom
Brian P. Gaucher	International Business Machines Corporation, United States of America
Yang Hao	Queen Mary, University of London, United Kingdom
Koichi Ito	Chiba University, Japan
Duixian Liu	International Business Machines Corporation, United States of America
Frank Pasveer	Philips Research, Netherlands
Xianming Qing	Institute for Infocomm Research, Singapore
Kazuyuki Saito	Chiba University, Japan
Terence S.P. See	Institute for Infocomm Research, Singapore

1

Introduction

Zhi Ning Chen
Institute for Infocomm Research, Singapore

Electronic devices are a part of modern life. We are constantly surrounded by the electromagnetic waves emitted from a variety of fixed and mobile wireless devices, such as fixed base stations for audio/video broadcasting, fixed wireless access points, fixed radio frequency identification (RFID) readers, as well as mobile terminals such as mobile phones, wireless access terminals on laptops, sensors worn on the body, RFID tags, and radio frequency/microwave thermal therapy probes in hospitals. Besides the fixed base stations, many wireless devices are expected to be portable for mobile applications. Mobile phones, laptops with wireless connection, wearable sensors, RFID tags, wireless universal serial bus (USB) dongles, and handheld microwave thermal therapy probes have been extensively used for communications, security, healthcare, medical treatment, and entertainment. Users of portable wireless devices always desire such devices to be of small volume, light weight, and low cost.

With the huge progress in very large scale integration (VLSI) technology, this dream has become a reality in the past two decades. For example, the mobile phone has seen a significant volume reduction from $6700\,cm^3$ to $200\,cm^3$ since 1979 [1]. However, with this dramatic reduction in overall size, the antennas used in such portable devices have become one of their biggest components. Therefore, much effort has been devoted to miniaturizing the size of antennas to meet the demand for devices with smaller volume and lighter weight.

In the past two decades, antenna researchers and engineers have achieved considerable reductions in the size of antennas installed in portable devices, although physical constraints have essentially limited such reductions. Today, almost all antennas for portable devices can be embedded in the devices. This creates a transparent usage model for the user, that is, the user never needs to be ware of the presence of the antenna. The appearance of the device is enhanced, and the possibility of accidental breakage is reduced.

Antennas for Portable Devices Zhi Ning Chen

Antennas for portable devices may be small in terms of [2]:

1. *electrical size*. The antenna can be physically bounded by a sphere having a radius equal to $\lambda_{freespace}/2\pi$. Planar inverted-F antennas with shorting pins or/and slots are typical examples of this category.
2. *physical size*. An antenna which is not electrically small may feature a substantial size reduction in one dimension or plane. Microstrip patch antennas with ultra low profiles belong to this category.
3. *function*. An antenna which is not electrically or physically small in size may possess additional functions without any increase in size. Dielectric resonator antennas operating in multiple modes fit this definition.

Therefore, the miniaturization of antennas for portable devices can be carried out in various ways because basically, the research and development of antenna technology are application-oriented.

With the rapid increase in the number of mobile portable devices, many technologies have been developed to miniaturize the antennas. The technologies can be broadly classified as follows:

1. *The design and optimization of antenna geometric/mechanical structures, in particular, the shape and orientation of the radiators, loading, as well as the feeding network.* This is a conventional approach and most often employed in antenna design. Inverted-F antennas, top-loaded dipole antennas, and slotted planar antennas all fall into this category.
2. *The use of non-conducting material.* Antennas loaded with ferrite or high-permittivity dielectric materials (for instance, ceramics) are examples of this type of technology, as is the dielectric resonant antenna.
3. *The application of special fabrication processes.* The fabrication of printed circuit boards and low-temperature co-fired ceramics have made co-planar and multiple-layer microstrip patch antennas popular. Such technologies are conducive to the mass production of miniaturized antennas at low cost.

This book aims to introduce the advanced progress in miniaturizing antennas for portable mobile devices. The portable mobile devices will include: mobile phone handsets; RFID tags; laptops with embedded wireless local area network (WLAN) access points; medical devices for microwave thermal therapy; sensors installed on or above the human body; and ultra-wideband (UWB) based high-data-rate wireless connectors such as the wireless USB dongle. All of these portable mobile devices are widely used. The antennas used in them have become a bottleneck in the miniaturization of portable devices in terms of performance, size, and cost. The increasing design challenges have made the antenna design for portable devices much more critical than before.

In this book, various challenging design issues will be addressed from a technology and application point of view. Authors from both academia and industry will present the latest concepts, procedures, and solutions for practical antenna designs for portable devices. Several case studies will be provided, together with detailed descriptions of the technologies and systems.

Figure 1.1 Handsets with embedded and external antennas.

Chapter 2 presents the antennas for the most popular wireless communication devices and handsets, delving into practical issues covering the radio frequency (RF) link budget, small antenna basis, measurement and simulation methods, and specific absorption rate (SAR). Handsets with embedded and external antennas are shown in Figure 1.1.

The term *handset* here covers almost all mobile devices such as mobile phones, camera phones, personal digital assistants, and any other handheld devices which are able to communicate through wireless networks or from device to device. The large number of antenna designs has been detailed in many references and published books. This chapter is mainly focused on the discussion of antennas operating in the environment of the handset and the influence of handset design on the potential RF performance. Much of the discussion will be of relevance to the industrial designer, the layout engineer, as well as the antenna engineer.

This chapter will treat the topics in the general order of the process by which the antenna designer will evaluate the target specifications from customers, the dimensions and configuration of the handset, and the local environment of the antenna relative to other components. After examining these factors, the antenna engineer will begin the design procedure by choosing potential electrical designs for the antenna by simulation and experiment, testing all the parameters of interest, optimizing the antenna performance before finalizing the design to be embedded into the handset devices.

In Chapter 3 a systematic description of antenna design issues related to the RFID system and tags is provided. The RFID is a technology which transmits data by using a mobile tag. The data will be read by an RFID reader and processed according to the needs of the particular application. The data transmitted by the tag may provide identification, location information, or specifics about the product tagged, such as price, colour, and date of purchase. RFID systems have been widely applied in tracking and access applications since the 1980s. Recently, RFID applications have increasingly captured the attention of academia and industry because of the growth in demand from sectors such as warehousing, libraries, retail, and car parks due to the great reduction in the cost of RFID systems, especially for tags with an antenna and microchip. Figure 1.2 shows RFID tag antennas operating at 13.56, 433, 869, and 915 MHz, developed in the Institute for Infocomm Research, Singapore.

This chapter will briefly introduce RFID systems in order to give readers a basic understanding of RFID operation and the requirements for RFID antennas, particularly tag antennas. Next, the RFID tag antenna design will be addressed. As the frequency used for RFID varies from very low (below 135 kHz) to millimetre wave (27.125 GHz), so will the antenna design. For near-field (inductively coupled) RFID systems, the antenna is made

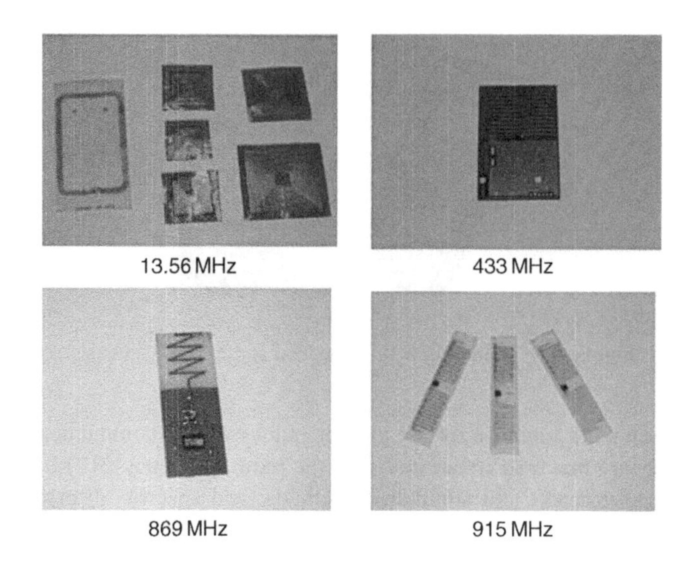

Figure 1.2 RFID tag antennas operating at 13.56, 433, 869, and 915 MHz.

up of a coil with a specified inductance for circuit resonance with an adequate quality factor. For far-field (wave radiation) RFID systems, various types of antennas, such as the dipole antenna, meander line antenna, and patch antenna, can be used. Generally, a tag antenna must have the following characteristics: small size, omnidirectional or hemispherical radiation coverage, good impedance match, typically linear polarization or dual polarization, robustness, and low cost.

This chapter also investigates the environment effect on RFID tag antennas. Tag antennas are always attached to specified objects, such as books, bottles, boxes, or containers. These objects may affect the performance of the tag antenna. The effects on the tag antennas will be severe when it is attached to metal objects or lossy materials. Some results are presented in the last part of this chapter.

Chapter 4 will discuss the integrated antenna design, test, and integration methodology for laptop computers as shown in Figure 1.3. A laptop has a much larger potential surface area for the antenna than a mobile phone. However, unlike the handsets of mobile phones, the laptop enclosure is intentionally designed to prevent electromagnetic emissions and, as a consequence, RF emissions. In addition, laptop users do not expect antenna protrusions as normally found on mobile phones. Two key parameters are proposed and discussed for laptop antenna design and evaluation: standing wave ratio (SWR) and average antenna gain. Though seemingly obvious, a novel averaging technique is developed and applied to yield a measurable, repeatable, and generalized metric.

The chapter covers three major topics. First, it discusses the antenna locations on laptops, particularly on the laptop display. Actual measurements are performed at different locations using an inverted-F antenna. The measurements indicate that the antenna location effects on the radiation patterns and SWR bandwidth. The second topic discusses link budget calculations. These calculations relate the antenna average gain value to wireless communication performance such as data rate or coverage distance. The third topic covers some practical antenna designs used in laptops for Bluetooth™ and WLAN. A PC card version of the wireless system is

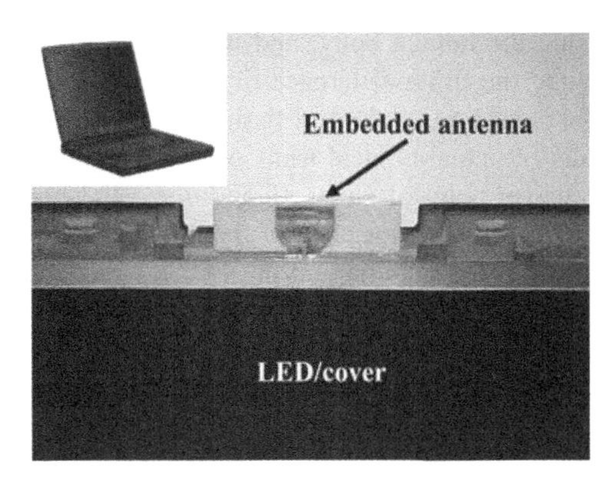

Figure 1.3 An antenna embedded into the cover of a laptop computer.

also discussed and compared with the integrated version. An integrated wireless system always outperforms the PC card version. This chapter emphasizes practicality by extensive measurements and using actual laptop antennas.

Chapter 5 introduces the antenna design for portable medical devices. This is the only chapter in the book which does not involve wireless communications but microwave-based applications. It exhibits the wide coverage of antenna technology for antenna researchers working on wireless communications as shown in Figure 1.4.

Recently, a variety of microwave-based medical applications have been widely investigated and reported. In particular, minimally invasive microwave thermal therapies using thin antennas are of great interest, among them the interstitial microwave hyperthermia and microwave coagulation therapy for medical treatment of cancer, cardiac catheter ablation for ventricular arrhythmia treatment, and thermal treatment of benign prostatic hypertrophy. The principle of the hyperthermic treatment for cancer is described, and some heating schemes using microwave techniques are explained. Next, a coaxial-slot antenna, which is a type of thin coaxial antenna, and array applicators comprised of several coaxial-slot antennas are also introduced. Moreover, some fundamental characteristics of the coaxial-slot antenna and the array applicators, such as the specific absorption rate, temperature distributions

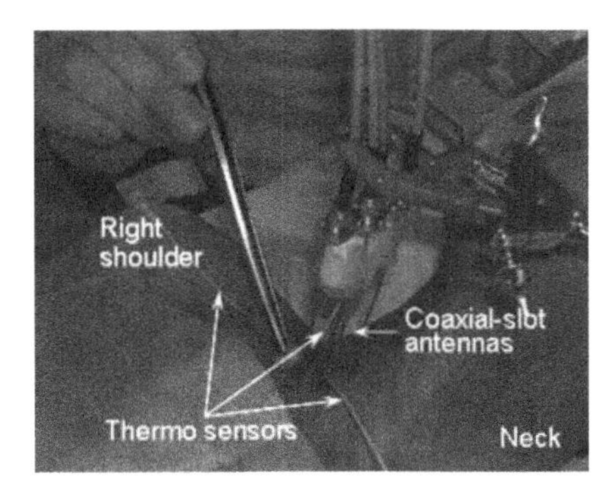

Figure 1.4 Microwave thermal therapy with coaxial slot antennas.

around the antennas inside the human body, and the current distributions on the antenna, are described by employing the finite-difference time domain (FDTD) calculations and the temperature computations inside the biological tissues by solving the bioheat transfer equation. Finally, some results of actual clinical trials using the proposed coaxial-slot antennas are explained from a technical point of view. In addition, other therapeutic applications of the coaxial-slot antennas such as the coagulation therapy for hepatocellular carcinoma, the hyperthermic treatment for brain tumours, and the intracavitary hyperthermia for bile duct carcinoma are introduced.

Chapter 6 briefly introduces the wireless personal area networks (WPAN) and the progression to body area networks (BAN), highlighting the properties and applications of such networks. Figure 1.5 shows the scenario where the antenna is installed on the human body (phantom) in simulation. The main characteristics of body-worn antennas, their design requirements, and theoretical considerations are discussed. The effects of antenna types on radio channels in body-centric networks are demonstrated. In order to give a clearer picture of the practical considerations required in antenna design for body-worn devices deployed in commercial applications, a case study is presented with a detailed analysis of the design and performance enhancement procedures to obtain the optimum antenna system for healthcare sensors.

Communication technologies are heading towards a future with user-specified information easily accessible whenever and wherever required. In order to ensure the smooth transition of information from surrounding networks and shared devices, there is a need for computing and communication equipment to be body-centric. The antenna is an essential part of the wireless body-centric network. Its complexity not only depends on the radio transceiver requirements but also on the propagation characteristics of the surrounding environment. For the long to short wave radio communications, conventional antennas have proven to be more than sufficient to provide the desired performance, minimizing the constraints on the cost and time spent on producing such antennas. On the other hand, for the communication devices today and in the future, the antenna is required to perform more than one task,

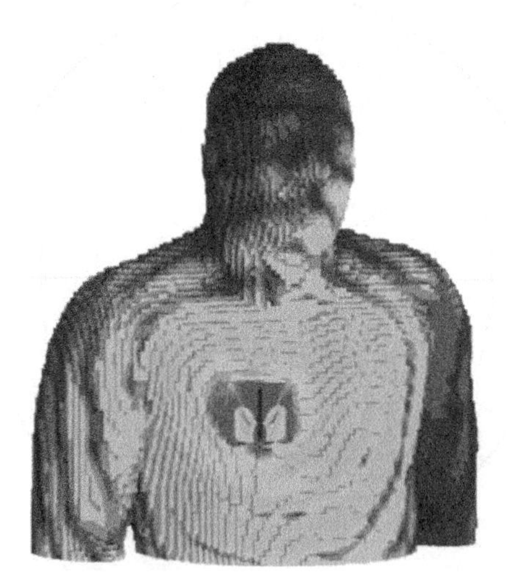

Figure 1.5 Wearable antenna on the human body (phantom in simulation).

Figure 1.6 Antenna embedded into a UWB-based wireless USB dongle.

or in other words, the antenna will be needed to operate at different frequencies so as to account for the increasing introduction of new technologies and services available to the user. Therefore, careful consideration is required for antennas applied in body-worn devices, which are often hidden, small in size, and light in weight.

In Chapter 7, the final chapter of this book, the UWB, an emerging technology for short-range high-data-rate wireless connections, high-accuracy image radar, and localization systems is introduced. Due to the extremely broad bandwidth and carrier-free features, antenna design is facing many challenges. The conventional design considerations are insufficient to evaluate and guide the design. Therefore, this chapter will begin with a discussion of the special design considerations for UWB antennas. The design considerations reflect the uniqueness of the UWB system requirements for the antennas. In accordance with these considerations, the antennas suitable for portable mobile UWB devices are presented. In particular, this chapter elaborates the design and state of the art of the planar UWB antennas. The latest developed UWB antennas will be reviewed with illustrations as well as simulated and measured data. Finally, a new concept for the design of small UWB antennas with reduced ground plane effect is introduced and applied to practical scenarios. Two versions of the small printed UWB antennas designed for wireless USB dongles installed on laptop computers are investigated in the case studies. Figure 1.6 shows an antenna embedded into a UWB-based wireless USB dongle.

As the design of antennas for portable devices is an area of rapidly growing research and development, this book is expected to provide readers with the fundamental issues and solutions to existing as well as forthcoming applications.

References

[1] Z.N. Chen and M.Y.W. Chia, *Broadband Planar Antennas: Design and Applications*. John Wiley & Sons, Ltd, Chichaster, 2006.

[2] K. Fujimoto, A. Henderson, K. Hirasawa, and J.R. James, *Small Antennas*. Letchworth: Research Studies Press, 1988.

2

Handset Antennas

Brian S. Collins
Antenova Ltd, United Kingdom

2.1 Introduction

The user's experience of a mobile communications system entirely depends on the performance of the bidirectional radio link between the base station and the handset. Each mobile network operator creates a system of linked base stations located to provide coverage of as large a physical area as possible, providing as much coverage and capacity as is appropriate for the expected traffic demand – or as much as technical considerations will permit. While the base station will typically be equipped with a high-gain antenna and a transmitter capable of delivering some tens of watts of radio frequency (RF) power, the handset relies on an antenna whose dimensions are severely constrained by those of the handset in which it is fitted, with a typical maximum effective radiated power of 1 watt. While the base station antenna is generally mounted in a clear location 10 m or more above ground level, the handset will be in the user's hand, perhaps held against the head, within 1.5 m of the ground.

Practical considerations of radio link performance result in the need to allow a substantial margin on the link budget, and the shortcomings of the handset form the single largest intrinsically reducible loss in the system. These shortcomings are not very noticeable when the handset is used in a well-served urban environment, but they become critical when the user is in an area of marginal network coverage or inside a building.

The design challenge posed by handset antennas is becoming more critical as networks evolve to offer a wider range of services. We now expect a pocket-sized mobile terminal to be able to deliver telephony (potentially video telephony), high-speed data services, location and navigation services, entertainment . . . and more to come in the future. Not only do some of the new services require higher data rates, but the increasing number of different facilities in the terminal puts great pressure on the available space for antennas. Handset designers expect that multiple antennas can be operated successfully in close proximity to components such as cameras, flash units, loudspeakers, batteries and the other hardware needed to support the growing capabilities of the terminal.

Antennas for Portable Devices Zhi Ning Chen

In this chapter the term *handset* will be used to cover mobile phones, camera phones, personal digital assistants (PDAs), entertainment terminals and any other pocket-size devices which must communicate with the network. To avoid repeated lists of frequencies the band terminology listed in Table 2.1 will be used in this chapter. The complexity of this list itself – which omits some major national assignments – emphasizes the varied demands made on the functionality of handsets.

This chapter deals only in outline with the large numbers of different designs for the antenna itself – these are described in detail in the accompanying references. The main emphasis is on the operation of the antenna in the environment of the handset and the influence of handset design on the potential RF performance that can be obtained. Much of the discussion is of importance to the industrial designer and the engineer laying out the electronic components of the handset, as well as to the antenna engineer.

The order of treatment of the topics follows the general order of the process by which the antenna designer will assess the design task – reviewing the target specification, the dimensions and configuration of the handset and the local environment of the antenna relative to other components. The antenna engineer will choose potential electrical designs for the antenna after first examining these factors.

Table 2.1 Frequency bands, nomenclatures and uses.

Frequency band	Short reference	Service
550–1600 kHz	MF radio	Radio broadcast
2–30 MHz	HF radio	Radio broadcast
88–108 MHz	Band II	Radio broadcast
174–240 MHz	Band III	T-DMB TV
450–470 MHz	450 MHz	Phone + data
470–750 MHz	Band IV/V	DVB-H TV
824–890 MHz	850 MHz	Phone + data
870 (880)–960 MHz	900 MHz	Phone + data
824–960 MHz (850 and 900 MHz)	Low bands	Phone + data
1575 MHz	GPS	Geolocation
1710–1880 MHz	1800 MHz	Phone + data
1850–1990 MHz	1900 MHz	Phone + data
1900–2170 MHz	2100 MHz	Phone + data
1710–2170 MHz (1800, 1900 and 2100 MHz)	High bands	Phone + data
2.4–2.485 MHz	2.4 GHz	WLAN
2.5–2.69 GHz	2.5 GHz	WiMAX™
3.4–3.6 GHz	3.6 GHz	WiMAX™
4.9–5.9 GHz	5 GHz	WLAN, WiMAX™

The list above includes most major world-wide assignments, but some other frequency bands are allocated in certain countries. Future bands for UMTS are not included.

Following the transfer of broadcast TV services to digital formats, it is expected that a significant amount of the present analog TV bands will be assigned for mobile services.

2.2 Performance Requirements

Before examining antenna design we should define handset performance parameters and examine the way in which these interact with network operation. A subset of these parameters is usually specified in connection with a target handset design.

Gain. The use of the simple term *gain* when applied to a handset antenna is not explicit and should generally be avoided – see *efficiency* and *mean effective gain* below.

Efficiency. The efficiency of a handset antenna is the ratio of the total power radiated by the antenna to the forward power available at its terminals (or those of its associated matching network). Some workers separately define *terminal efficiency* as indicating the ratio of radiated power to the net power delivered to the antenna (forward power – reflected power), and *total efficiency* as meaning the definition here adopted, but these terms will not be used in this chapter.

Efficiency may be measured either with the antenna driven from an external signal source (its *passive efficiency*) or with the antenna driven by the RF output of the phone (its *active efficiency*). In active measurements it is difficult to determine the forward power, so the better active parameter is a measurement of the total radiated power (TRP) – which is what matters in network performance.

Bandwidth. The bandwidth of an antenna is the frequency range over which some specified set of parameters is maintained. The objective of handset antenna design is that the bandwidth is sufficient to cover the frequency bands over which the handset is intended to operate.

Radiation patterns. It is relatively uncommon for the specification for a handset antenna to include any reference to its radiation patterns, although these are commonly measured during the development of the antenna. The reason for this lack of specification is partly that the designer has only limited ability to control the patterns, but also that the handset will be operated in contact with the hand (and sometimes the head) of the user, so any measurement of the patterns is of limited significance. Radiation patterns are usually measured in the three principal planes of the physical handset.

Polarization. The radiation from a handset is regarded as having randomly-oriented elliptical polarization. Measurements of radiated power are usually made separately for linear orthogonally polarized signal components. This means that the full 3D radiation characteristics of the handset are characterized by six separate patterns (three cuts and two polarization components). For most purposes the energy contained in orthogonal linear polarizations is added vectorially – as, for example, in efficiency or TRP measurements.

Mean effective gain (MEG). This is calculated by averaging the measured gain at sufficient points on a (typically spherical) surface around the handset. If the antenna were lossless, then the mean gain would be 0 dBi, so the MEG is effectively the same as $10\log_{10}\eta$, where η is the efficiency.

Total radiated power. This is the total power flowing from the handset when it is transmitting. To measure this the handset is controlled by a base station simulator and the outgoing power (summed in orthogonal polarizations) sampled at points over a closed surface surrounding the handset.

Total isotropic sensitivity (TIS). The sensitivity is defined as being the input signal power which gives rise to a specific frame error rate or residual bit error rate. The sensitivity is sampled in orthogonal polarizations at points spread over a surface surrounding the handset.

The TIS and TRP together determine the effectiveness of the handset as a piece of radio equipment, in particular the maximum range at which the handset can operate from a base station with some given level of performance. It is essential that the TIS and TRP are properly related to one another. The link budgets for the up- and down-links (to and from the base station) are based on specific assumptions about handset performance assuming that efficiency is maintained across the whole of both the transmit and receive frequency bands.

Input return loss and voltage standing wave ratio (VSWR). Input matching can be described either by return loss or VSWR, the two terms being easily converted:

- $VSWR = (1+\rho_v)/(1+\rho_v)$, where ρ_v is the modulus of the voltage reflection coefficient – the ratio of the reflected wave to the forward wave expressed in volts.
- *Return loss* $= 20\log_{10}\rho_v$. (The – sign should be omitted, as it is unnecessary and leads to confusion in such phrases as 'a return loss greater than $-8\,\text{dB}$'.)

In this chapter the less specific terms *match/matching* can be taken as referring to either.

- The *power reflection coefficient* $= \rho_v^2$ so the power delivered to the load is $(1-\rho_v^2)$ and the corresponding *reflection loss* $= 10\log_{10}(1-\rho_v^2)$.

The input match of a handset antenna is one of its most important parameters. As we shall see, the small size of the handset and its antenna create fundamental problems in obtaining a low-input VSWR over the required frequency bands. The main effect of high VSWR is to cause input reflection loss which reduces the efficiency of the handset; in general, the efficiency target takes precedence over VSWR which is not regarded as the primary parameter. It is of essential interest to the antenna designer, but it is efficiency which determines network operation.

Passive test. In a passive test a small coaxial cable or microstrip line is connected between the antenna and an input connector, allowing the designer to measure the VSWR, radiation patterns and efficiency of the antenna mounted in place on the handset (or perhaps for initial evaluation on a representative dummy of the handset). Passive testing is used during initial design while the antenna configuration is optimized and an input matching circuit is devised.

Active test. In an active test. no external connection is made to the handset. A base station simulator is used to set up a call to a complete operating handset in an anechoic chamber and measurements are made to establish the TRP and TIS. These parameters determine the performance that will be experienced by a user in network service. If users experience poor call quality they will usually complain about poor network coverage; to avoid this, many network operators establish standards of TRP/TIS performance which must be met by handsets before they are permitted to be used on their network. Standard test methods are described in [1].

Sometimes a handset that appears to perform well in passive tests is shown to be substandard in active tests. There are several factors which contribute to the differences between active and passive measurements:

- In a TRP measurement the output of the power amplifier (PA) is fed through internal connecting transmission lines and a switch or diplexer. These may be mismatched or may be more lossy than expected.
- In a passive measurement the antenna is fed from a well-matched 50-ohm source. In an active TRP measurement the power delivered by the PA will depend on the complex impedance presented to it by the transmission line connecting it to the switch/diplexer and antenna. The load-line characteristic of the PA will determine how much power it will deliver into this impedance, which is likely to change very significantly over each operating band.
- In a TIS measurement the sensitivity of the receiver is reduced by any noise sources within the handset because the noise may mask low-level signals. Displays and cameras, and their associated feed circuits, often generate noise, especially in the low bands.

Active measurements are important because they represent the behavior of the handset in use; passive measurements are simpler to understand. The difference between them is a very important diagnostic tool for rectifying unexpected problems.

Free-space, in-hand and head-position measurements. During the antenna development process the measurements described above are typically made with the handset in an isolated test fixture made from low-density polystyrene foam. In operation the handset may be held away from the head (for example, when texting or accessing Web-based services) or against the head as in normal phone operation. To simulate these scenarios a handset is tested in conjunction with physical models of lossy hands and heads – known as phantoms. There are often major differences between performance in the presence of the phantom and in a free-space environment.

The effects of head and hand are sometimes referred to as *detuning*, but this term is not really very helpful; whether the resonant frequency of the antenna changes or not, power is deposited in the phantom. The input match may actually improve when the handset is placed against it, but this simply confirms that less power is being reflected from the antenna. Specifying detuning by reference to the change of the frequency of optimum antenna match is not helpful; it neither indicates the fall in efficiency in the presence of the phantom, nor the frequency of optimum efficiency. It is more helpful to refer to the change in efficiency when the handset is held, averaged across the relevant frequency bands.

Specific absorption rate (SAR). A handset placed alongside the user's body will deposit energy in the tissue penetrated by electromagnetic fields. To study possible effects on body tissues we must examine the rate at which energy is deposited in a given volume of tissue. This is the *specific absorption rate*, whose units are watts per kilogram of tissue. To control the possibility of high local peaks, the maximum permitted SAR is specified as applying to any 1 g or 10 g of tissue.

It is important to distinguish between limits for exposure to electromagnetic fields and maximum permitted SAR levels. Limits for exposure to fields are quoted in terms of the power (in watts per square metre) carried by a plane wave (or by specifying maximum electric or magnetic fields). SAR limits are more complex and relate to the power absorbed by the user's body.

There is no single world-wide standard limit for SAR, and some current standards are shown in Table 2.2 [2–11].

In use, the handset is positioned so that its near field penetrates the user's body. The body is not electrically homogeneous – bone, brain, skin, and other tissues have different

Table 2.2 SAR limits for the general public specified by various administrations.

	Australia	Europe	USA	Japan	Taiwan	China
Measurement method	ASA ARPANSA	(ICNIRP) EN50360	ANSI C95.1b:2004	TTC/MPTC ARIB		
Whole body	0.08 W/kg	0.08 W/kg	0.08 W/kg	0.04 W/kg	0.08 W/kg	
Spatial peak	2 W/kg	2.0 W/kg	1.6 W/kg	2 W/kg	1.6 W/kg	1 W/kg
Averaged over	10 g cube	10 g cube	1 g cube	10 g cube	1 g cube	10 g
Averaged for	6 min	6 min	30 min	6 min	30 min	

densities, dielectric constants, dielectric loss factors and complex shapes. This is a situation which has to be simplified to provide handset designers with engineering guidelines with which they can work, so for regulatory purposes a standard physical phantom head is used in which the internal organs are represented by a homogeneous fluid with defined electrical properties. With a handset positioned beside the phantom and with its transmitter switched on, the fields are probed inside the phantom. They are translated into SAR values and the pattern of energy deposition is mapped to determine the regions with the highest SAR averaged over 1 g and 10 g samples. Simulations are often carried out using this 'standard head', but more realistic information is obtained using high-resolution computer models based on anatomical data.

Extensive investigation of possible health effects of RF energy absorbed from mobile phones has been carried out in many countries. Current results suggest that any effects are very small, at least over the time period for which mobile handsets have been in widespread use. Those interested should consult the websites of the major national occupational health administrations and medical journals. The responsibility of the antenna designer is to ensure that the user is exposed to the lowest values of SAR consistent with the transmission of a radio signal with the power demanded by the network.

Hearing aid compatibility. Handsets operating with time-division multiplex protocols such as GSM emit short pulses of radio energy. A hearing aid contains a small-signal audio amplifier and if this is presented with a high-level pulsed radio signal the result of any non-linearity in the amplifier will be the generation of an unpleasant buzzing sound. Some administrations place networks under a responsibility to provide some proportion of their handsets which are designed to minimize these interactions.

2.3 Electrically Small Antennas

The dimensions of handset antennas are very small compared with the operating wavelength, particularly in the low bands. Not only is the antenna small, but the length of the handset to which it is attached – typically between 80 and 100 mm – is also only a fraction of a wavelength long. A typical handset antenna is less than 4 ml in volume (about one thousandth of a cubic wavelength) and a 90 mm chassis is only 0.27λ long at 915 MHz.

The operation of electrically small antennas is dictated by fundamental relationships which relate their minimum Q-factor to the volume of the smallest sphere in which they can be enclosed, often referred to as the Chu-Harrington limit [12, 13]. The Q relates stored energy and dissipated energy, and a small antenna intrinsically has a very reactive input impedance with an associated very narrow bandwidth. We can compensate for the input reactance by adding an opposite reactance, but the combination will have a higher Q and less bandwidth. We can trade efficiency for bandwidth, but we want to achieve the highest possible efficiency at the same time as enough bandwidth to cover the mobile bands – perhaps several bands. Whatever ingenuity we apply, it is often impossible to obtain the combination of properties we need from such a small device.

A simple small antenna is shown in Figure 2.1, where a short monopole is fed against a groundplane. This antenna looks capacitive all the way from DC to the frequency at which it is almost $\lambda/4$ long. The input impedance has the form $Zin = R + jX$, where R is small and X is very large. The bandwidth will be limited by the Q of the device, where $Q = X/R$. If the antenna is a very small fraction of a wavelength long, it is necessary to excite a very large current in it to persuade it to radiate any significant power; put another way, its radiation resistance is very small so it must carry a large current to radiate the required power. Unfortunately the radiation resistance may be comparable with the loss resistance in its conductors and the equivalent loss resistance of any insulating components needed to support it. We are therefore confronted with a very small bandwidth and a problem with efficiency – any current will create losses as well as radiation. The efficiency η will be limited to a value given by $R_r/(R_l + R_r)$where R_l is the equivalent loss resistance and R_r is the radiation resistance. To feed energy into the antenna we will need to match it to a transmission line, and the matching circuit will contribute further losses.

Figure 2.1(a) shows a short vertical radiator over ground – for the moment we can regard this as perfect ground. The current at the top of the radiator is zero and it rises linearly to some maximum value at the bottom (it is approximately linear because although the distribution is approximately sinusoidal, $\sin\theta \approx \theta$ when θ is small). We can improve matters by extending a horizontal conductor from the top of the antenna (Figure 2.1(b)); this occupies no more height but the current zero is now moved to the ends of the horizontal sections and a larger and almost constant current flows in the vertical section. We have increased the radiation resistance (R_r) and at the same time reduced the capacitive reactance X_c at the feedpoint, so the Q of the antenna has fallen. Figure 2.1(c) shows an alternative configuration with similar characteristics, known as an inverted-L antenna. In both cases the top conductor contributes little radiation because of the proximity of its anti-phase image in the groundplane.

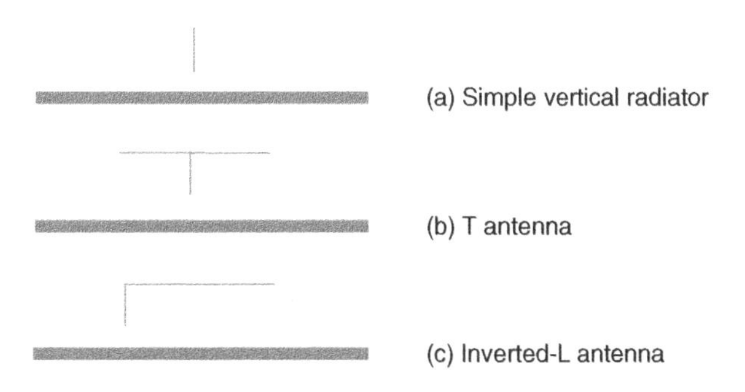

Figure 2.1 Short radiators over ground.

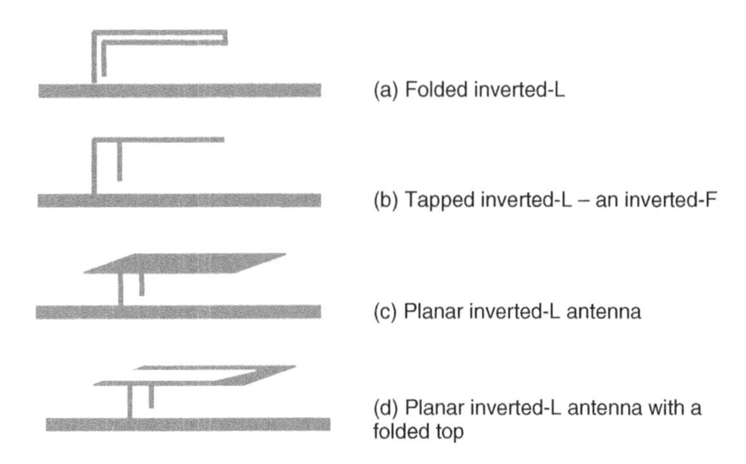

Figure 2.2 Derivatives of an inverted-L.

To further increase the value of R_r we can fold the antenna as in Figure 2.2(a), or tap it in the manner shown in Figure 2.2(b) – an inverted-F antenna. This will be naturally resonant when the total length of the upper limb is around $\lambda/4$, and by selecting the position of the feedpoint the input impedance can be chosen to be close to 50 ohms.

We can replace the wire top of the inverted-L with a plate (Figure 2.2(c)) and slot the plate to make the loading more compact (Figure 2.2(d)). Unfortunately we have still not overcome the constraint created by the small volume of the antenna and we need another trick to allow us to solve our problem. An important feature of all these configurations is that they are unbalanced. If we conceive the ground as an infinite perfect conductor we can envisage an image of the antenna in the groundplane and calculate the radiation pattern by summing the contributions of the antenna and its image.

When we build one of these antennas on a handset, the groundplane is only around $\lambda/4$ long – about the same length as one half of a dipole. What we have created is a kind of curiously asymmetrical dipole; one limb comprises the groundplane of the handset, while the other limb is the F-structure we have fed against it. What properties might we expect of this configuration?

Polarization. The polarization of the inverted-F antenna (Figure 2.2(c)) is vertical – orthogonal to the groundplane. We can envisage this from the direction in which we apply the feed voltage, the current in the vertical radiating leg and the alignment of the E-field between the top and the ground. By contrast, our asymmetric dipole is polarized in the direction of its long axis, along which most of the radiating current flows.

Radiation patterns. The inverted-F antenna would have an omnidirectional pattern in the plane of the ground, while the asymmetric dipole would be omnidirectional in the plane bisecting the groundplane.

If we now examine the behavior of a typical handset we see that it really does have these properties. The antenna has very little relationship to the prototypes from which we derived it. The polarization is aligned with the long axis of the phone, and its radiation pattern in the low bands looks very much like that of a half-wave dipole aligned with the groundplane (see Figure 2.11 below).

Bandwidth. The derivation we have followed makes it unsurprising that we can obtain a far greater impedance bandwidth than would have been possible from the tiny structure we usually refer to as the antenna (and which we can now recognize as being some kind of coupling structure whose main purpose is to allow us to excite currents in the groundplane). Not surprisingly the largest bandwidth will be obtained when the phone is of a resonant length, as in this event the impedance presented to the currents flowing into the groundplane will change less rapidly with frequency [14].

High-band performance. In the high bands the antenna is electrically larger and we could expect that it might operate more independently of the groundplane. In fact the polarization usually remains along the groundplane and the radiation pattern simply looks like that of a long dipole driven from a point off-center (see Figure 2.12 below). A small antenna can provide adequate high-band performance, and we shall later examine the possibility of making a balanced antenna operating substantially independently of the groundplane.

The chassis of the handset. What has been referred to as the groundplane comprises all those parts of the handset that are connected to the groundplane, including the battery, display, case metallization and screening cans. For a two-part handset (clamshell or slide-phone) it will comprise the grounded parts of both components.

Losses. An ideal antenna will radiate all the energy supplied to it. In practice losses are created by:

- *Reflection caused by the mismatch between the antenna and its feedline.* The reflection loss is a major cause of inefficiency; it increases if the antenna VSWR rises when the handset is held or placed against the head.
- *Absorption by circuits and other components inside the handset.* RF energy may be coupled into the drive circuits for loudspeakers, cameras and other components if they are close to the antenna and exposed to RF fields. This coupled energy will not contribute to radiation from the handset.
- *Absorption* by flexi-circuits connecting various handset components. Although these are not close to the antenna they can contribute losses by coupling energy into internal circuits.
- *User effects.* The user's hand and head change the antenna VSWR, absorb RF energy, and may block the potential propagation path between the handset and the base station.
- *Dissipation within the antenna.* Dissipation of RF energy within the antenna is relatively much less important than most of the other effects.

The demands on mobile phone performance have increased rapidly over the last few years. The economics of manufacture makes it very desirable to make handsets that cover several of the increasing number of world frequency bands. For high-end products both economics and user expectations require them to cover as many bands as possible. Currently at least five bands are assigned for world-wide mobile services (850, 900, 1800, 1900 and 2100 MHz), so many antennas must cover 824–960 MHz and 1710–2170 MHz with high efficiency. Not only must the bandwidth of the antenna be very wide, but modern large color displays are power-hungry and place heavy demands on battery life. When transmitting data using high-order modulation schemes such as EDGE (enhanced data rate for GSM evolution) and HSDPA (high-speed downlink packet access), it is very important that handset antenna gain and efficiency are as high as possible. If the received signal level is too low, the base station will raise the handset power level and request retransmission of blocks of lost data; this will consume additional network resources (additional coding is added, so the time taken to transmit a given amount of revenue-earning data is extended) and demand longer transmission times at high power from the handset, discharging the battery much faster than would have been necessary had the antenna performed better.

Additional pressure is placed on the antenna designer by the shrinking size of handsets, the increased competition for physical space in the handset – the user wants a camera and a music player, not an antenna – and the power demands of the latest hardware and games. The handset may provide other services that require antennas – for example, GPS position fixing, Bluetooth™ or wireless local area network (WLAN) connectivity, and radio or TV entertainment services. Antennas for these services compete for physical space and it is necessary to avoid unwanted interaction between the electronics supporting the different services.

2.4 Classes of Handset Antennas

Large numbers of alternative handset antenna designs can be found in the technical literature and a useful summary is provided in [14]. There are relatively few basic designs, but each has many variants. A convenient method for reviewing the basic designs is to examine their history over the period of development of modern mobile radio systems. Designers should be aware that many configurations are the subject of current patents.

Whip antennas. A quarter-wavelength whip or blade mounted on a large handset provides efficiency which still forms the standard by which other antennas are judged. Unfortunately low-band whips are inconvenient: they typically have to be extended or folded up when the phone is in use and the moving mechanical parts are costly and become worn or broken. Pull-out whips need careful attention to the design – many of these antennas can be pulled out of the handset by a sharp tug and cannot be refitted correctly without dismantling the handset. Hinged blades are vulnerable to damage in both stowed and operating positions.

Meanders and coils. To make whips more acceptable to users, the simple straight conductor is wound into a helix or meandered so the quarter-wave conductor is contained in a short housing, often designed to be flexible.

Dual-band whips and coils. The progressive introduction of a second tier of mobile services in the high bands quickly led to requirements for dual-band handsets. These allowed users to roam between networks operating on different bands, created the possibility of overlay/underlay dual-band network configurations and provided economies of scale in handset manufacture. The commonest early designs comprised whips fed by a coupling structure, but these have been replaced in most markets by dual-band concentric helix-whip and non-uniform helical structures [15], both of which were externally similar to their single-band predecessors. These remain standard external antennas but in many markets users increasingly choose handsets with internal antennas.

Early internal antennas. One of the earliest forms of internal antenna was a meandering conductor etched on the main printed circuit board (PCB), often configured as a form of T or inverted-L antenna. The addition of shunt-feeding to the inverted-L created the inverted-F antenna (IFA) which has become a classic standard form of internal antenna. In the planar inverted-F antenna (PIFA) the upper loading wire of the conventional inverted-F becomes a flat plate (Figure 2.2(c)).

Dual-band internal antennas. The frequency assignments for the low and high bands are about an octave apart, so it is not easy to provide an acceptable input VSWR using a single internal element. The standard solution is to use two radiating elements fed in parallel at their common point. This principle can be applied to monopoles and to PIFAs [16]. In both instances the short (high-band) element creates a capacitance in parallel with the lower impedance of the resonant (low-band) element, while at the high band the long element has a high impedance and most of the power is radiated by the short element which is approximately a quarter-wavelength long.

An alternative hybrid antenna is shown in Figure 2.22(c). below the whole length of the conductor operates on the low band as a folded-up monopole, while at the high band the antenna acts as a half-slot. The input impedances in both bands depend on the same dimensions, making this format tricky to optimize.

Triple-, quad- and penta-band antennas. The growth of world-wide mobile services has seen a progressive increase in the number of frequency bands that must be supported by a handset. For a quad-band or penta-band antenna, the low-band response must range over 826–960 MHz (15.3%) and that of the high band over 1710–2170 MHz (24%). These bandwidths far exceed those of the early dual-band antennas.

Multiple antennas. Techniques such as dual-antenna interference cancellation (DAIC) require the provision of a second receiving antenna [17]. The challenge is to find room for this second antenna and ensure that neither antenna is blocked by the user's hand. Use of DAIC on a single band is relatively simple but extension of this technique to multiple frequency bands requires a second broadband antenna.

Multiple-input, multiple-output (MIMO) schemes. These exploit multipath transmission to enhance the available data rate. Multiple signal samples are transmitted and the data stream is reassembled after being received by multiple independent receiving antennas [18].

Additional services. At the upper end of the market, handsets are becoming ubiquitous terminals for communications, information and entertainment. This is driving requirements to add antennas capable of supporting GPS, WLAN, Bluetooth™ and DVB-H, VHF and later medium/high frequency digital radio, Band II analog FM, DAB (Digital Audio Broadcasting) and DRM (Digital Radio Mondiale). The antenna designer must not only create new designs capable of providing these facilities but also manage the interactions that can limit their usefulness. This represents a major challenge.

A common characteristic of the antennas described above is that they are unbalanced. In each case the antenna is driven from a single terminal on the handset PCB.

There are two different approaches to placing an antenna in a handset – the groundplane can be left in place under the antenna or removed (Figure 2.3). If the groundplane is left in place the most critical dimension is the height h available above the groundplane. Designs with no groundplane under the antenna suffer less restriction on the thickness of the handset, but the PCB length must be extended to accommodate the antenna and no components can be mounted on the opposite face of the board. While the size of on-groundplane designs can be compared in terms of the volume occupied by the antenna, it is not easy to compare on- and off-groundplane designs in this way. This can lead to an impression that off-groundplane designs are smaller, but the volume they effectively deny to other components may be large, and the additional length they require may be unacceptable.

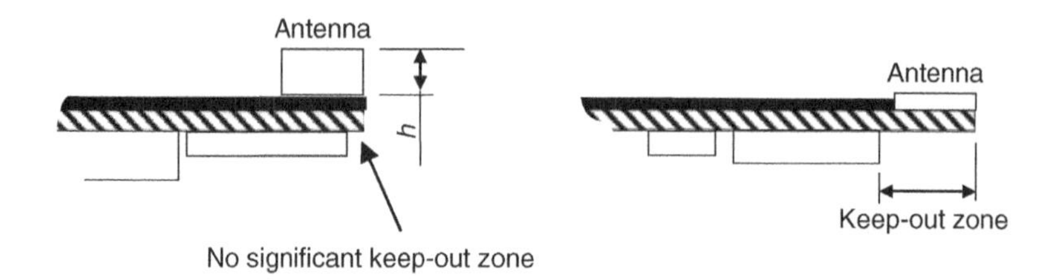

Figure 2.3 On the left the antenna is mounted over the groundplane; on the right he groundplane has been completely removed under the antenna but components can no longer be mounted underneath the end of the PCB.

2.5 The Quest for Efficiency and Extended Bandwidth

In the quest for increased operating bandwidth we are constrained by two main parameters, the dimensions of the handset chassis and the permitted size of the antenna. As we noted in Section 2.3, the behavior of small unbalanced antennas is strongly dependent on the dimensions of the groundplane. Figure 2.4 shows the typical relationship between the available impedance bandwidth and the length of the groundplane (see also [14]). The absolute bandwidth depends on the design of the antenna and the width of the chassis – it is generally slightly greater if the chassis is wider, and the length for optimum efficiency is reduced. In the example shown, the VSWR bandwidth available with a chassis length of 120 mm is double that for a length of 90 mm.

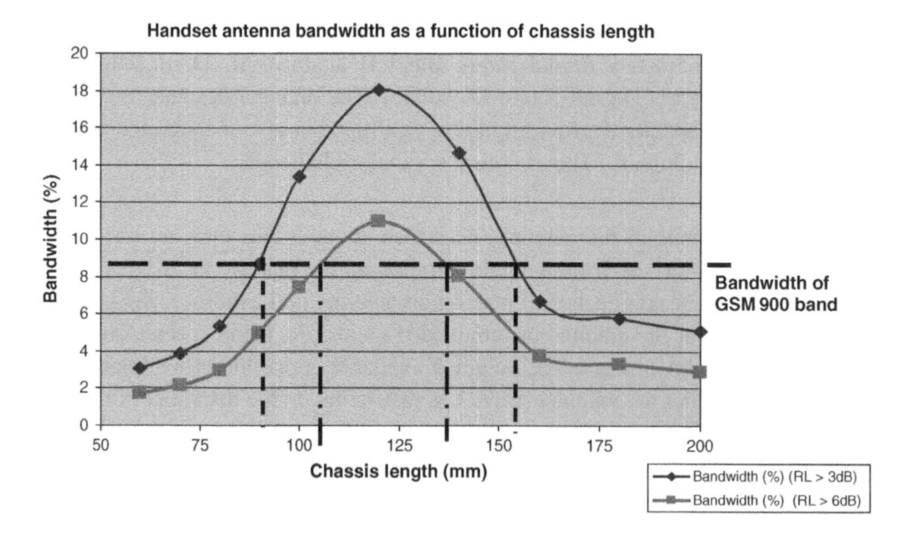

Figure 2.4 Typical relationship between antenna impedance bandwidth of a 900 MHz PIFA antenna mounted on one end of a handset chassis and the length of the chassis.

2.5.1 Handset Geometries

The relationship in Figure 2.4 applies to a single-component handset, often referred to as a bar (or candy-bar) phone. Matters are more complicated when the handset has a variable configuration. Clamshell phones comprise two components joined by a hinge, so the antenna must operate efficiently in open and closed configurations – some variants have a complex hinge allowing two axes of rotation which effectively adds a third operating configuration. Slider phones comprise two separate components placed with their large faces together, connected with a slide mechanism. These are used in open and closed configurations.

Other geometries have appeared, but none has been adopted on a significant scale. These include handsets with the two components hinged on the long side like a small diary, and handsets which can be opened along either the long or the short edge (three operating states).

The requirement to operate with full efficiency in both open and closed configurations was not so significant with early handsets because they were normally opened for use. Lower efficiency was acceptable in the closed condition; in this state they only needed to respond to network control messages and ringing – both of which are well protected against poor efficiency by lower code rates. Modern handsets must retain the greatest possible efficiency when closed because many are capable of use for voice calls when open or closed. Large incoming data volumes may be handled when the handset is closed, possibly when the handset is in the user's pocket, purse or belt pouch.

2.5.2 Antenna Position in the Handset

For each handset geometry there are several possible antenna positions. Each geometry and position creates a different set of challenges for the antenna designer in terms of the available shape and volume, the proximity to other components likely to interact with the antenna, and the ability of the antenna to excite radiating currents in the chassis.

Barphones almost universally have their antennas located at the upper end of the handset, above or behind the display. This position uses the whole length of the chassis to achieve maximum bandwidth. If the handset is more than about 90 mm long and has the right 'feel' in the hand, the user will hold the lower part of the phone and the antenna will not be covered when the handset is held to the ear. Shorter barphones tend to be held with the hand covering most of the rear surface, so the antenna may be completely covered by the user's hand. Some handsets have a sticker suggesting: 'Keep your fingers away from the antenna', but this is likely to be quickly taken off by the user and the message forgotten.

Clamshell phones do not have a universal position for the antenna and three different locations are used (Figure 2.5):

(a) *Top of the flip.* Although occasionally used, this is not a very satisfactory position from the antenna performance point of view.

- The flip is usually thin – often only 5 mm, including the thickness of the case.
- The area round the antenna may not be well grounded.
- The antenna competes for space with the loudspeaker.
- The PA is usually positioned on the main PCB so an interconnecting coaxial cable is required, usually with at least one demountable connector. This is an expensive

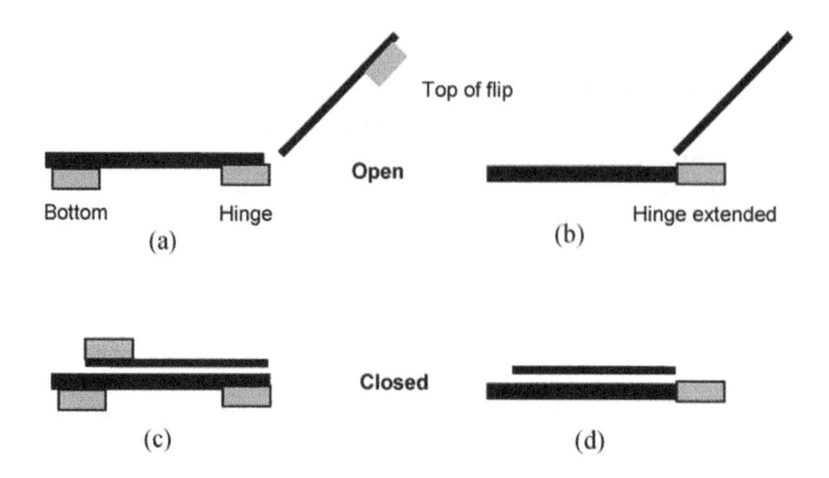

Figure 2.5 Typical antenna positions in a clamshell handset. A variant of the hinge position allows the lower part of the handset that contains the antenna to extend beyond the hinge (right).

arrangement that complicates the mechanical design of the hinge which must accommodate both a flexible PCB (FPCB) driving the screen and a coaxial cable.

- If the groundplane is removed the antenna is very close to the user's ear, so the SAR may be high.

(b) *End of the main component of the handset, adjacent to the hinge.* This is the usual position. The antenna is usually clear of the loudspeaker, but the position suffers a number of disadvantages.

- When held to the ear in the open position, the handset is often held near the hinge and the user's hand covers the antenna.
- When closed, the antenna lies at one end of the handset but when open the antenna position is close to its mid-point. This change in relative position leads to a large change in impedance characteristics when the phone is opened and closed.
- The hinge accommodates flexible connections between the display, camera and processor. The flexi-circuit is excited by RF fields close to the antenna, leading to loss of RF energy, and the high-frequency digital signals in the flexi-circuit radiate noise over a wide spectrum, desensitizing the receiver, particularly in the low bands.

It will be seen from Figure 2.5(d) that when the lower component of the handset is extended past the hinge this position is very similar to that of a typical short helical external antenna in a clamshell handset.

(c) *Lower end of the main component of the handset.* This position is generally clear of hand cover when the handset is open and in use for voice calls. Other advantages of the lower end position are:

- The antenna is well-separated from the FPCB at the hinge.
- The antenna does not have to share space with the speaker.

- The antenna is not close to the head or to any hearing aid worn by the user – only the (inevitable) radiation fields interact with the user's head, not the local stored-energy fields associated with the antenna.
- The antenna is positioned at the end of the handset in both open and closed states – this makes the change in antenna impedance between the two states more manageable.

Slider phones typically have the configurations and antenna positions shown in Figure 2.6. The slider configuration is relatively uncommon, so the design can be regarded as rather less mature than the barphone and clamshell. The lower component of the handset usually contains the keyboard and RF components while the upper component contains the camera and display. The two typical antenna positions are:

(a) *Top end of the lower component – under the display when the handset is closed.* This is the most common position. The groundplane usually extends over the antenna, limiting the extent to which the local fields of the antenna interact with the upper component when the handset is closed. Interaction with the speaker is limited because it is usually housed in the upper component. Slider phones can only be made thin if both components are thin, so there is always great pressure on the available height for the antenna. The antenna is at the end of the handset in the closed position but is about a third of the way down the handset when it is open. This creates a large difference between the open and closed antenna input impedances.
(b) *Bottom of the lower component (under the keypad).* Although this is a less common position, it has the advantage that the antenna is at the end of the handset in both open and closed positions. The antenna is also in a low-noise area of the handset, well separated from the potentially noisy camera and display.

2.5.3 The Effect of the User

There is strong interaction in terms of handset efficiency between antenna position and user grip – the way users typically hold their handsets while making calls or using the handset for interactive data, Web browsing, playing games and writing text messages. Modes of grip which cover the antenna with the hand are likely to have high hand losses compared with those which leave the antenna uncovered. Careful observation of users clearly shows that many common assumptions in this respect are not accurate. A sample of several hundred Japanese users of clamshell handsets showed that almost all used their handsets to access data (perhaps checking the times of their trains or letting their families know they were on their way home) by hooking their index finger round the upper end of the handset body (where the antenna is usually located) and operating the keypad with the thumb of the same

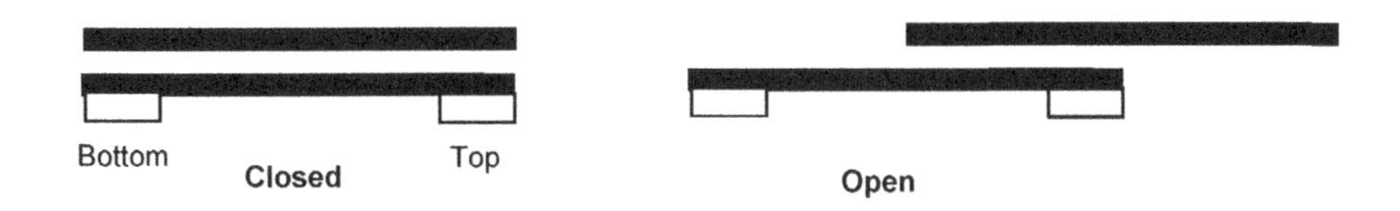

Figure 2.6 Typical antenna positions on slider phones.

hand. The natural grip for a handset depends on its shape and feel, and the effect of the user on the antenna depends strongly on its position relative to the hand. These are features determined by the industrial design (ID) of the handset and not by the antenna designer. This implies that by the time the antenna designer receives a prototype handset some very important limits have already been set on its potential performance. Users will hold the handset in the manner that feels natural to them; the industrial designer must understand and use this to try to ensure that the antenna remains uncovered when the phone is in use.

During voice calls users may modify their grip in the event that the perceived audio volume is too low or there is a high level of local acoustic noise. Their grip may also be changed if the perceived audio quality is poor. The handset is often pressed closer to the ear and the user's hand may be cupped round the top of the handset in an effort to hear more clearly. This often results in covering the antenna and further reducing the signal. In data modes the user will not generally have much feedback about signal quality and no significant feedback mechanism will apply.

2.5.4 Antenna Volume

There is an unavoidable connection between the physical volume of the antenna and the bandwidth that can be obtained. This may be regarded as an expression of the Chu-Harrington limit, but it is probably better seen as describing the effectiveness of the antenna structure in driving radiating currents in the chassis of the handset. The relationship may be seen in two ways: for a given antenna volume there is a minimum chassis length that is necessary to provide a certain bandwidth; for a given chassis length there is a certain minimum antenna volume that can provide the required bandwidth. Handsets that have both small length and a small available volume present a particular challenge to the antenna designer's ingenuity. Once the ID has been fixed, the maximum efficiency that can be obtained has also been fixed; suboptimal antenna and circuit design will lead to the achievement of some lower efficiency, but the maximum was determined by the ID. The relationships between dimensions and bandwidth for a barphone are indicated in Figures 2.7–2.9.

2.5.5 Impedance Behavior of a Typical Antenna in the Low Band

In discussing the optimization of the impedance bandwidth of an antenna it is useful to be able to refer to the behavior of the input impedance by some convenient shorthand terms. We will define these by reference to a typical standard PIFA (Figure 2.10). If the feed position is close to the short circuit the input impedance has the characteristic behavior shown in Figure 2.10(a). The impedance plot remains close to the edge of the Smith chart at most frequencies, with a small circle at the frequency at which the antenna is resonant. This will be referred to as an under-coupled response. As the input position is moved away from the short circuit, the size of the circle indicating the resonant frequency grows until at some point it passes through the center of the Smith chart ($50 + j0$ ohms), a situation which we will refer to as critical coupling (Figure 2.10(b)). As the feedpoint is moved further from the short circuit the size of the resonant circle continues to grow and we will refer to this state as being over-coupled (Figure 2.10(c)). We can separately control the coupling and the resonant frequency by using two parameters, the distance between the feedpoint and the

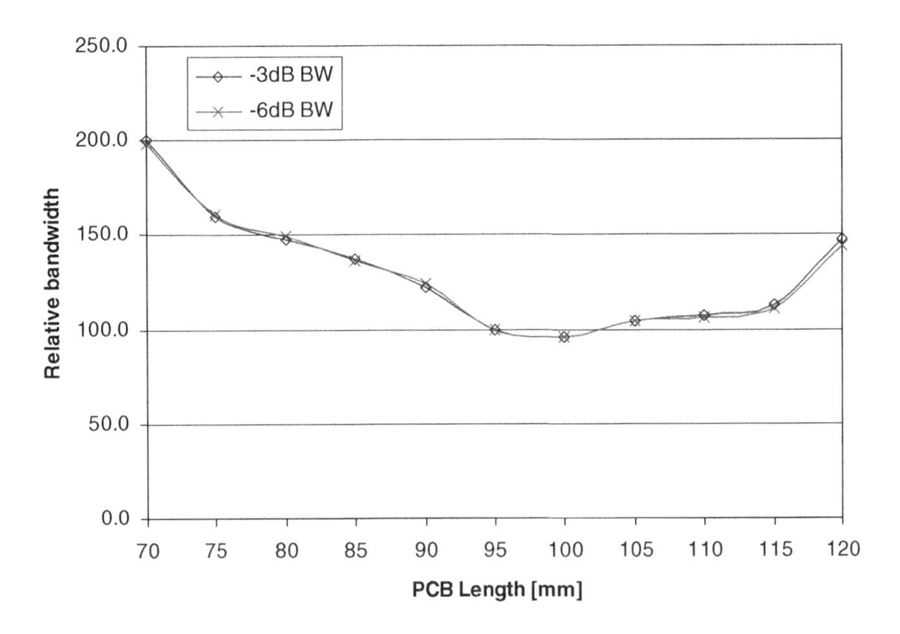

Figure 2.7 Relationship between length and bandwidth at 1850 MHz. For the low-band relationship see Figure 2.4.

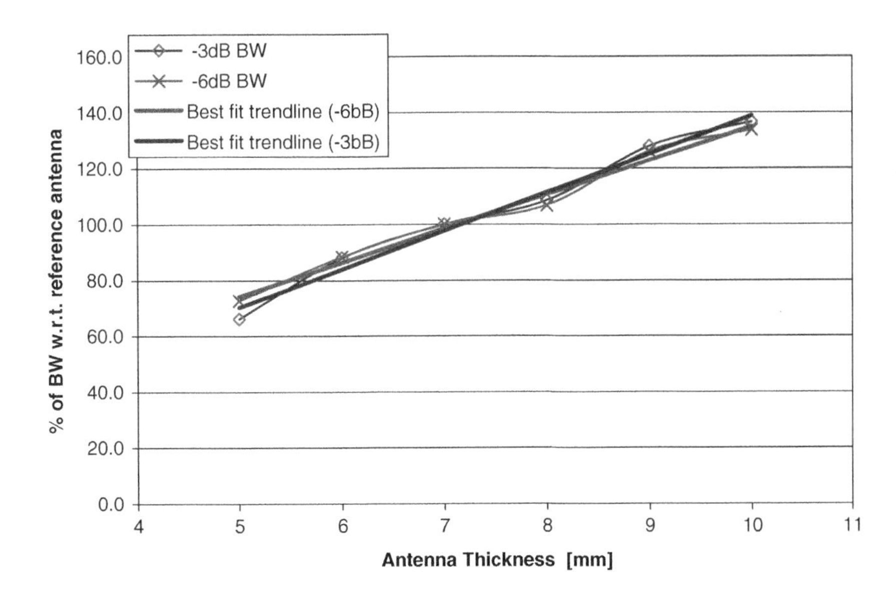

Figure 2.8 Relationship between bandwidth and antenna height at 890 MHz.

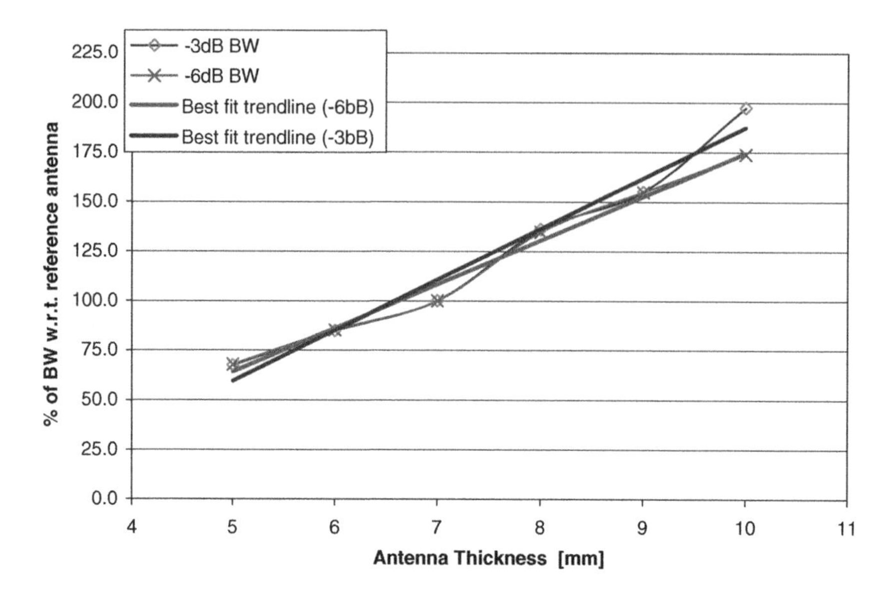

Figure 2.9 Relationship between bandwidth and antenna height at 1850 MHz.

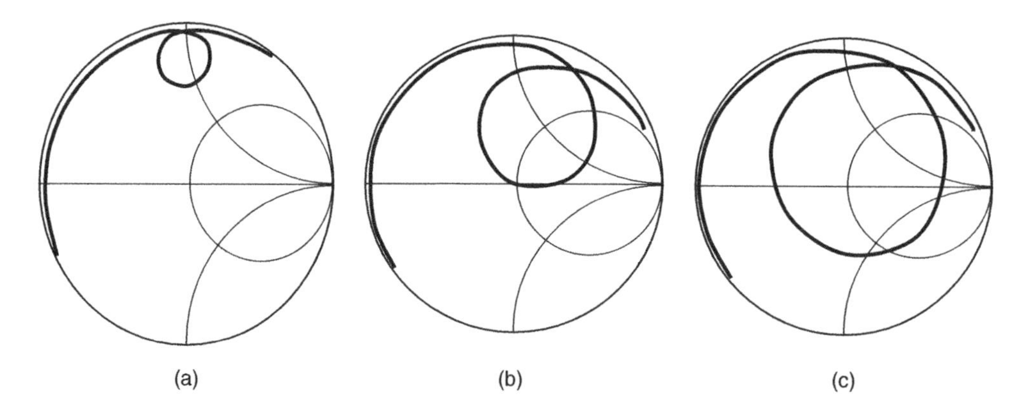

Figure 2.10 Typical feed positions and impedance plots for a PIFA: (a) under-coupled; (b) critically coupled; and (c) over-coupled.

short circuit and the overall length of the top of the antenna – this is true whether the top is in a straight line or is convoluted in some way.

We will return to a discussion of the construction of the antenna in Section 2.6, but the dominant influence of the handset configuration and dimensions suggests that we should examine these first. When we examine a handset to see whether it will be possible to provide the required antenna functionality, the most important considerations are the size and geometry of the handset and the volume and position available for the antenna. If these have already been determined, the antenna designer will need to use considerable diplomacy in persuading others that the necessary performance may be simply impossible to achieve without changes to the design of the handset.

2.5.6 Fields and Currents on Handsets

One of the most useful features of computer programs for electromagnetic simulation is that they help in the visualization of radiating currents and fields. They allow us to develop an understanding of the mechanisms underlying the behavior of antennas on handsets and to develop strategies for the optimization of bandwidth and efficiency.

Figure 2.11 is a simulation result for a conventional PIFA placed at the end of a simple rectangular groundplane 100 mm long [19]. It shows the electric field that exists on a plane slightly displaced from the axis of the groundplane, which can be seen immediately below the simulation plane. The displacement of the simulation plane avoids displaying the local fields around the antenna and allows us to see the fields that are contributing to radiation – without the clue of the input connector (below the groundplane on the right) it would be difficult to identify the location of the antenna, so this simulation emphasizes the way in which the groundplane contributes a dipole-mode field. Unsurprisingly the corresponding far-field radiation pattern is very similar to that of a half-wave dipole (Figure 2.11(b)).

It is more surprising that the same relationships hold at the high bands, but again the unbalanced feed for the antenna results in dominant fields being produced by the chassis (Figure 2.12). The radiation patterns are similar to those of a long dipole with an asymmetric

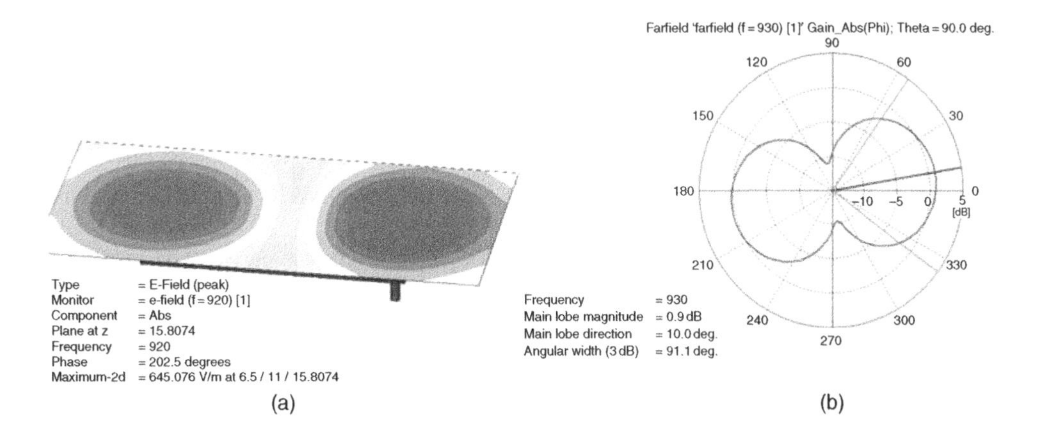

Figure 2.11 Simulations of the E-field and radiation patterns of a typical barphone at 920 MHz.

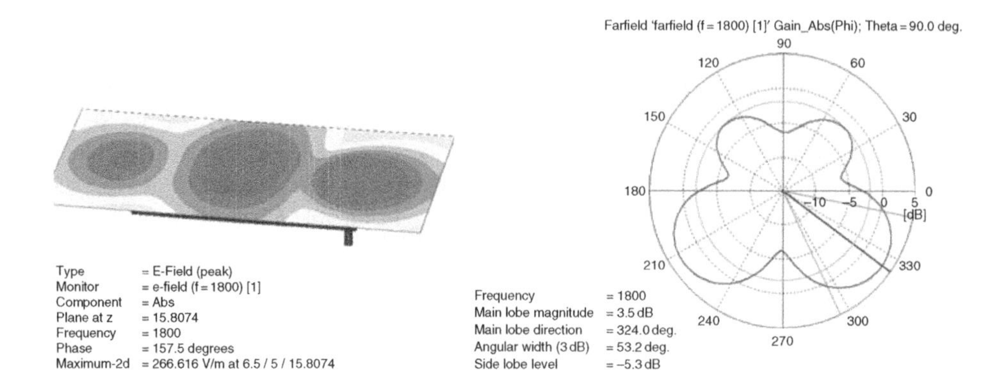

Figure 2.12 Simulations of the E-field and radition patterns of a typical barphone at 1800 MHz.

feed. At both bands almost identical field amplitudes and distributions are seen on both faces of the chassis – again emphasizing its central role in the radiation process.

Not only do the simulated fields match the simulated (and measured) radiation patterns, but it is quickly found that the impedance bandwidth available from an antenna depends critically on the extent to which it excites currents in the groundplane. For a real handset the currents will flow in the groundplane and all the grounded hardware connected to it, a configuration which we will refer to collectively as the chassis of the handset. Repeated simulations using different antenna and chassis parameters clearly show that any configuration in which chassis currents are weak is likely to have a narrow impedance bandwidth. If the E-field is computed on the groundplane itself, this effect is not seen very clearly because most simulation programs rescale their output to accommodate the strong fields around the antenna. These are intense local fields – for example, between the antenna and the groundplane – but their effect at a large distance from the handset is not very significant. The universal rule is seen to be that the larger the current the antenna can establish in the chassis, the more effective it is as an antenna! If a simulation shows that a low-band antenna stimulates only small currents in the chassis, its impedance inevitably changes rapidly with frequency. The antenna itself has too small a size to be able to operate as a wideband antenna – the role of the device we refer to as the 'antenna' proves to be that of a coupling device, stimulating radiating currents in the handset chassis. In view of what we know of the bandwidth limitations of small antennas, this explanation makes perfect sense!

This conclusion has several important consequences:

- It is best to design and position the antenna with the objective of stimulating chassis currents.
- For a given *radiated* power, there is a necessary value of chassis current and resulting local E- and H-field. Any local effect of this *radiating* current is an inevitable consequence of the absolute power radiated.

These are very significant conclusions. We have noted that the radiating fields on the opposite side of the chassis to the antenna are very similar to those on the face containing the antenna, so undesirable effects caused by electromagnetic fields in the region of the user's head can be reduced only by radiating less power or by reducing the interaction between the *local* fields surrounding the antenna. The most obvious precautions are to place the antenna on the opposite side of the groundplane to the user and to position the antenna so it will not be close to the user's head when the handset is in use. Interference with a hearing aid worn by the user will be reduced by the same precautions.

2.5.7 Managing the Length–Bandwidth Relationship

The relationship between dimensions and the ability of the antenna to excite chassis currents, essential for low-band performance, now has to be understood in the context of the wide variety of handset geometries and antenna locations encountered in practice. To emphasize that the radiation efficiency is governed by a complex interaction between the antenna and the handset, the term *handset efficiency* is used here in preference to the more usual term *antenna efficiency.*

2.5.7.1 Barphones

Most barphones are shorter than the length that provides the largest possible bandwidth (about 115 mm) so, from the point of view of both bandwidth and the convenience of laying out the components of the rest of the handset, it is usually advisable to mount the antenna at the end of the handset.

The provision of high efficiency across both the low bands is almost always challenging, but when the chassis is less than about 80 mm long it rapidly becomes difficult to provide a satisfactory return loss and efficiency across either the 850 MHz or the 900 MHz band. In this situation it is useful to position the antenna feedpoint at one corner of the main PCB (or, to put it another way, to drive the main PCB from one corner). This effectively increases the chassis length to the dimension of its diagonal. It also creates the typical skew in the axis of the radiation pattern seen in Figure 2.11.

With very short handsets it may be necessary to create some form of artificial extension to the length of the chassis, for example by adding a flange or slotted flange at the end remote from the antenna (Figure 2.13). This is a well-known technique seen in commercial handsets. Adding a flange may be compared with capacitively loading the ends of a short dipole. Additional measures can be taken to make the added section appear to be electrically longer than its physical length, but this may be at the expense of frequency bandwidth – it may allow the handset to operate better at the bottom of the band, but may inhibit its performance at other frequencies. Methods involving inductive notches cut in the groundplane are also possible, but they may limit unacceptably the layout of other components on the PCB – already difficult on a very short board.

The result of adding length in this way will not be as effective as if the chassis had the same overall length in one plane, but the advantage obtained may be sufficient to provide a small but critical increase in low-band efficiency. Adding further inductive or capacitive loading within the extended structure is also sometimes useful, but this is a process of diminishing returns – if the loading is itself resonant, then its effective bandwidth is reduced and its interaction with the resonant behavior of the whole chassis may not produce the expected result. At frequencies lying between various resonances the efficiency of the handset may be lower than expected.

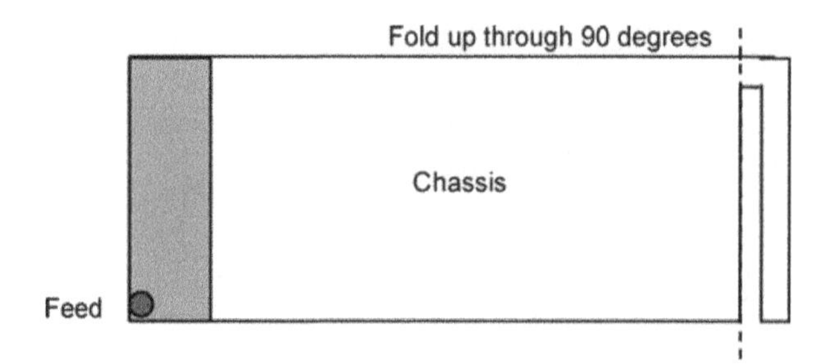

Figure 2.13 A short chassis can be extended by fitting a conducting extension, folded (along the dashed line) against the far end of the case.

2.5.7.2 Clamshell Phones

A typical clamshell phone is around 80 mm long when closed and 140 – 160 mm long when open. The relationship between the open and closed lengths depends on whether the flip and the body of the handset are of equal length (the flip is often shorter than the body) and also the position of the hinge (see Figure 2.5). Referring to Figure 2.4 we see that when closed the chassis length is less than that required for optimum bandwidth, and when open it is too long. In both configurations the length is far from the optimum and bandwidth is restricted. A measurement of the antenna impedance in both open and closed positions will generally show a very large change in the antenna coupling (Figure 2.10) as well as a change in the resonant frequency. It is difficult to achieve optimum return loss (and efficiency) performance in both open and closed states – to achieve this objective we need to identify design parameters that, as far as possible, separately control the antenna impedance when the handset is open and closed.

The reason for this impedance change is clear: the configuration of the (radiating) chassis has changed and the position of the antenna relative to the positions of current maximum on the chassis has also changed. We can improve matters to some extent by adopting a compromise antenna design (in which the coupling and resonant frequency take values on either side of the optimum in the two states). Typically performance is weighted toward the open state on the basis that voice calls and Web browsing will only take place when the handset is open, but this may not apply if the handset can be used for voice calls or for transmitting pictures when it is closed. Increased functionality has not only increased pressure on handset efficiency in general, but demands similar efficiency in all states of the handset.

The Open Clamshell

Figure 2.14 shows a simplified view of a clamshell handset with the antenna placed at the lower end of the main component. The two components are connected by a flexi-circuit whose function is to provide DC and data connectivity to the display, camera, speaker and other components mounted in the flip. The physical form of the connection varies, but it typically consists of several layers of FPCB with at least one layer having a continuous ground conductor on one face. The points of connection in the handset components are usually close to their adjacent ends, but are sometimes as far as 25 mm from the ends. The flexi-circuit is shaped to fit through a channel provided in the hinge and is usually looped around the hinge axis to ensure that the movement of the hinge causes it to bend in the

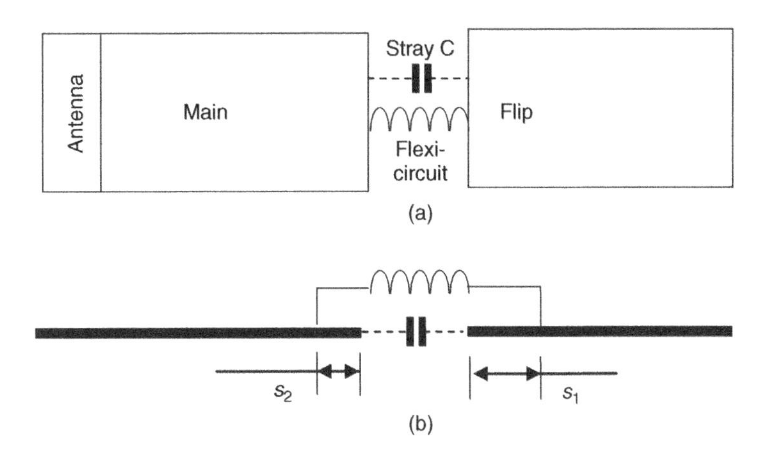

Figure 2.14 (a) The effective RF configuration of an open clamshell handset. (b) The overhangs of the components beyond the connection points.

desired plane. Any tendency to twist or bend in the plane of the circuit must be avoided or it will rapidly fail in use. When opened flat, a typical hinge FPCB is between about 50 mm and 80 mm long, measured along the ground conductor, and about 6 mm wide.

From an RF standpoint the flexi-circuit provides an inductive connection between the two components with an inductance determined by its length and configuration. The net reactance between the two components depends on the values of the series connecting inductance (L_s) and the shunt capacitance (C_p) which is created by the geometries of the hinge and flexi-circuit. If the two reactances are parallel-resonant in the lower band, the effective length of the open handset will be that of the main component; if they are non-resonant, the effective length will be that of the entire length of the open handset loaded by the series reactance between them. Figure 2.15 shows a simulation of the surface currents in

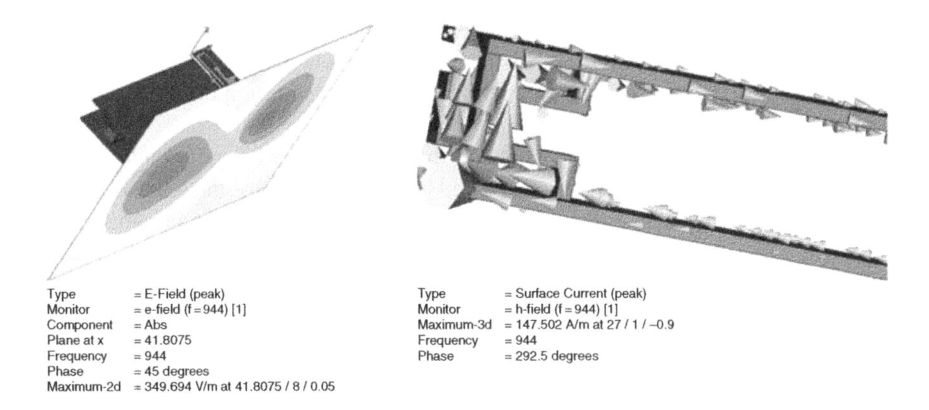

Figure 2.15 A closed clamshell handset with the connecting flexi-circuit modeled as a simple meandered link showing (a) the E-field on a plane displaced to fall just outside the handset and (b) the currents in the chassis.

a simplified clamshell handset. The connection supports a current almost as large as that flowing in the PIFA antenna; this increases the importance of the value of the reactance in which it is flowing (if the connection lay at a current minimum its value would be less significant). Fortunately we can manage the effective electrical length of the open handset by adjusting the net reactance across the hinge. This parameter, vital to the RF performance of the handset, is only adjusted by close co-operation between the antenna engineer and the industrial and mechanical designers. We can understand the significance of both L_s and C_p by examining some extreme cases:

1. If L_s is very small it requires a large value of C_p to bring it into resonance and the rate of change of reactance with frequency around resonance is large. The exact value of C_p is also very critical.
2. If L_s is very large it may already be self-resonant or it will require only a very small value of C_p to bring it to resonance. The net reactance is likely to be capacitive and is difficult to change.
3. If C_p is very large the net reactance is small and capacitive whatever the value of L_s.
4. If C_p is very small the reactance is dominated by L_s and the effective handset length is too long.

The best situation is that in 4, because we can gain control of the connecting reactance by deliberately increasing C_p to a value chosen to be sufficiently close to resonance to deliver the greatest available bandwidth. In practice it will be found that this occurs when the net reactance approaches resonance from the capacitive direction. Depending on the effective value of L_s the optimum value of C_p may be extremely critical (to within a few picofarads). In general terms L_s is determined by the overall length of the connecting FPCB (measured along the RF current path) and the position and alignment of the ends of the connection relative to the ends of the components (s_1 and s_2 in Figure 2.14). C_p is determined by the geometry of conducting components in the area of the hinge and the dimensions s_1 and s_2. If the hinge design results in the two ends of the flexi-circuit passing close to each other with facing flat faces, perhaps close to the exit point from the hinge, this will create a very large value of C_p which will be difficult to reduce. As s_1 and s_2 are increased the length of the connection (and its inductance) inevitably increases; the regions in which the connections pass over the groundplanes form transmission line stubs that further increase L_s.

The antenna designer can generally simulate the connection with sufficient accuracy to gain some insight into the likely net reactance and the RF operation of the handset. The ideal solution is to allow some later optimization, perhaps controlling the hinge capacitance by adjusting the configuration of the conducting surfaces around it. An all-metal handset case is likely to cause severe problems!

The Closed Clamshell

The radiating currents on a closed clamshell handset are seen in Figure 2.15(a) where the antenna position is at one end. Once more the dominant field mode, seen in a plane displaced so it falls just outside the handset, is a dipole mode with high fields and minimum currents at the ends and maximum current in the center. If we move the simulation plane we also see a current flowing along the folded chassis in the mode typical of a short-circuit stub. The antenna impedance and its bandwidth will be determined by the impedances encountered by the radiating (chassis) current mode, the transmission line current mode and the antenna currents. If the net connection impedance was optimized for the open position, we must use other methods to adjust the closed antenna impedance with the objective of ensuring minimal change between the open and closed states. This suggests that we should examine the effect of the transmission line current in the stub formed between the components. This has a length determined by the geometry of the components, and a characteristic impedance determined by the width of the chassis and the distance between the components in the closed state. The capacitance across the open end of this stub has the greatest effect on the antenna impedance and by optimizing the shapes and spacing of conducting components in this region we can optimize the closed antenna impedance without significantly affecting the open impedance.

This optimization is amenable to simulation, with physical adjustments made at the prototype stage to achieve optimum geometry.

The radiating modes for a hinge-position antenna are similar to those for the end-position antenna. The main difference is in the effects of s_1 and s_2, which modify the antenna impedance in both states. The transmission line current may have a smaller value, but the impedance it encounters is likely to vary more rapidly with frequency because the transmission line stub is excited near its short-circuited end, producing a higher value of loaded Q. These effects are best studied using simulation software, identifying the separate and combined effects of the parameters involved.

Fortunately the effects of chassis length on antenna impedance are less marked (although still significant) at the high bands. At regular intervals in the optimization of low-band performance it is worth checking that high-band performance has not been compromised. This is specially important when operation is needed over the whole range from 1710–2170 MHz, when it may be difficult to recover high-band performance if the impedance plot has spread too far across the Smith chart.

2.5.7.3 Slider Handsets

A slider handset typically has the configuration shown in Figure 2.6 The components are physically connected together by a slide mechanism, usually containing both metal and plastic parts, while electrical connection is made by a U-shaped flexi-circuit connected to each component in the area that overlaps when the handset is open. This flexi-circuit is typically 100 mm long and 15 mm wide and carries a large number of very fine traces (perhaps 50) carrying digital signals to the display and camera, and audio signals to the speaker. In a typical configuration ground conductors are provided along each edge of the flexi-circuit, but these are very narrow and are not really of any use as RF grounds – anyway, with a length of 100 mm they are ineffective irrespective of their width. The position of the 180° U-bend moves along the flexi as the slide operates, so each station along the central part of the flexi experiences a 180° flexure around a radius controlled by the handset design every time the handset is opened or closed. Handset designers resist the suggestion of providing a wide ground conductor because the small clearance between the components requires that the flexi-circuit is bent with a very small radius and the repeated flexure causes early failure of the conductor. The antenna excites a potential difference between the handset components, and currents flow through the capacitance between them which is in parallel

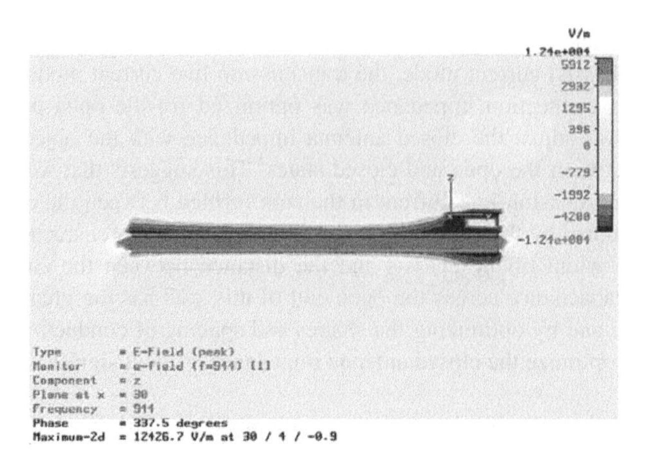

Figure 2.16 A closed slider phone showing the very high field between the components.

with the effective reactance of the flexi-circuit. The lack of screening of the signal tracks in the flexi results in RF currents flowing into other handset circuits where they contribute only to losses and not to useful radiation. Figure 2.16 is a simulation result showing the very high E-field existing between the components. Not only is this field trapped (raising the Q of the handset) but it drives currents through the lossy flexi-circuit.

One method of reducing the RF currents in the flexi is to ensure that a large capacitance exists between the components in both open and closed states [20]. This is easier to achieve when the handset is open, when there is a larger facing area; in the closed state it may require the provision of additional conductive areas on both components and connection of these areas directly to the internal RF circuit grounds. (This is a parameter that suits an all-metal handset.)

Given the provision of adequate inter-component capacitance, the main effect on antenna impedance is that the length of the chassis changes between states. This is to some extent offset by the fact that both components of slider phones are usually relatively long so the severest effects of a short chassis are avoided. Antenna impedance is also affected by the change in its relative position on the chassis when the handset is opened. This affects models where the antenna is placed under the display, where the impedance responds to changes in both length and relative antenna position. The under-keypad position is at the bottom of the handset in both states, so the antenna impedance responds only to the length change.

Slider phones are relatively uncommon and will perhaps continue to be so as demand increases for extra keypad functionality and larger displays.

2.5.7.4 Other Handset Configurations

A small number of handsets provide a hinge along the long axis of the handset. These are usually not a major problem as they are long enough in the closed state to provide reasonable antenna bandwidth. Some clamshells handsets have hinges that allow their flips to be rotated while open. As well as creating a further geometry in which the handset must operate efficiently, the more complex hinge geometry usually requires a longer flexi-circuit.

The use of more metal parts in the hinge to provide mechanical strength can result in a high capacitance across the ends of the flexi, especially if the flexi is wound round a metal pin. The loss of control of the inter-component capacitance almost inevitably leads to lower RF efficiency as the antenna designer struggles to balance the conflicting demands of the three operating states. Another variation is a compound hinge allowing operation in both long-edge and short-edge opening – this provides the same potential challenges as the three-way hinge, with a long flexi and metal hinge parts.

The continuing trend towards the integration of user functionality – mobile phone, camera, PDA, audio and video player, and radio/TV capabilities – will drive the market to create new hybrid device formats. These will be subject to huge constraints on dimensions and weight. All the skill and ingenuity of the antenna designer will be needed to integrate the necessary multiple antennas into a single compact device. New physical device formats will appear and they will continue to influence the extent to which the desired RF performance can be reliably achieved.

The transmission characteristics of signals at lower frequencies – in particular their reduced diffraction loss round obstacles – make them attractive in rural areas and in other situations where coverage is limited by transmission loss rather than capacity. The design principles described in this chapter are applicable to any device in which antenna space is limited, especially when the longest dimension of the device is small compared with the operating wavelength.

2.5.8 The Effect on RF Efficiency of Other Components of the Handset

A number of the matters discussed below seem to be unrelated to antenna design, but it is the antenna designer who is usually regarded as being responsible for the RF efficiency of the handset and is probably best able and equipped to advise on these matters. If the potential problems are not appreciated at an early stage of the design it becomes increasingly difficult to persuade others of the need for change and the opportunity to correct matters is lost. The handset will have poor efficiency and the antenna engineer will be seen as being responsible.

2.5.8.1 Loudspeakers

One of the most difficult aspects of design with barphones is the intensive use of surface area and volume in the region at the top of the handset. It is common for the loudspeaker to be placed close to or underneath the antenna. This always reduces the efficiency of the antenna, although the extent of the reduction can be controlled by the positioning of the speaker, choice of the speaker configuration, and appropriate isolation and decoupling to prevent RF currents flowing into the audio circuit. Successful designs have been produced by using the same physical structure to function as the acoustic resonator and the carrier for the antenna – the loudspeaker is integrated into this assembly and is connected to pads on the supporting PCB by spring pins. This design requires co-operation between the antenna designer and the acoustic engineer who will specify the required volume, sealing and venting of the resonator. Loudspeakers mounted close to antennas must have direct low-profile spring connections (not wires) and be connected through inductors in series with both terminals. Some further increase in antenna efficiency may be produced if the terminals are directly connected by a bypass capacitor (typical value 10 – 30 pF). Speaker designs with some measure of external screening to prevent direct inductive coupling of the RF fields to the speech coils are more successful than designs with no screening.

The speaker is normally mounted against the groundplane, keeping as much clearance between the speaker and the antenna as possible. An alternative strategy is to raise the speaker above the groundplane and connect it with leads running down the antenna short-circuit.

2.5.8.2 Cameras

Careful attention must be given to the screening and grounding of cameras when they are mounted in close proximity to antennas. Coupling of antenna fields into cameras and their drive circuits will result in loss of efficiency; coupling of camera drive circuit currents into the antenna may lead to increased receiver noise levels – a consideration that is particularly important if the camera is to be used for video calls when the camera and receiver will operate together. A screen wrapped round a camera module may have a length resonant in the high bands unless the positions for its grounding to the main PCB are chosen with care – multiple ground points will almost certainly be required.

A camera connected with a flexi-circuit and sited close to an antenna can be very destructive of antenna performance unless both the camera and the connecting flexi are well screened. Checking the antenna impedance with the camera in various positions will allow the effectiveness of the screening to be checked – and active measurements will confirm this.

RF currents, inevitably present in the hinge area of a clamshell handset at low band, will couple into a poorly screened camera and its connections, reducing handset efficiency. Cameras in the hinge itself are the most difficult to deal with, as their flexi-circuit may be integral with the inter-component flexi; current densities in the ground conductor of this flexi can be very large – not only is it in the high-current region of the middle of the handset, but its restricted width increases the current density.

2.5.8.3 Vibrators

These are sometimes placed near antennas, but they often do not form good neighbors, perhaps because of inadequate internal screening round the motor and its windings. The off-center weight may not have a well-defined rest position.

2.5.8.4 Batteries

The battery occupies a significant proportion of the surface area of a handset – particularly in a closed clamshell phone – and any disruption it causes the surface currents is likely to result in changes to the antenna impedance and bandwidth. The case of a NiMH battery is DC positive, while the groundplane of the handset is usually DC negative. This means that the battery case – in which RF currents are likely to be induced – is connected directly to the internal DC supply rail of the handset. From an RF point of view this is highly undesirable, so precautions must be taken to adequately decouple the battery case. Decoupling usually takes the form of series inductance in the DC connection and a bypass capacitor fitted between the battery case and the handset ground. The case is grounded in this way at one end, so the rest of the case forms a transmission line stub; the impedance seen at the open-circuit end opposes the flow of currents in the surface of the handset chassis (Figure 2.17).

Figure 2.17 The battery creates a transmission line stub creating an impedance in series with currents flowing along the surface of the chassis.

This impedance will have least effect if it occurs at a point where the surface current is smallest, so it is better for the battery connections to be provided at the inboard end and the open circuit end to lie close to the end of the handset.

Further problems can arise if RF currents are allowed to flow into the static discharge or charge control circuits; this possibility must be avoided by adequate isolation and decoupling.

If the spacing between the antenna and the battery is too small the battery intersects the local fields round the antenna. This causes strong excitation of currents in the battery case and effectively restricts the volume available to the antenna. If the handset layout provides inadequate clearance between the antenna and the battery it may be better to reduce the antenna dimensions to increase the inadequate clearance.

In almost every instance the efficiency of a handset will be found to decrease when the battery is fitted. This effect is almost avoidable if the matters indicated here are properly managed.

2.5.8.5 Electromagnetic Compatibility (EMC) Shielding

In early handsets the entire inside surface of the case was covered with conductive paint; the external antenna projected through a hole in this conductive enclosure whose purpose was to provide an EMC (RF interference) shield for the handset. In many recent designs, shielding is provided by metal cans placed over the susceptible components, but many handset designers still use conductive paint as the last line of defense. It is the outermost conductive surface that carries radiating currents on the chassis, so where EMC paint is used it forms the effective radiating surface of the handset. EMC coating materials are of two types, lossy and highly conductive. The paint should have high conductivity because any ohmic losses decrease the power that will be radiated. The impact of surface currents on antenna impedance and bandwidth mean that any EMC coating must be in place during RF testing early in the design process or unexpected changes in RF performance may occur when the coating is applied. The coating should form a closed box (avoiding the creation of any resonant structure that can be excited by the radiating currents) and connections between the coating and the PCB ground made with low-resistance gaskets. These both increase the shielding effectiveness of the coating and also reduce losses where radiating currents flow from the coating to ground.

2.5.8.6 The Antenna Feed Circuit

It is important that low insertion loss and good impedance matching is maintained between the PA (and the receiver input) and the antenna matching circuit. Any losses reduce the power reaching the antenna. Frequency-dependent mismatch in the connecting lines will make it difficult to obtain consistency of output power (TRP) over the operating frequency band. The traces forming the RF connections must not pass over other internal lines unless an intervening ground layer is provided and adequate vias must be provided along their length to ensure that they do not excite unwanted modes between different groundplanes. As much consideration must be given to the return (ground) path as to the outgoing stripline or microstrip.

2.5.8.7 The Groundplane

An adequate groundplane must be provided over the whole length of the PCB. If this is not done, RF currents will flow in other components such as the display and keypad and their associated connections. These currents will encounter resistance and will consequently cause RF power loss. The design of the bonding between handset components and the groundplane needs to recognize the probability that RF surface currents will be present and they should be provided with a low-resistance path in which to flow. Conductors forming loops will couple to surface currents; where loops are formed deliberately – as in the interconnecting flexi-circuit in a clamshell handset – they must be kept to minimum size, be stable in mechanical configuration and be screened and/or decoupled. An interconnecting flexi that is capable of being installed in more than one precise configuration will give rise to antenna impedance changes; unless the engineers diagnosing the problem appreciate these interactions they are likely to look for inconsistencies in the manufacture of the antenna when the problem lies elsewhere.

2.5.9 Specific Absorption Rate

In the previous discussion of the optimization of the RF performance of a handset no mention has been made of acceptable levels of SAR. This is because on the whole the achievement of levels of SAR that fall within the international limits is not onerous as long as the handset design and antenna placement avoid the user's head being exposed to the local fields in the immediate vicinity of the antenna. In general it is observed that exposure to magnetic (rather than electric) fields gives rise to large values of SAR. It might be expected that a very efficient handset gives rise to higher levels of SAR than a less efficient one, but the relationship between efficiency and SAR is not simple. The main low-band radiating currents flow in the handset chassis so the SAR associated with the radiating currents for a given radiated power will be similar for most handsets. Loss mechanisms may result in the need for higher antenna currents to sustain the required total radiated power, causing increased SAR in the user's head.

The general reduction in mean radiated power by handsets since the adoption of digital modulation systems has reduced the difficulty in complying with SAR limits, and the achievement of low SAR is not a significant factor in handset antenna optimization unless, as sometimes happens, network operators or handset manufacturers adopt lower limits.

If SAR is measured on an early mock-up of a handset (with no plastic case) misleadingly high values may be measured in some of the standard handset positions because the bare PCB is likely to be much closer to the head than when it is housed in its case. Compliance with the relevant SAR criteria is usually mandatory and in many countries handset manufacturers are obliged by law to publish the results of SAR tests for each type of handset placed on the market.

2.5.10 Hearing Aid Compliance

The generation of unpleasant buzzing sounds caused by the interaction of pulsed RF signals with the small-signal circuits of a hearing aid is of obvious concern to hearing aid users and the manufacturers of handsets and hearing aids. The design considerations parallel those relating to SAR, but the achievement of low interference levels is a function both of the handset and the hearing aid. Just as for SAR compliance, the best that the antenna designer can do is to ensure that the hearing aid is not excited by the local fields surrounding the antenna. The hearing aid manufacturer in turn must ensure that designs are sufficiently well screened and linear, so that the threshold for the production of objectionable buzzing lies below the field strength which can be expected to occur close to a well-designed handset in which the chassis radiates the required nominal RF power. Handset designs in which the antenna is located well away from the head – such as clamshell phones with hinge- or bottom-mounted antennas – are likely to be most successful in this respect, as are slider phones with bottom-mounted antennas. The key reference documents are [21], [22].

Direct interference from the RF output from a handset is only one cause of hearing aid interference; other causes include electromagnetic interference from displays and backlights, and direct audio feedback. A scoring system is used to indicate the extent of compatibility to the purchaser. Network operators in the USA are obliged to ensure that, for each network protocol they use, a mandatory proportion of handsets complies with specified levels of compatibility.

2.5.11 Economic Considerations

A handset is usually shaped to provide a pleasing external appearance, so the antenna must be correspondingly profiled to fit the available space – most handset antennas require some degree of profiling in three dimensions. The differences in the shapes of various handsets and the variations in the configuration of the other major components mean that, both mechanically and electrically, handset antennas are unique in design for each handset. One result of this is that the fixed costs of development, often referred to as non-recurring engineering costs (NRE), and the cost of unique production tooling may be a very significant part of the cost of each antenna, especially when the production run is short. NRE may equal or exceed the production cost of simple low-complexity antennas for volumes less than 250 000 units. A complex and well-optimized design – perhaps a quad-band design integrating a speaker – will have a higher NRE and, even with production volumes over a million units, the NRE will form a significant part of the overall cost of the antenna.

An apparently attractive method by which the varying input impedance of compound handset designs could be overcome is the use of switched or self-tuning matching circuits, perhaps using micro-electromechanical systems (MEMS) components. These methods have not been adopted for high-volume products, although they may find application in the future. Other technologies such as the use of ferroelectric devices or plasma antennas have also not yet achieved an attractive price–performance ratio.

2.6 Practical Design

2.6.1 Simulations

The emphasis placed on the interaction of the handset with the antenna makes it clear that to provide real accuracy, any computer simulations need to include aspects of the design of the handset as well as of the antenna itself. A complete model is not practicable; it would be excessively complex and would run too slowly to allow optimization of the very large number of relevant parameters. In practice, when the antenna designer starts work there may be no detailed design for other aspects of the handset, so a complete model is not possible. The most practical approach is to model the antenna on a representative groundplane of the right dimensions, with any major components modeled as simple blocks with an appropriate conductivity. This will give some idea of the possibility of achieving the required impedance bandwidth – and in consequence some indication of the reflection losses to be expected.

Simulations are useful for studying the effects of varying parameters such as the clearance to batteries, but components such as speakers and cameras cannot really be simulated with sufficient accuracy to be useful. What is useful is an indication of the antenna performance if the nearby components are perfectly screened, as this will help the designer to identify the effects of imperfect screening when the whole handset is available.

The design may begin from scratch with some new innovation, but it is more likely to start close to a previous design that has proved effective on a previous project, modified as required to fit the constraints of the new project.

One of the features of the design which is fixed very early will be the position of the feed connection(s) on the handset PCB and the corresponding position of any ground connection(s). These positions will be used by the handset designer in the layout of other components and it is often difficult to make changes once the project is under way. If it is likely that an antenna matching circuit will be required, its position and topography also need early agreement. The principle here is to make sure that choice is maintained for as long as possible; the number of component positions can always be reduced later, but it may not be possible to increase their number if inadequate provision was made at an early stage.

If, when the antenna design is transferred to a real prototype, the impedance matching is significantly better than predicted, this immediately raises the possibility that unexpected loss mechanisms are at work. Losses always reduce the antenna Q and increase its impedance bandwidth. Factors external to the antenna are probably involved if the efficiency is much lower than would be expected from its reflection loss (calculated from its measured return loss).

Tracing the cause of unexpected losses is a slow and time-consuming process. It is usually tackled by deconstructing the handset and measuring the efficiency as each component is removed, or by the reverse process of building the handset up from a stripped-down PCB. It is particularly difficult to identify the causes of excess loss when the loss mechanism is associated with interactions between other handset components; removing components one by one may produce different conclusions according to the order in which they are removed.

2.6.2 Materials and Construction

As we have seen, the most significant causes of reduced efficiency are the mismatch at the antenna terminals and effects related to the design of the rest of the handset. Rather counterintuitively, this means that although the antenna is electrically small the ohmic and dielectric losses in the antenna itself are not really very significant. The radiating element may be made from copper, but stainless steel, brass, phosphor bronze, nickel silver and other materials can be used almost as effectively. The choice is likely be made on the basis of cost, which will also influence the form of construction. Possible constructions include the use of a printed circuit on a rigid or flexible substrate, electroplating or other method of metal coating applied directly to the carrier, or pressed metal.

While high-impact polystyrene would be an ideal electrical choice for the dielectric support for the radiating element (usually referred to as the carrier) because of its low dielectric loss factor, in practice the much more lossy polycarbonate or polycarbonate/acrylonitrile-butadiene-styrene (PC-ABS) co-polymers are almost equally good. PC-ABS is commonly used for handset cases, so it is often used for the carrier. The carrier is almost always injection molded from a tool designed specifically for the target handset. The radiating element may be glued, heat-staked or clipped in to the carrier, or the carrier may be molded over the radiator.

The carrier is usually provided with integral hooks, pins or other features to engage with the main PCB in the handset, or with some features on the case. The first method allows the antenna to be securely located on the PCB while the whole assembly is placed in the case, the second requires that the antenna is placed in the case and the PCB then inserted. Electrical connections for the feed(s) and ground connection(s) are usually provided by some kind of compliant contacts which can be in the form of metal springs, sprung pins or compliant conductive gasket materials. In choosing a form of connection the main considerations will be the stability of contact over both long and short term, its reproducibility in production and the associated cost of materials and assembly. Designs suitable for automated assembly are usually preferred.

All these methods and forms of construction will be seen in practice. Additional important variants employ ceramic dielectric materials either for loading or exciting a conducting radiating element [23], or with integral conductors as in low-temperature co-fired ceramic antennas [24].

2.6.3 Recycling

Electronic equipment is the subject of world-wide regulations prohibiting the use of toxic materials and encouraging the practice of design for easy recycling. Designers must be aware of these requirements and able to demonstrate compliance with them [10, 11].

2.6.4 Building the Prototype

Early prototype antennas will generally be made by processes quite different from those intended for the finished production antenna. It is important to consider the possible impact of the technique to be used for later mass production and to approximate the final form of the antenna as closely as possible.

2.6.5 *Measurement*

2.6.5.1 Performance

Most of the measurements on the prototype will be performed by connecting the antenna to the test equipment by means of an external coaxial cable. The cable between the antenna tuning network and the point of exit from the handset should be of minimum diameter (usually around 1 mm). Its outer conductor is soldered at regular intervals (say, every 15 mm) to the groundplane. It should exit the handset close to the mid-point of the long edge of the groundplane, and the terminating connector (often type SMA) should project as little as possible. The reason for these important precautions is clear. We have already seen that the bandwidth available from an antenna is strongly dependent on the length of the chassis; if the connection between the handset and the measurement system will allow groundplane currents to flow along its outer conductor, the bandwidth will be enhanced by the presence of the cable. When the cable is removed the performance will be significantly changed – often for the worse. This effect is less significant at the high bands, where performance is less dependent on length.

It is usual to connect a quarter-wavelength sleeve choke (often referred to as a *balun*) directly to the connector on the handset (Figure 2.18) to prevent currents flowing in the outer conductor of the connecting cable [25]. Ferrite beads are lossy at the frequencies of the mobile radio bands, their use may create errors in efficiency measurements, so a choke is to be preferred. Positioning the point of connection at the center of the PCB places it at the point of minimum voltage (minimum impedance) so the projecting connector has less influence on the measurements than it would if the cable were positioned at the end. It is often convenient to place the feed cable on the reverse side of the PCB to avoid any local influences on the fields surrounding the antenna.

Input impedance measurements are usually made using a vector network analyzer (VNA). Two methods are available to allow accurate calibration to a reference plane at the input to the antenna matching circuit (P2 in Figure 2.18).

- The VNA is calibrated to the end of the balun (P1). It is then connected to the feed cable while the cable is still open-circuit at the antenna matching circuit P2 and the *electrical delay* parameter of the VNA is adjusted to coalesce the trace to a point at the open-circuit point on the Smith chart. The value of electrical delay is recorded so the system can be recalibrated later without having to detach the measurement cable.

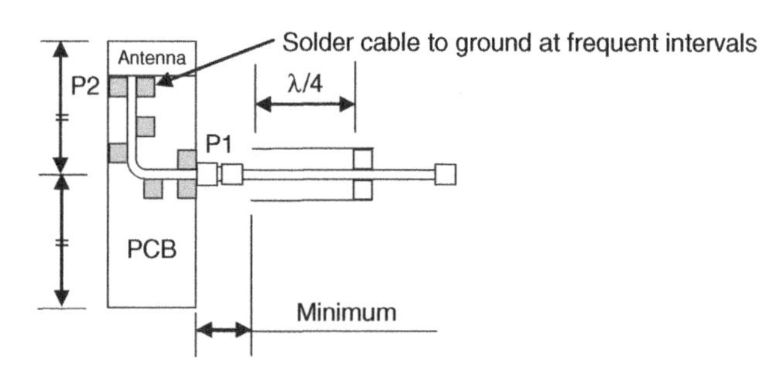

Figure 2.18 Handset with measurement cable and choke.

- Two identical feed cables are cut and terminated at one end. One is then fitted to the handset as shown, and the other is reserved for use as a calibrating cable.

The same internal cable and external balun are used for input impedance, radiation pattern and efficiency measurements. The handset case is modified to allow the cable to pass through it when it is refitted to the handset.

Note that it is wise to discharge a battery before putting it into a cabled-up handset. The operation of fitting the measurement cable may short-circuit some internal components and give rise to a risk of fire when the battery is refitted. (The alternative possible precaution of insulating the terminals may not produce a result typical of the correctly fitted battery.)

2.6.5.2 Impedance Measurements

Impedance is measured by calibrating the VNA to the measurement plane as described above, placing the handset on an insulating support such as a block of low-density polystyrene foam and connecting the measurement cable through the balun. *Always* measure the complex impedance (on a Smith chart) because it contains more information than the return loss or VSWR measurement (display both if you miss the more familiar format).

2.6.5.3 Radiation Patterns

Radiation pattern measurements are made in three dimensions in an anechoic environment. Several methods are available, and the choice between them is made on the basis of the speed of measurement, cost of personnel time and capital investment required. In order of increasing capital cost they include:

1. a single-axis far-field chamber, using simple fixtures to locate the handset as it is rotated in steps about the second axis (preferably the long axis of the chassis);
2. a two-axis far-field chamber (usually elevation-over-azimuth);
3. a spherical chamber with one axis through the handset and a second axis to rotate the illuminating horn around an orthogonal axis, operating in either far-field or near-field modes;
4. a chamber with a vertical axis through the handset, surrounded by a circle of fixed switched sampling antennas.

Radiation patterns are measured using two orthogonal polarizations, and it is normal to add together the powers measured in each polarization by RMS addition of the measured fields in each direction in space. The use of wide-band dual-polar illuminating antennas allows measurements to be made in all relevant frequency bands and in both polarizations during a single rotation of the principal measurement axis; this increases the productivity of the measurement facility.

2.6.5.4 Gain

Gain measurements are made using the same facility as the radiation pattern measurements, but accurate absolute calibration of the facility is needed, generally carried out by the use of a calibrated standard-gain antenna – usually a horn or a standard dipole. Great care must be taken in the design of cables and connections within the measurement system and careful attention paid to mechanical stability and the protection of vulnerable components from damage. The effects of temperature change on system calibration must be assessed; they may be reduced by careful system design or by limiting the extent to which the ambient temperature is able to vary.

2.6.5.5 Efficiency

Efficiency is measured by integrating the total power flux (or gain) measured over the whole spherical surface containing the device under test. The details of computation vary according to the distribution of the measurement points over the (virtual) measurement surface.

2.6.5.6 Specific Absorption Rate

SAR is measured by placing the handset under test next to a plastic phantom head filled with a sugar-saline solution with similar dielectric properties to brain tissue. A probe is moved inside the phantom and field levels are measured as its position is varied. Industrial-grade robotic control is used in high-accuracy systems capable of absolute measurements for certification purposes, while comparative tests can be made with less costly hand-operated equipment.

2.6.5.7 Hearing Aid Compatibility

This is evaluated by measuring the axial and radial magnetic fields in the vicinity of the user's ear [21].

2.6.6 Design Optimization

This begins with the adjustment of the antenna and matching circuit to achieve a low input VSWR over the working bands followed by measurement of the efficiency of the antenna in place on the handset. Efficiency is usually optimized by investigating and modifying the interactions between the antenna and other handset components and by reference to simulations to assist with an understanding of the fields and loss processes which may be occurring.

Optimization is not easily reducible to a simple procedural algorithm. It requires clear understanding of the possible mechanisms at play, the development of insight into the operation of the antenna and its interaction with the handset, experience, lots of patience and a certain amount of luck.

2.7 Starting Points for Design and Optimization

The design of an antenna for a particular handset is constrained by the available dimensions. These include any keep-out areas over components located under the antenna that might need access – for example a test port connector, or a loudspeaker sufficiently thick to make it unlikely that the antenna can extend over it. The design may begin with one of the familiar canonic antenna models but the geometry will be modified to fit the available space.

2.7.1 External Antennas

The design of an external antenna is relatively straightforward. Extensive references to dual-band external helical antennas are provided by Ying [15] and Haapala [26] (see Figure 2.19), and design procedures for printed spirals are provided by Huang [27]. An alternative but less common form of 3D branched monopole is proposed by Sun [28] (see Figure 2.20).

Figure 2.19 Non-uniform spiral antennas in cylindrical format [15] and concentric whip and spiral [26].

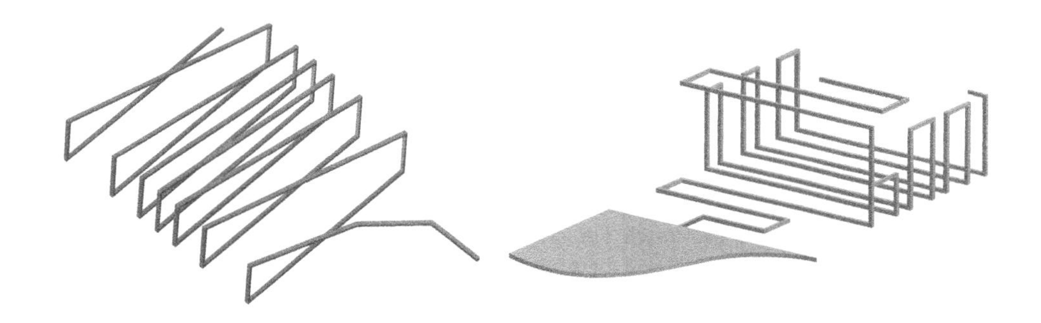

Figure 2.20 Non-uniform spiral in flattened format [27] and 3D branched monopole [28].

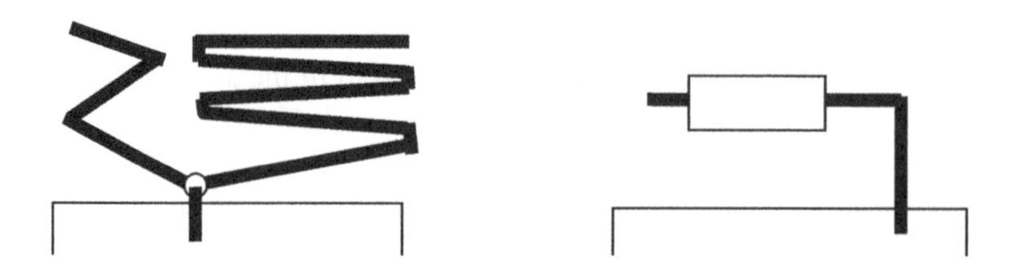

Figure 2.21 A 2D branched monopole [29] and a hybrid dielectric-loaded/DRA [30].

2.7.1.1 Off-Groundplane Antennas

These often take the form of branched monopoles [29] (Figure 2.21), one or both of which may be loaded with a dielectric pellet. The shape and dimensions of these are very variable and they can be modified to suit the available space. An alternative format comprises a single element which operates as a loaded monopole at one frequency and a dielectric resonator antenna (DRA) at a higher frequency [30].

2.7.1.2 On-Groundplane Antennas

The almost universal format for on-groundplane antennas is some form of PIFA. These exist in a wide variety of shapes and configurations [31]. The following paragraphs provide general descriptions of a number of subclasses of PIFA. For all of them the basic relationships between bandwidth, efficiency, antenna dimensions and chassis size apply. The more complex forms have been created in an effort to increase the number of bands covered and to squeeze the highest bandwidth and efficiency from a given geometry and environment. In most cases the capacitive top of the antenna can be meandered, folded or convoluted to reduce the maximum linear dimensions of the antenna – the exceptions to this being those designs which themselves seek to optimize the geometry of the radiating element.

Simple Single-Band PIFAs

Because of requirements for multi-band operation, PIFAs are now most frequently used as antennas for Bluetooth™, Zigbee and WLAN. To reduce their dimensions they are typically dielectrically loaded, often with meandered conductors, and are produced in the form of surface-mounted devices, using printed-circuit or LTCC techniques. Short-range protocols demand less antenna efficiency than is needed for mobile phone applications – the low power levels at which they operate allows some compensation for losses by increased power, so more severe compression of dimensions is often accepted.

Multi-Band PIFAs

Early designs [16] have been elaborated to increase the available bandwidth as the number of assigned mobile bands has increased. The simple two-pronged radiator is usually folded so the low-band radiator either encloses the high-band radiator or lies next to it (Figure 2.22). The choice between these configurations lies in the relative performance needed in the two band groups – the radiator with the open circuit end on the outer edge generally performs better, so Figure 2.22(a) has better high-band performance than Figure 2.22(b) which may be preferable when the chassis is short or the height is restricted. It is often possible to reverse

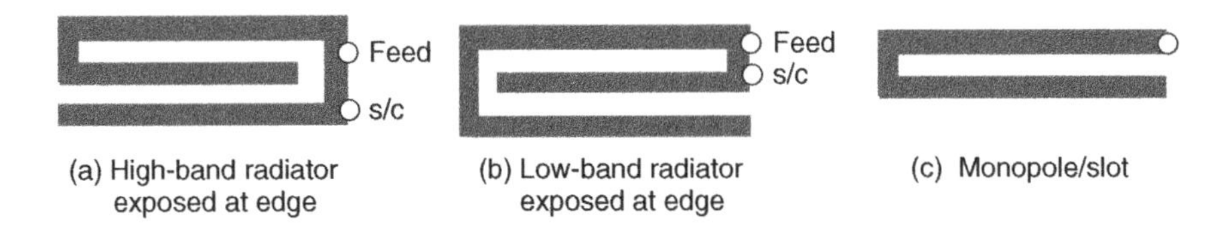

Figure 2.22 Typical multi-band configurations.

the positions of the feed and short-circuit pin, again causing some change in the relative performance in the two bands.

Extended PIFAs

It is common to use an external matching circuit to optimize the impedance characteristic of the antenna. The matching circuit is usually placed close to the feedpoint on the PCB; alternatively it can be placed on the antenna. Some configurations make use of matching components placed in series with the ground pin. Further variants use multiple ground pins, perhaps with matching components in one, or different components in each. Further variations are possible in which the antenna is provided with multiple feeds.

PIFAs with Parasitic Radiators

The impedance bandwidth if a PIFA can be modified by placing parasitic radiating elements alongside, above or below it.

Dielectric-Excited PIFAs

Kingsley and O'Keefe [23, 32] devised a class of PIFA antenna in which the radiating element is capacitively excited by means of a small ceramic pellet placed under the element, fed on a metallized face. This configuration provides a number of additional parameters for optimization and has provided some useful enhancements in efficiency over extended bandwidths (Figure 2.23). The radiator for this antenna can be bifurcated in the manner shown in Figure 2.22 for operation in both high- and low-bands, and this form of design has proved to be capable of providing high efficiencies concurrently over all five GSM/UMTS bands with a single feedpoint.

2.7.2 Balanced Antennas

In much of the discussion of handset antennas we have assumed that the antenna is unbalanced – a device with a single terminal, fed against ground. We have also seen that this configuration results in a strong interaction between antenna performance and the size of the groundplane, as well as allowing unwanted interactions with the user's body in the

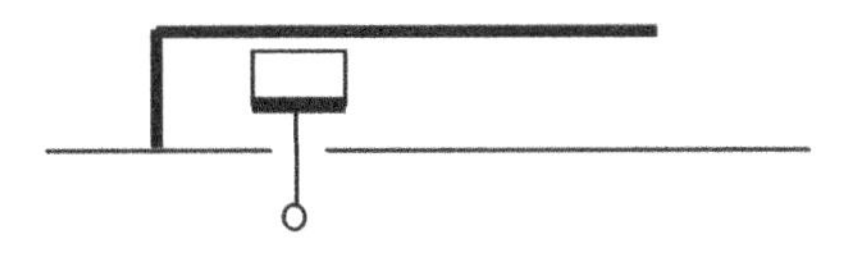

Figure 2.23 Dielectric-excited PIFA (patent applied for, Antenova Ltd, 2006).

form of detuning and the direct absorption of energy transferred to the user's hand by the groundplane currents.

Unfortunately in the low bands this is an inevitable state of affairs. The use of an electrically very small antenna (say 3.5 ml at a frequency where the wavelength is 300 mm) results in a very narrow bandwidth and great susceptibility to small losses. As we have seen, the more current we can induce onto the groundplane the more effectively the handset will operate. In the high bands this dependence is much less important. Measured in cubic wavelengths (the relevant dimensions) the antenna is at least eight times larger (2^3) so it is feasible to realize a small balanced antenna having the bandwidth needed.

The use of a balanced antenna has many benefits:

- The dimensions of the handset have almost no affect on antenna performance.
- There is almost no effect on antenna impedance when the user grasps the groundplane.
- There are no currents over most of the chassis, so hand/head loss is much reduced.
- A balanced antenna does not need to be placed on the end of the handset because its operation is independent of groundplane excitation.
- Multiple balanced antennas on a handset have less coupling than multiple unbalanced antennas because they do not excite a common groundplane mode.
- A balanced antenna can be directly interfaced to a balanced or differential amplifier.

These features create significant advantages, but, as we have seen, operation on the low bands continues to require unbalanced operation to realize bandwidth through the excitation of radiating currents in the groundplane. One solution is to use an antenna which operates in an unbalanced mode at lower frequencies and a balanced mode at higher frequencies – such an antenna [33] is shown in Figure 2.24. It can be integrated with a differential amplifier to realize additional benefits in terms of the reduced total complexity of the RF circuits of a handset.

2.7.3 *Antennas for Other Services*

It is increasingly common for handsets to incorporate services requiring additional antennas, including provision for GPS, Bluetooth™ or WLAN. The radio circuits for these services are usually contained in separate RF integrated circuits. The antennas are usually separate

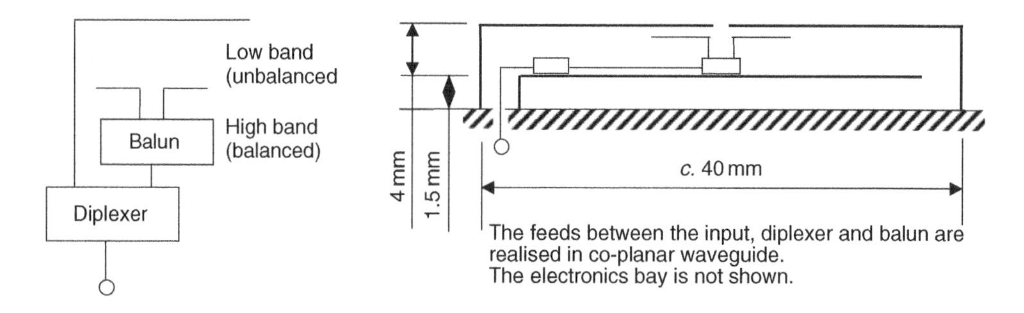

Figure 2.24 Antenna providing balanced operation at high band and unbalanced operation at low band [33].

from the main antenna for mobile phone services, and the design and layout of the handset must provide enough isolation to avoid unwanted interactions such as intermodulation and receiver blocking. The secondary antennas are usually of simple form and their efficiency is less critical than that of the main antenna. Typical formats include simple meandered monopoles and dielectric loaded or LTCC PIFAs and monopoles, usually located either close to the main antenna or part-way down the edge of the main PCB. The main challenge posed by these secondary antennas is usually the requirement for isolation combined with an inadequate volume for the entire antenna complement of the handset. System designers should not expect that more than 12–15 dB isolation can be obtained between different antennas sharing the positions discussed above, with more isolation available if the antennas can be separated or isolated by polarization. The isolation in a given environment can be investigated using simulations at an early stage in the design of the handset, but an exact reproduction of the simulated results cannot be expected in practice because of the limited accuracy of the model.

GPS signals are transmitted from satellites using circular polarization. The link budget for handheld devices used outside allows a linearly polarized antenna to be used (with a consequent 3 dB penalty relative to a circularly polarized antenna). Indoors the signal can be critically low, but multiple reflections result in random polarization at the receiver, so in this limiting case there is very little penalty in the use of a linearly polarized antenna [34].

2.7.4 Dual-Antenna Interference Cancellation

To satisfy the need for increased data throughput and higher reliability of the radio link to handsets, it will become increasingly common to equip handsets with a second receiver, especially for high-speed code-division multiple access (CDMA) and EDGE services. Providing two separate receiving antennas, typically at opposite ends of the handset, allows the implementation of two-branch diversity and/or the cancellation of interference by null steering and signal processing – known as *dual-antenna interference cancellation* (DAIC). An alternative system using one antenna and co-detection of wanted and unwanted signals, known as *single-antenna interference cancellation* (SAIC), can be applied to GSM-based systems but creates no new requirements to the antenna designer.

The antenna requirement for DAIC is relatively easy to satisfy because the additional antenna only needs to cover the receive band 2110–2170 MHz. At this frequency the handset is a significant fraction of a wavelength long and it is possible to provide sufficient isolation and decorrelation of the signals from the main antenna and the secondary receiving antenna. The reduced bandwidth for the secondary antenna will allow the use of small antenna formats such as those mentioned above for WLAN. What must be avoided is reducing antenna performance so far that the link-budget benefits of DAIC are lost because they are traded for smaller antennas.

2.7.5 Multiple Input, Multiple Output

A further step in the achievement of higher data rates is the adoption of MIMO in handsets. This can function remarkably well if the base station antenna has dual-polar antennas [35], but higher and more reliable performance can be provided by transmitting from antennas with a large physical separation which compensates for the small separation of the antennas on the handset [36]. The separation can be provided by transmitting from separate base station locations, so it may be easier to provide in a picocell/microcell environment than in larger cells. The advent of multiple antennas for mobile phone bands in laptop computers will create new possibilities in this area, and increased user expectation is then likely to create demand for enhanced services to handsets and PDAs.

2.7.6 Antennas for Lower-Frequency Bands – TV and Radio Services

The design of effective antennas for entertainment services is a highly significant challenge, given a device of the dimensions of a mobile handset or PDA. Conventional portable radio and TV sets had generally unreliable performance even when used in the primary coverage areas of standard broadcast stations – a situation which has continued even since the advent of digital services. Users are likely to wish for coverage in trains, cars and offices as well as outdoors where signal coverage is much easier to provide. The technical challenges of achieving sufficient antenna performance will dominate the range and quality of the available services and for this reason will critically impact the economics of service provision.

Services that should be considered for inclusion include the following:

- AM radio in the MF and HF bands (550–1605 kHz and 3–30 MHz). These are currently analog (AM, with some use of stereo), but are progressively moving to digital signal formats.
- VHF radio (88–108 MHz in most countries) is moving to digital signal formats in the same frequency band or in the DAB band 174–230 MHz.
- TV broadcasts are transitioning to digital formats, mostly in the band 470 – 860 MHz, but for mobile services this may be restricted to 470–750 MHz to reduce front-end filtering problems associated with coexistence with low-band mobile radio services.
- A number of countries are establishing radio and TV services in other frequency bands, in particular in the L-band.
- New datacasting services which may be added to the resources offered by the existing media on the frequencies listed above, or may appear on new systems in new bands.

Work carried out on portable digital television receivers has shown the benefit of polarization diversity for this application [37]. It can be expected that diversity will be even more desirable on mobile devices, where the user will be unable to place the device in the best part of the room, or to orient its antenna in the optimum manner for signal reception. A successful mobile device will need to make optimum use of any signal available whatever its orientation or polarization.

An antenna can be situated in a number of locations; the following possibilities are listed in order of probable gain (or effective height) beginning with the lowest:

1. *Internal antenna* within the equipment housing, where it is subject to the dimensional constraints of a housing. At lower frequencies a ferrite antenna works well; at frequencies below at least the middle of the HF band the environment is electrically very noisy so even a small antenna is externally noise limited.

   For operation in the VHF and UHF bands the antenna may take the form of a very compressed T or a PIFA; in both cases these can be tuned by a mixture of capacitive loading of the antenna and an adjustable tuning network in the feed line. The VHF or UHF antenna will project from the end of the chassis, so if mobile phone functionality is also needed it may be desirable to make sure that there is an effective low impedance path from the chassis to this antenna at the mobile phone low bands so the VHF/UHF antenna is used as part of the low-band groundplane – this is probably most easily arranged if it is a PIFA. Given the narrow operating bandwidth of the antenna, auto-tuning may be effective in compensating the detuning effects of the user's body. With some degree of intelligence an auto-tune system could learn to associate specific multiplex channel frequencies with the user's location; there is also no reason why channel-searching and learning of tuning states cannot take place off-line when the device is first turned on in an unknown location as indicated by the network ID, the cell ID or a GPS position fix.

   Internal antennas will be electrically extremely small and will have a very high Q-factor and a very small radiation resistance with either a very high capacitance in series or very low inductance in parallel. To provide the maximum signal-to-noise ratio from the input stage of the receiver the antenna should be noise-matched to the input impedance of the amplifier. The high Q of the antenna means that this can be achieved over only a very narrow frequency band unless some form of tuning is provided – either the user must be provided with a tuning knob (not very acceptable from an image point of view and probably not practicable), or some form of automatic tuning is needed. Excluding the local transmit signals requires that a filter is provided between the antenna and the amplifier. This results in a schematic such as Figure 2.25.
2. *External antenna* designed to function as a stand or other functional external part of the handset; this could be a monopole or a loop.

   An external antenna is obviously more acceptable if it appears to the user to have some other useful function. Care must be taken to make sure the antenna is conveniently

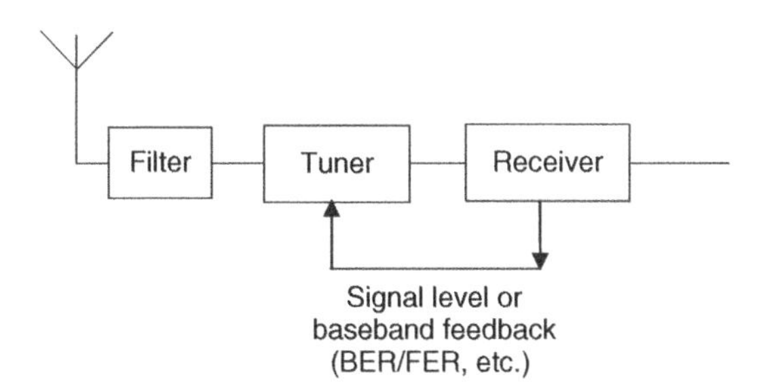

Figure 2.25 Block diagram of the tuning arrangement for an internal antenna for TV reception.

deployed in the intended manner in any mode of use of the device. Antennas doing duty as prop-stands may be appropriate when the device is used on a table top, but would be inconvenient for laptop use while watching the news on a train.

Although external antennas are less constrained in dimensions than internal antennas, the grade of service will be enhanced by the addition of self-tuning functionality as described for internal antennas.

3. *An external whip antenna* – usually a telescopic pull-out or fold-up model – is often used on portable radio sets. Whips have been used on some early mobile TV terminals, but they are not liked by users and would not be popular in a crowded train. This suggests that devices with external whips may need some other form of antenna for use if it is not convenient to deploy them – the other antenna may function as a diversity antenna when both are in use.
4. *The cable connecting the earpiece/headphones* has been used as an antenna on small radio sets and some handsets with radio functionality. The usual arrangement is unbalanced and suffers from the fact that the long headset has only a small counterpoise (the ground plane of the radio) and in consequence performance drops sharply unless the radio is held in the user's hand or placed on a metal surface. By designing a special-purpose antenna/headset it is possible to obtain much better performance as an antenna, and even to provide some measure of polarization diversity [38]. This improved functionality in headset mode probably matches operational requirements quite well. A user may be content to use a whip or internal antenna while using loudspeaker mode at home, but would probably wear headphones when traveling by train with others; this matches the radio requirement in which static operation may allow some choice of location and antenna position, while a high-velocity user has no choice of location or orientation, and Doppler shift may result in an unacceptable error rate unless more signal is available.

The quality and reliability of reception is improved by adding a second receiver channel. Diversity combining methods can be chosen to suit cost and performance requirements, using for example simple switched diversity at the receiver input, or much more complex but better-performing maximal ratio combining. Diversity systems using dual receivers are not only more expensive but consume more power than single-receiver solutions. The payback comes in the stability of reception in difficult environments such as fast-moving vehicles.

2.8 The RF Performance of Typical Handsets

The standard of performance available from handsets is of great importance to network planners and operators. The following figures provide some indication of what can be expected of a well-optimized antenna implemented in a successful handset. Figure 2.26 shows the measured free-space efficiency of a penta-band antenna installed in a well-designed commercial handset. At the other end of the performance scale, the measured free-space efficiency of some handsets falls below 15%, at some frequencies. Typical specifications require mean handset efficiency of around 50%, with a minimum in any band somewhere above 40%.

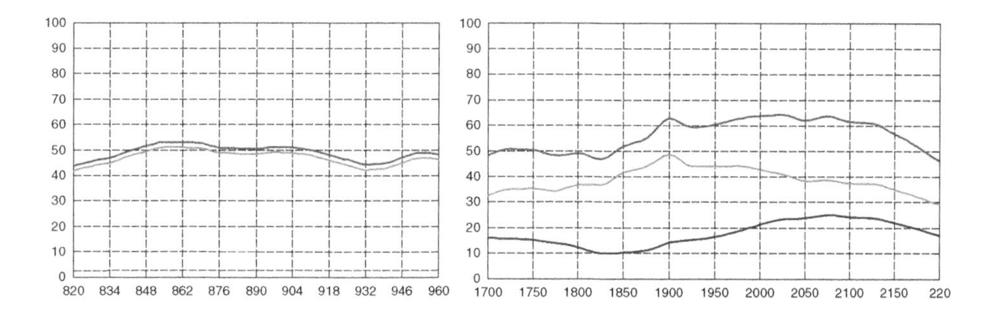

Figure 2.26 Measured performance efficiency of a penta-band antenna installed in a commercial handset (Antenova Ltd).

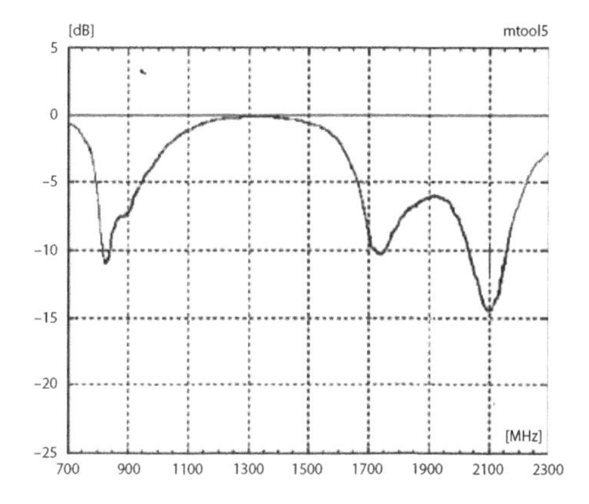

Figure 2.27 Input return loss of the penta-band antenna in Figure 2.26.

The input return loss of the antenna in the same handset is shown in Figure 2.27. The result is typical of many antennas; the choice of the components of the antenna matching network has been made on the basis of optimizing the efficiency of the antenna taken together with the matching circuit; the result of the losses in typical matching components is often, as in this case, that the component values chosen to provide optimum total efficiency do not necessarily provide the optimum input VSWR.

Typical handset radiation patterns in both low and high bands are shown in Figure 2.28. These measured patterns are of almost exactly the same form as the simulated patterns shown in Figures 2.11 and Figure 2.12, confirming the simulated field distributions.

The SAR distribution shown in Figures 2.29 and 2.30 has been simulated using a high-resolution head model and gives a good impression of the way in which SAR is distributed in the neighborhood of the handset.

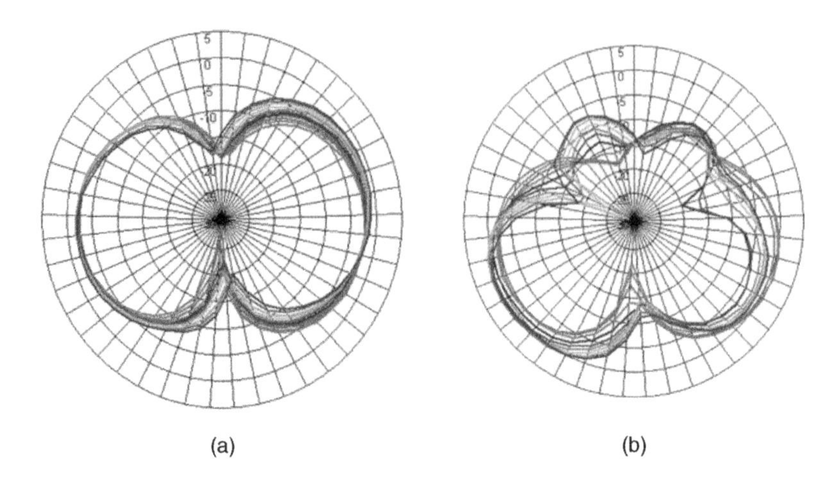

Figure 2.28 Typical radiation patterns for a handset antenna plotted at a number of frequencies across (a) 824–960 MHz and (b) 1710–1990 MHz. These are very similar to the simulations in Figures 2.11 and Figure 2.12.

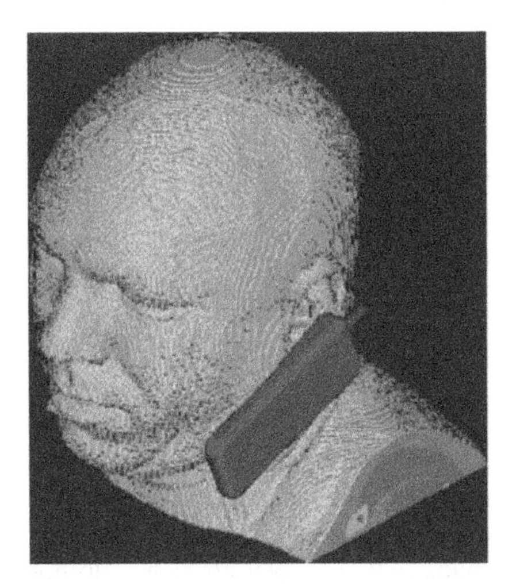

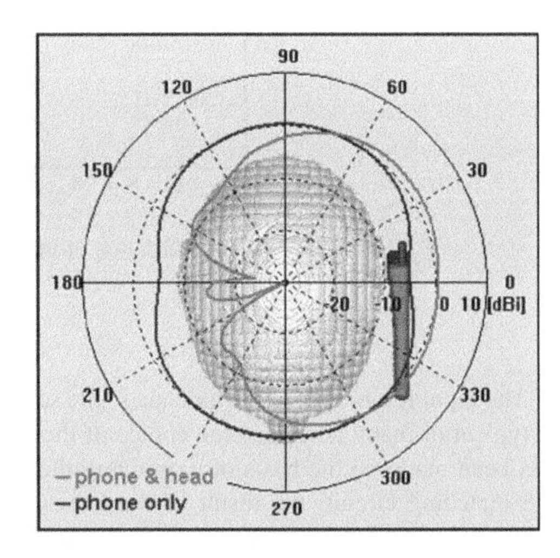

Figure 2.29 A simulation showing the huge effect of the user's head on the radiation pattern of a handset. In the high bands (as here) much of the energy impinging on the head is reflected, so the gain of the antenna rises in the direction away from the head. At 900 MHz much of the energy is absorbed. The azimuth pattern with no head differs from that of Figure 2.12(b) because of the oblique angle of the handset. (Reproduced with permission of CST GmbH).

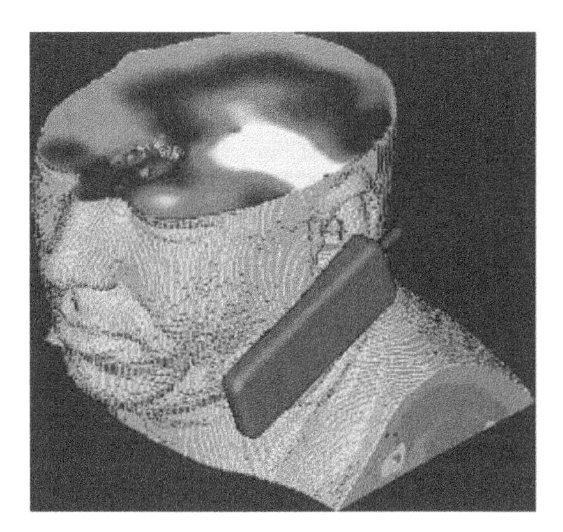
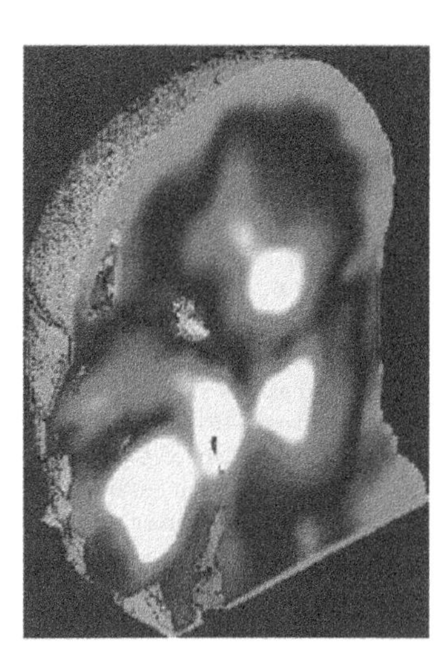

Figure 2.30 Simulations showing the complexity of typical SAR distributions. Even with an external antenna, the vertical section through the middle of the head (right) shows the characteristic triple hotspots (shown as light areas) along the axis of the handset seen in Figure 2.12(a). (Reproduced by permission of CST GmbH).

2.9 Conclusion

The art and science of the handset antenna are still in a state of rapid change. Market demands for new services, higher data rates, and additional functionality in a small, light and inexpensive piece of electronics place very onerous demands on the design of every part of the handset. The performance of the antenna is crucial to the communications function of the handset and its efficiency influences both battery life and the quality of the user's experience.

The development of new services and radio techniques is likely to increase the number of antennas fitted to hands. Optimization of their design will demand an increasingly integrated approach to the design of the handset.

References

[1] *Method of Measurement for Radiated RF Power and Receiver Performance*, Cellular Telecommunications & Internet Association, CTIA Certification Department, Washington, DC.

[2] *Evaluating Compliance with FCC Guidelines for Human Exposure to Radiofrequency Electromagnetic Fields*, Federal Communications Commission, *OET Bulletin* 65, ed. 97-01; supp. C, ed. 01-01. Washington, DC: Office of Engineering and Technology, FCC, 2001.

[3] International Commission on Non-Ionising Radiation Protection, Guidelines for limiting exposure in time-varying electric, magnetic, and electromagnetic fields (up to 300 GHz). *Health Physics*, 74 (1998), 494–522.

[4] *Product Standard to Demonstrate the Compliance of Mobile Phones with the Basic Restrictions Related to Human Exposure to Electromagnetic Fields (300 MHz–3 GHz)*, EN 50360:2001, CENELEC, Brussels, 2001.

[5] *Basic Standard for the Measurement of Specific Absorption Rate Related to Human Exposure to Electromagnetic Fields from Mobile Phones (300 MHz to 3 GHz)*, EN 50361:2001, CENELEC, Brussels, 2001.

[6] *Radiocommunications (Electromagnetic Radiation – Human Exposure) Standard 2003*, Australian Communications Authority, Melbourne, March 2003.

[7] *ARPANSA Radiation Protection Standard No. 3: Maximum Exposure Levels to Radio-Frequency Fields – 3kHz to 300GHz*, Australian Radiation Protection and Nuclear Safety Agency, Sydney, 2003.

[8] *Specific Absorption Rate Test Method Using Phantom Model of Human Head*, ACA EMR Standard Schedule 1, Australian Communications Authority, Melbourne, 2001.

[9] *Measurement Method for Devices 20 cm or Less from the Human Body: Information for Documenting Compliance*, ACA EMR Standard Schedule 2, Australian Communications Authority, Melbourne, 2003.

[10] *Recommended Practice for Determining the Peak Spatial-Average Specific Absorption Rate (SAR) in the Human Head from Wireless Communications Devices: Measurement Techniques*, IEEE 1528:2003, Institute of Electrical and Electronics Engineers, New York, 2003.

[11] *Human Exposure to Radio Frequency Fields from Hand-Held and Body-Mounted Wireless Communication Devices – Human Models, Instrumentation, and Procedures – Part 1: Procedure to Determine the Specific Absorption Rate (SAR) for Hand-Held Devices Used in Close Proximity to the Ear (Frequency Range of 300 MHz to 3 GHz)*, IEC 62209-1, International Electrotechnical Commission, Geneva, 2005.

[12] L.J. Chu, Physical limitations of omnidirectional antennas. *Journal of Applied Physics*, 19 (1948), 1163–1175.

[13] G.A. Thiele, P.L. Detweiller, and R.P. Penno, On the lower bound of the radiation Q for electrically small antennas. IEEE *Transactions on Antennas and Propagation*, 51 (2003), 1263–1269.

[14] P. Vainikainen, J. Ollikainen, O. Kivekäs, and I. Kelander, Resonator-based analysis of the combination of mobile handset antenna and chassis, *IEEE Transactions on Antennas and Propagation*, 50 (2002), 1433–1444.

[15] Z. Ying, Ericsson, 1996, US Patent 6212102 (WO9815028).

[16] Z. Liu, and P.S. Hall, Dual-band antenna for hand held portable telephones, *Electronic Letters*, 32 (1996), 609–610.

[17] D. Cairns, T. Fulghum, and R. Bexten, Experimental evaluation of interference cancellation for dual-antenna UMTS handset. *IEEE 62nd VTC Fall 2005*, Vol. 2, pp 877–881.

[18] M.A. Jensen and J.W. Wallace, A review of antennas and propagation for MIMO wireless communications. *IEEE Transactions on Antennas and Propagation*, 52 (2004), 2810–2824.

[19] B.S. Collins, Improving the performance of clamshell handsets. *IEEE International Workshop on Antenna Technology*, IWAT 2006, White Plains NY, March 2006.

[20] Mobile telephone handset with capacitive radio frequency path between first and second conductive components thereof, Patent Application WO 2005/112405 (Antenova Ltd).

[21] ANSI, *Methods of Measurement of Compatibility between Wireless Communications Devices and Hearing Aids* (ANSI C63.19–2005).

[22] HAC Report and Order (FCC 03-168), 2003.

[23] S.P. Kingsley *et al.*, A hybrid ceramic quadband antenna for handset applications. *6th IEEE Circuits & Systems Symposium on Emerging Technologies*, Shanghai, May 31 – June 2, 2004, pp 773–774.

[24] S. Holzwarth, J. Kassner, R. Kulke and D. Heberling, Planar antenna arrays on LTCC-multilayer technology. *IEE ICAP 2001*, Vol. 2, pp 710 – 714.

[25] B.S. Collins and S.A. Saario, The use of baluns for measurements on antennas mounted on small groundplanes. *IEEE International Workshop on Antenna Technology*, IWAT 05, Singapore, Jun 2005, pp 266–269.

[26] P. Haapala, P. Vainikainen and P. Eratuuli, Dual frequency helical antennas for handsets, 1996. *IEEE 46th VTC May 1996*, Vol. 1, pp 336–338.

[27] T-F Huang, Methodology of external dual-band printed helix design. *IEEE APS International Symposium, 2005*, Vol. 1A, pp 458–461.

[28] B. Sun, Q. Liu and H. Xie, Compact monopole antenna for GSM-DCS operation of mobile handsets. *Electronic Letters*, 39 (2003), 1562–1563.

[29] D. Liu, A multi-branch monopole antenna for dual-band cellular applications. *IEEE APS 1999*, Vol. 3, pp 1578–1581.

[30] Z. Wang *et al*, Optimisation of a broadband dielectric antenna. *LAPC*, Loughborough University, April 4–6, 2005, Paper 41.

[31] K.-L. Wong, *Planar Antennas for Wireless Communications*. Hoboken, NJ: Wiley Interscience, 2003.

[32] Hybrid antenna using parasitic excitation of conducting antennas by dielectric antennas. Patent application WO 2004/114462 (Antenova Ltd).

[33] B.S. Collins *et al.*, A multi-band hybrid balanced antenna. *IEEE International Workshop on Antenna Technology*, IWAT 2006, White Plains NY, March 2006.

[34] V. Pathak *et al.*, Mobile handset system performance comparison of a linearly polarized GPS internal antenna with a circularly polarized antenna. *IEEE APS 2003*, Vol. 3, pp 666–669.

[35] B.S. Collins, Polarization diversity antennas for compact base stations. *Microwave Journal*, 43 (2000), 76–88.

[36] I. Sarris, A. Doufexi, and A,R. Nix, High-performance antenna array architectures for line-of-sight MIMO communications. *Proceedings of the Loughborough Antennas & Propagation Conference*, LAPC'06, pp 20–24.
[37] Y. Lévy, DVB-T – a fresh look at single and diversity receivers for mobile and portable reception, *EBU Techinical Review*, No. 298, European Broadcasting Union, Geneva, Apr. 2004.
[38] UK patent applied for, Antenova Ltd.

3

RFID Tag Antennas

Xianming Qing and Zhi Ning Chen
Institute for Infocomm Research, Singapore

3.1 Introduction

Radio frequency identification (RFID), which was developed around World War II, provides wireless identification and tracking capability and is a technology, more robust than the bar code [1]. The purpose of an RFID system is to enable data to be transmitted by a mobile device, called a tag, which is read by an RFID reader and processed according to the needs of a particular application. The data transmitted by the tag may be used to provide identification or location information, or specifics about the product tagged, such as price, color, and date of purchase. RFID technology was first used in tracking and access applications during the 1980s. RFID has been popular because of its ability to track moving objects and its low-cost implementation. As the technology is refined, more pervasive and possibly invasive uses for RFID tags are likely to appear.

In a typical passive RFID system, each individual object is equipped with a small and inexpensive tag (transponder) which comprises an antenna and an application-specific integrated circuit (ASIC, or microchip) that is given a unique electronic product code [2]. The reader (interrogator) emits a signal to activate the tag, which passes through the electromagnetic zone generated by a reader antenna and decodes the data encoded in the tag's microchip. The data are then passed to a host computer for processing. The querying signal coming from the reader must have enough power to activate the tag microchip, perform data processing, and transmit back a modulated string over the required reading distance. Since the maximum output power of the reader is constrained to local regulated effective isotropically radiated power (EIRP), the reading distance achieved is dependent on the performance of the tag, that is, the microchip selected and the characteristics of the antenna.

This chapter introduces the methodology of designing and evaluating RFID tag antennas from a system performance point of view. RFID fundamentals such as system configuration, classification, regulation, and standardization are first introduced. Furthermore, tag antenna design issues will be addressed. Since the frequency available for RFID applications varies

Antennas for Portable Devices Zhi Ning Chen

from very low (below 135 kHz) to microwave (up to 24.125 GHz), antenna designs will vary with specific applications. In RFID applications with an operating frequency lower than 100 MHz, coils have been used for power transfer and information communications. The main design consideration is to configure coil antennas to achieve a specified inductance for circuit resonance and adequate quality within a limited area. In systems with operating frequency higher than 100 MHz, a variety of antennas such as the dipole antenna, meander line antenna, patch antenna, and slot antenna can be used. The important considerations are to achieve the required radiation characteristics and impedance matching between the antenna and the microchip. Practical designs are provided to demonstrate the method of designing a tag antenna with a prior selected microchip. In Section 3.4 a critical issue, namely the effect of the environment on RFID tags, will be discussed because the tag is often attached to a specific object in practical applications. The performance of the RFID tag is strongly dependent on the properties of the attached object. The effect of materials such as metal and water on the tag antenna characteristics and tag performance will be investigated and discussed.

3.2 RFID Fundamentals

3.2.1 RFID System Configuration

A typical RFID system, as shown in Figure 3.1, is composed of an interrogator or reader, a reader antenna, a tag (transponder), a host computer, and a software system. The tag is capable of storing a significant amount of information (several kilobytes) and can be affixed to a wide variety of items for the purpose of identification and tracking. Each tag has a manufacturer-installed unique identification code as well as an additional available memory. The reader is capable of storing and transferring to a host computer system a wide variety of item information, which may include stock number, current location, and status. The reader emits a signal at a certain frequency. When an RFID tag passes within the reading range of the reader, the tag is detected and interrogated for its content information. This process is conducted by field coupling or electromagnetic wave capture via the antenna.

An RFID *reader*, also called an interrogator, is a device that can read from and write data to compatible RFID tags. Thus, a reader also doubles up as a writer. The act of writing the tag data by a reader is called creating a tag. The process of creating a tag and uniquely associating it with an object is called commissioning the tag. Similarly, decommissioning a tag means to disassociate the tag from a tagged object and optionally destroying it. The time during which a reader emits RF energy to read tags is called the duty cycle, and this is internationally regulated. The reader is the central nervous system of the entire RFID hardware system – establishing communication with and control of the reader are the most important tasks for the other parts which seek integration with it. A reader comprises the following main devices /components:

- transmitter
- receiver
- microprocessor
- memory
- input/output channels for external sensors, actuators

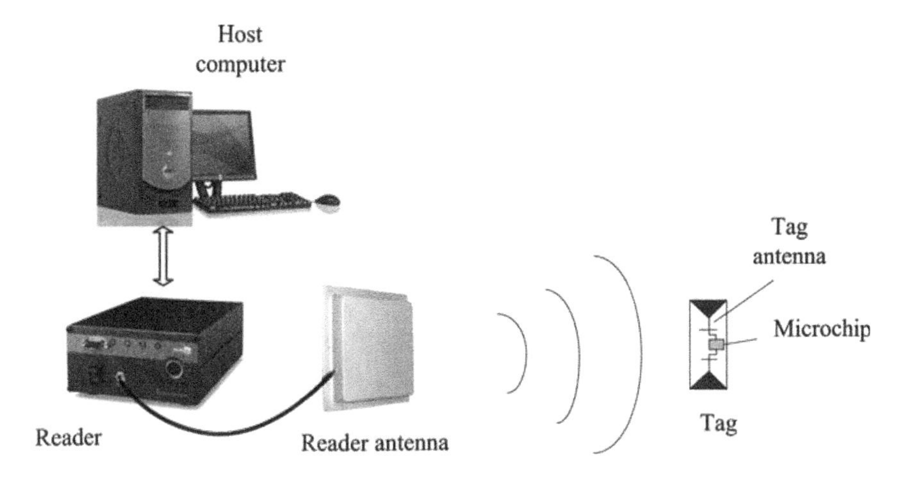

Figure 3.1 Configuration of a typical RFID system.

- controller (which may reside as an external component)
- communication interface
- power supply.

A reader communicates with a tag through the *reader antenna*. One reader can support more than one antenna by using a multiplexer. The antenna is also called the reader's coupling element because it creates an electromagnetic field to couple with the tag at low frequencies. An antenna broadcasts the RF signals from a reader transmitter to its surroundings and receives the response from the tags. Therefore, the proper positioning of the reader antenna is essential for high reading accuracy, although the reader has to be located close to the antenna to minimize the loss due to the RF cable. In addition, portable readers might have built-in antennas, where the antenna will be positioned with the reader simultaneously.

An RFID *tag*, or transponder, is a device that can store and transmit data to a reader in a contactless manner through radio waves. Some tags, known as 'chipless', do not use an integrated circuit (IC) chip. Most tags categorized as 'chip' tags comprise two main components: a small silicon microchip which contains a unique identification (ID) number, and an antenna that can send and receive radio waves. These two components can be tiny. The antenna consists of a flat metallic conductive track (see Figure 3.2), and the microchip is potentially less than half a millimeter in size. These two components are usually attached to a flat plastic tag that can be fixed to a physical item. These tags can be quite small and thin, and are, easily embedded into packaging, plastic cards, tickets, clothing labels, pallets, and books.

The *host computer system* is an all-encompassing term for the hardware and software components that are separate from RFID hardware such as the reader, reader antenna, and tag. The host computer system mainly comprises the following components:

- edge interface/system
- middleware
- enterprise back-end interface
- enterprise back end.

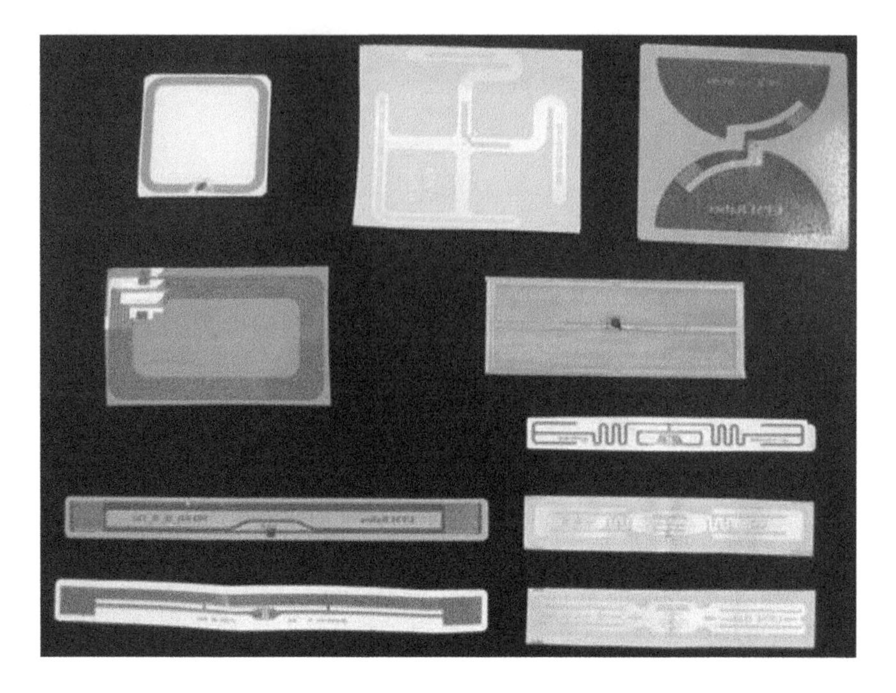

Figure 3.2 Various RFID tag configurations.

3.2.2 Classification of RFID Systems

Over the several stages of RFID system development, many types of RFID systems have emerged. The systems distinguish from each other by system usage, operating frequency, reading distance, protocol, power supplied to the tag, and the procedure for sending data from the tag to the reader, and soon [3]. From an antenna design point of view, RFID systems can be categorized as 'near-field' RFID and 'far-field' RFID in terms of the method of transmitting power from the reader to the tag. The other classification is based on the method of powering up the tags: RFID systems can be classified as 'passive', 'active' and 'semi-active'.

3.2.2.1 Near-field and Far-field RFID

There are two different approaches to transfer the power from an RFID reader to an RFID tag, namely inductive/capacitive coupling and electromagnetic (EM) wave capturing. These two approaches exploit the EM properties associated with an RF antenna – the near-zone and the far-zone fields. Both approaches can transfer enough power to a remote tag to sustain its operation–typically between 10 μW to 1 mW, depending on the tag type.

Near-field RFID

In near-field RFID systems, the power transfer from a reader to a tag as well as the communication between the reader and the tag can be achieved by inductive coupling through the interaction with magnetic fields, or by capacitive coupling through the interaction with electric fields. A near-field RFID system uses the most straightforward approach to implement a passive RFID system. This is the reason why it was the first approach taken and has resulted in many subsequent standards, such as ISO 15693 and 14443, and a variety of proprietary solutions.

The main limitation of the near-field RFID system is its limited reading distance. For inductive coupling RFID systems, the energy available for induction is a function of distance from the coil antenna. The magnetic field decreases at a rate of $1/r^3$, where r is the separation between the tag and reader coil antenna along a center line perpendicular to the coil's plane [2, 4]. The reading distance of a conventionally designed near-field RFID system is typically less than 1.5 m [5]. The further limitation is related to the orientation of the magnetic field. Along the boresight of the reader coil antenna the intensity of the magnetic field component perpendicular to the coil antenna plane is strong, whereas the magnetic field component parallel to the coil antenna plane is very weak or even zero. Therefore, if the tag is positioned parallel to the magnetic field from the reader coil antenna, the tag cannot be detected by the reader because no magnetic flux goes through the tag.

Far-field RFID

In far-field RFID systems, the power transfer from reader to tag as well as the communication between reader and tag are achieved by transmitting and receiving EM waves. The reader emits energy through a reader antenna, and some of the energy is then reflected from the tag and detected by the reader. The amplitude of the reflected energy from the tag can be influenced by altering the load connected to the tag antenna. By changing the antenna's load over time, the tag can reflect back more or less of the incoming signal in a pattern that encodes the tag's ID.

Far-field RFID systems operate at frequencies greater than 100 MHz, typically in an ultra high-frequency (UHF) band such as 868 MHz, 915 MHz or 955 MHz or a microwave frequency band such as 2.4 GHz or 5.8 GHz. The reading distance of such systems is determined by the intensity of energy received by the tag and the sensitivity of the reader's receiver to the reflected signals from the tag. The energy required to power a tag at a given frequency continues to decrease (currently as low as a few microwatts). The readers have been developed with improved sensitivity so that they can detect weak signals with power levels of the order of –80 dBm at a reasonable cost. A typical far-field reader can successfully interrogate tags 3–5 m away. The maximum reading distance of some specific far-field RFID systems can be up to 10 m or more.

3.2.2.2 Active, Semi-active and Passive RFID

According to the method of powering up the RFID tag, RFID systems can be categorized into three groups: active, semi-active, and passive.

Active RFID

In active RFID systems, tags have an on-board power source (for example, a battery) and electronics for performing specialized tasks. An active tag uses its on-board power supply to support microchip operation and transmit data to a reader. It does not need the power emitted from the reader for data transmission. The on-board electronics contain microprocessors, sensors, and input/output ports, and so on. In an active RFID system, the tag always communicates first, followed by the reader. As the presence of a reader is unnecessary for the data transmission, an active tag can broadcast its data to surroundings even in the absence of a reader. This type of active tag, which continuously transmits data with or without the presence of a reader, is also called a *transmitter*. Another type of active tag enters into a sleep or low-power state in the absence of interrogation by a reader. A reader wakes the tag from its sleep state by issuing an appropriate command. The ability to enter into a sleep state conserves battery power, and therefore this type of tag generally has a longer battery life than an active transmitter tag. In addition, the amount of induced RF noise in its environment is reduced because the tag transmits only when interrogated. This type of active tag is called a *transmitter/receiver*. The reading distance of an active RFID system can be 30 m or more.

An active tag mainly consists of the following components:

- microchip
- antenna
- on-board power supply
- on-board electronics.

Semi-active RFID

The tag in the semi-active RFID system has an on-board power source (for example, a battery) and electronics for performing specialized tasks. The on-board power supply provides the tag with energy for its operation. A semi-active tag is also called a battery-assisted tag. However, a semi-active tag uses the power emitted from the reader to transmit its data. In tag-to-reader communication for this type of tag, the reader always communicates first, followed by the tag. The advantage of using a semi-active tag over a passive tag is that, unlike a passive tag, a semi-active tag does not use the reader's signal to excite itself. It can be read over a longer distance than a passive tag. Since no time is required to energize a semi-active tag, such a tag could be in the reading zone of a reader for substantially less time for proper reading (unlike a passive tag). Therefore, even if the tagged object is moving at a high speed, its tag data can still be read if a semi-active tag is used. Finally, a semi-active tag might offer good readability for the tagging of RF-opaque and RF-absorbent materials. The presence of these materials might prevent a passive tag from being properly excited, resulting in the failure to transmit its data. However, this is not an issue with a semi-active tag. The reading distance of a semi-active RFID system can reach up to about 30 m.

Passive RFID

In passive RFID systems, the RFID tag has no on-board power source, and instead uses the power emitted from the reader to energize itself and transmit its stored data to the reader. Compared to active or semi-active tags, passive tags are consequently simpler in structure, lighter in weight, less expensive, more generally resistant to harsh environmental conditions, and offer a virtually unlimited operational lifetime. The trade-off is that passive tags have shorter reading distances than active tags and require higher-power readers. They are also constrained in their capacity to store data and their ability to perform well in electromagnetically noisy environments. The reading distance of passive RFID systems can be up to 10 m.

In tag-to-reader communication for this type of tag, the reader always communicates first, followed by the tag. The presence of a reader is mandatory for such a tag to transmit its data. A passive tag mainly consists of a microchip and an antenna.

3.2.3 Principles of Operation

As mentioned in Section 3.2.2.1, the energy coupling mechanism between the reader and the tag in near-field and far-field RFID systems is quite different due to the different EM field behaviors in each zone. In this section, the principles of power transfer from a reader to a tag and the procedure for communication between the tags and the reader, in both the near-field and far-field RFID systems, will be discussed.

3.2.3.1 Near-field Coupling

Electromagnetic field in the near zone is reactive and quasi-static in nature. Electric field is decoupled from magnetic fields, and which one will dominate determined by the type of antenna employed: the electric field dominates when a dipole antenna is used, whereas the magnetic field dominates in the case of a small-loop antenna. The coupling between tag and reader antennas may be achieved through interaction with the electric or magnetic fields. In near-field RFID systems, inductive coupling systems are much more widely available than capacitive coupling systems.

Inductive Coupling
In an inductive coupling RFID system (Figure 3.3) the reader coil generates strong magnetic field which penetrates the coil of the tag. The coils in both the reader and tag function as antennas. The tag usually comprises a microchip and a coil antenna. When a small part of the emitted field from the reader coil antenna penetrates the coil antenna of the tag, a voltage U_i is generated in the tag coil antenna by induction. This voltage is rectified and serves as the power supply for the microchip. A capacitor C_r is connected in parallel with the reader coil antenna, the capacitance being selected such that it works with the inductance of the coil antenna to form a parallel resonant circuit with a resonant frequency that corresponds to the transmission frequency of the reader. Very large currents will be generated in the coil antenna of the reader by resonance set-up in the parallel circuit, which can be used to generate the magnetic field for the operation of the remote tag.

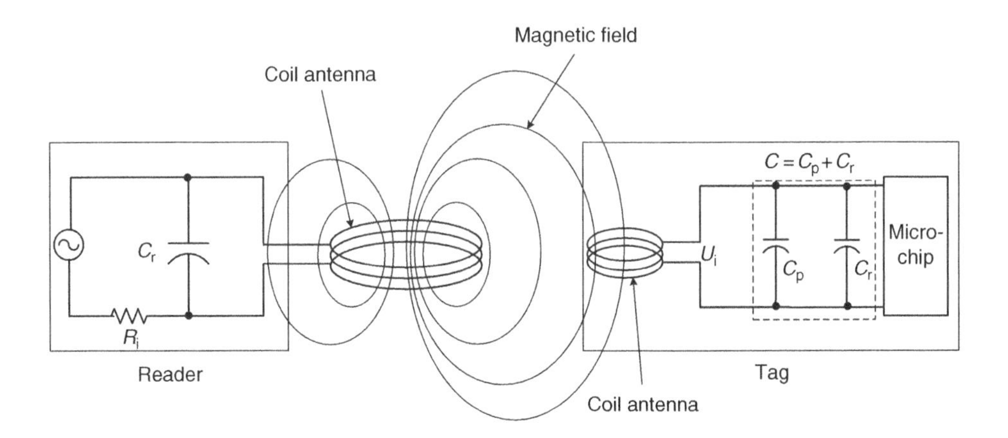

Figure 3.3 Power transfer and communication between reader and tag in an inductive coupling RFID system.

The efficiency of the power transfer between the coil antenna of the reader and tag is proportional to the operating frequency, the number of windings, the area enclosed by the coil antenna, the angle of the two coils relative to each other, and the distance between the two coils.

Capacitive Coupling

In capacitive coupling RFID systems, the antennas create and interact with quasi-static electric field. In these systems, it is the distribution of the charges rather than currents that determines the field strength and hence the coupling strength. As the coupling strength is dependent on the amount of accumulated charges, systems based on capacitive coupling are much less used than inductive coupling systems [6].

A dipole is a suitable antenna for capacitive coupling systems since the electric field dominates the magnetic field. Just as inductive coupling systems require resonant circuits for maximum coupling, so do capacitive coupling systems. Since the antenna has its own capacitance, inductance is added in parallel to the tag and in series to the reader. In addition, as in inductive coupling systems, the tag can communicate with the reader by varying its impedance.

Load Modulation

Tag-to-reader communication in inductive and capacitive coupling RFID systems is realized through the varying the tag's load impedance. When a tag is placed within the reading zone of a reader, the tag draws energy from the magnetic field or electric field which is generated by the reader antenna. The resulting feedback by the tag on the reader antenna brings about the change in the magnetic field or electric field. This change is subsequently detected by the reader. The data can be transferred from the tag to the reader if the timing within which the load resistor or capacitor is switched on and off is controlled by data. This type of data transfer is called load modulation.

The tag varies its load impedance by switching the load resistor or capacitor on and off, giving rise to resistive load modulation and capacitive load modulation [2]. In resistive load modulation, a resistance is switched on and off, ether in time with the data stream, or at some higher frequency (subcarrier). In capacitive load modulation, a capacitance is switched on and off. These changes are detected by the reader as a combination of amplitude and phase modulation.

3.2.3.2 Far-field Coupling

The tags in far-field RFID systems capture the EM waves radiating from the reader antenna. The tag antenna receives the energy and develops an alternating potential difference that appears across the ports of the microchip. A diode can rectify this potential and link it to a capacitor, which will result in an accumulation of energy in order to power up its electronics. The tags are located beyond the near-field zone of the reader, so that the information cannot be transmitted back to the reader by using load modulation. Therefore, in far-field RFID systems the communication between the reader and tags is realized by using a technology known as backscattering.

As shown in Figure 3.4, the reader antenna emits energy which will be received by the tag, and some of the energy is then reflected from the tag and detected by the reader. The variation of the tag's load (microchip) impedance causes the intended impedance mismatch

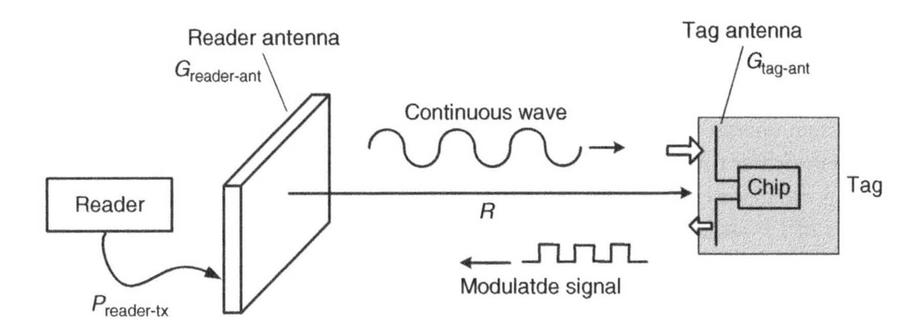

Figure 3.4 Power /communication mechanism for far-field RFID systems.

between the tag antenna and the load. The varying impedance mismatch results in the variation of the reflected signals. In other words, the amplitude of the reflected energy from the tag can be influenced by altering the load of the tag antenna. By changing the tag antenna's load over time, the tag can reflect back more or less of the incoming signal in a pattern that encodes the tag's ID. This type of communication is called 'backscattering modulation'.

3.2.4 Frequencies, Regulations and Standardization

3.2.4.1 Frequencies and Regulations

The range and scalability of an RFID system are heavily dependent on the radio frequency which the system utilizes. The operating frequency can significantly affect reading distance, data exchange speed, interoperability, antenna size and type, and surface penetration. The need to ensure that RFID systems coexist with existing radio systems such as the mobile, marine and aeronautical radio systems significantly restricts the range of operating frequencies available for RFID systems; it is only possible to use the frequency ranges that have been reserved specially for the Industrial, Scientific and Medical (ISM) frequency ranges. In addition to the ISM frequencies, the entire frequency ranges below 135 kHz (in North and South America) and 400 kHz (in Japan) are also available for RFID applications. Figure 3.5 shows the main frequency spectrum available for RFID applications.

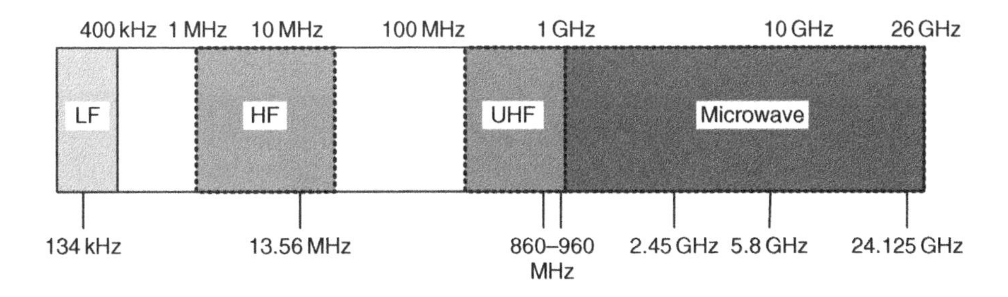

Figure 3.5 Main frequency ranges available for RFID applications.

RFID systems are generally distinguished by four frequency ranges: low, high, ultra high and microwave.

Frequencies between 30 kHz and 400 kHz are considered as low frequency (LF). A typical LF RFID system operates at 125 kHz or 134.2 kHz. Such systems generally use passive tags, have low data-transfer rates from the tag to the reader, and are suitable for applications where the operating environment contains metals, liquids, dirt, snow, or mud (a very important characteristic of LF systems). Active LF tags are also available in RFID applications.

The high frequency (HF) band ranges from 3 MHz to 30 MHz, in particular with 13.56 MHz being the typical frequency used for HF RFID systems. A typical HF RFID system uses passive tags, has a low data-transfer rate from the tag to the reader, and offers fair performance in the presence of metals and liquids.

The ultra high frequency (UHF) band ranges from 300 MHz to 1 GHz. A typical passive UHF RFID system operates at 915 MHz in the United States and at 868 MHz in Europe. A typical active UHF RFID system operates at 315 MHz or 433 MHz. A UHF RFID system can therefore use both active and passive tags and has a high data-transfer rate between the tag and the reader. However, the UHF frequency range for RFID applications has not been accepted worldwide.

The microwave frequency (MWF) band ranges upward from 1 GHz. A typical MWF RFID system operates either at 2.45 GHz, 5.8 GHz, or 24.125 GHz, of which 2.45 GHz is the most commonly used frequency and is accepted worldwide. The MWF RFID system can use both semi-active and passive tags, and has the highest data-transfer rate between the tag and the reader. Since the antenna size is inversely proportional to the frequency, the antenna of a passive tag operating in the MWF range can be much smaller than those of the RFID systems operating in other frequency range.

The various RFID operating frequency bands and associated characteristics are summarized in Table 3.1 [7]. The frequencies and power limitations of UHF RFID systems in different countries/regions are shown in Table 3.2 [8].

3.2.4.2 Standardization

The number and adoption of standards for RFID technology and its associated industries are quite complicated and involve a number of bodies. The effort of standardization is ongoing. The existing standards have been produced to cover four key areas of RFID applications, namely air interface standards (for basic tag-to-reader data communication), data content and encoding (numbering schemes), conformance (testing of RFID systems), and interoperability between applications and RFID systems [9].

There are several standards bodies involved in the development and definition of RFID technologies, including the International Organization for Standardization (ISO), EPCglobal Inc., the European Telecommunications Standards Institute (ETSI) and the Federal Communications Commission (FCC). Standards for RFID technology have been drawn up in a number of areas, including the following [2, 9, 10]:

- *Animal identification.* ISO 11784 ('Radio frequency identification of animals – Code structure'); ISO 11785 ('Radio frequency identification of animals – Technical concept'); and ISO 14223/1 ('Radio frequency identification of animals – Advance transponders').

- *Contactless smart cards.* ISO 10536 ('Contactless integrated circuit(s) cards – Close-coupled cards'), which specifies the structure and operating parameters of contactless close coupling smart cards, with range limited to about 1 cm; ISO 14443 ('Identification cards – Contactless integrated circuit(s) cards – Proximity cards'), which describes the operating method and operating parameters of contactless proximity coupling smart cards with range around 7–15 cm; and ISO 15693 ('Identification cards – Contactless integrated circuit(s) cards – Vicinity cards'), which describes the method of functioning and operating parameters of contactless vicinity coupling smart cards, with range up to 1 m.
- *Item management.* The ISO 18000 series ('Information technology – Radio frequency identification for item management'), which specifies the parameters for air interface communications below 135 kHz, at 13.56 MHz, 433 MHz, 860–960 MHz, and 2.45 GHz; and Class 1 Generation 2 UHF Air Interface Protocol Standard Version 1.0.9 [10], which defines the physical and logical requirements for a passive-backscatter, interrogator-talks-first (ITF) RFID system operating in the 860–960 MHz frequency range.

Table 3.1 Near-field and far-field RFID systems and associated characteristics.

	Near-field RFID		Far-field RFID	
	LF	HF	UHF	MWF
Frequency	30–400 kHz	3–30 MHz	300 MHz–3 GHz	1–30 GHz
Typical RFID frequency	125–134 kHz	13.56 MHz	433 MHz or 865 – 956 MHz	2.45 GHz
Approximate reading distance	less than 0.5 m	Up to 1.5 m	433 MHz: up to 100 m (active tag) 865–956 MHz: 0.5–5 m	Up to 10 m
Typical data-transfer rate	less than 1 kbit/s	Approximately 25 kbit/s	30 kbit/s	Up to 100 kbit/s
Characteristics	Short range, low data-transfer rate, penetrates water but not metal	Higher ranges, reasonable data rate (similar to GSM phone), penetrates water but not metal	Long ranges, high data-transfer rate, cannot penetrate water or metals	Long range, high data-transfer rate, cannot penetrate water or metal
Typical applications	Animal ID, Car, immobilizer	Smart labels, contactless travel cards, access and security	Specialist animal tracking Logistics	Moving vehicle toll

Table 3.2 Frequencies and power limitations of UHF RFID system in different countries/regions.

Country/region	Status[a]	Frequency[b] (MHz)	Power[c]	Technique[d]	Comments
Africa					
South Africa	OK	865.6–867.6	2 W ERP	LBT	Regulations should be in place in 2006.
South Africa		917–921	4 W EIRP	FHSS	Indoor.
Tunisia	IP	865.6–867.6	2 W ERP	LBT	Plans being adopted.
Asia-Pacific					
Australia	OK	920–926	4 W EIRP		4 W EIRP available through license managed by GS1 Australia.
China	IP	917–922	2 W ERP		Provisional allocation. Temporary license required.
Hong Kong, China	OK	865–868	2 W ERP		
		920–925	4 W EIRP		
India	OK	865–867	4 W ERP		Approved in May 2005.
Indonesia	IP				Band 923–925 MHz being considered.
Japan	OK	952–954	4 W EIRP	LBT	License required for using 952–954 MHz at 4 W EIRP. Is available for unlicensed use at 20 mW EIRP.
Korea	OK	908.5–910	4 W EIRP	LBT	Approved July 2004.
		910–914	4 W EIRP	FHSS	Approved July 2004.
Malaysia	OK	866–869			Allocation under consideration. 868 MHz is available at 50 mW power.
		919–923	2 W ERP		Unlicensed use allowed up to 2 W ERP. Use up to 4 W ERP allowed under license.
Singapore	OK	866–869	0.5 W ERP		
		923–925	2 W ERP		
Taiwan	OK	922–928	1 W ERP	FHSS	Indoor.
		922–928	0.5 W ERP	FHSS	Outdoor.
New Zealand	OK	864–868	4 W EIRP		
Europe					
Finland	OK	865.6–867.6	2 W ERP	LBT	New regulations in place since 3 February 2005.

Table 3.2 Continued

France	IP	865.6–867.6	2 W ERP	LBT	
Germany	OK	865.6–867.6	2 W ERP	LBT	New regulations in place since 22 December 2004.
Italy	IP	865.6–867.6	2 W ERP	LBT	Conflict with band allocated to tactical relays military applications.
Netherlands	OK	865.6–867.6	2 W ERP	LBT	
Russian Federation	IP	865.6–867.6	2 W ERP	LBT	Licensed use only.
Spain	OK	865.6–867.6	2 W ERP	LBT	New regulations will be in place by January 2007. Temporary licenses available.
Sweden	OK	865.6–867.6	2 W ERP	LBT	New regulations approved 13 December 2005. In the law since 1 January 2006.
Switzerland	OK	865.6–867.6	2 W ERP	LBT	
United Kingdom	OK	865.6–867.6	2 W ERP	LBT	New regulations in place since 31 January 2006.
North America					
Canada	OK	902–928	4 W EIRP	FHSS	
United States	OK	902–928	4 W EIRP	FHSS	
South America					
Argentina	OK	902–928	4 W EIRP	FHSS	
Brazil	OK	902–907.5	4 W EIRP	FHSS	
		915–928	4 W EIRP	FHSS	
Chile	OK	902–928	4 W EIRP	FHSS	

[a] OK indicates regulations are in place or will be in place shortly. IP indicates appropriate regulations expected within 6–12 months.
[b] Indicates the frequency band(s) authorized in the country for RFID applications. The objective is to get a band available in the 860–960 MHz spectrum.
[c] Indicates the maximum power available to RFID applications. The power is expressed either as EIRP (effective isotropic radiated power) or ERP (effective radiated power). Note that 2 watts ERP is equivalent to 3.2 watts EIRP.
[d] Indicates the reader to tag communication technique. FHSS stands for frequency hopping spread spectrum and LBT stands for listen before talk.

3.3 Design Considerations for RFID Tag Antennas

The most important aspect of an RFID system's performance is the reading distance – the maximum distance at which an RFID reader can detect the load modulated or backscattered signal from the tag. For a specific application with prior selected reader (including reader antenna), the reading distance depends on the performance of the tag. As stated previously, a typical passive RFID tag consists of an antenna and a microchip. The characteristics of the microchip are quantified by IC manufacturers and cannot be modified by users. The key challenge for tag antenna design is to maximize the reading distance with a prior selected microchip under the various constraints (such as limited antenna size, specific antenna impedance and radiation pattern, and cost).

Generally, the requirements for RFID tag antennas with prior selected microchips can be summarized as follows [11]:

- good impedance matching for receiving maximum signals from the reader to power up the microchip;
- small enough to being attached to or embedded into the specified object;
- insensitive to the attached object to keep performance consistent;
- required radiation patterns (omnidirectional, directional or hemispherical);
- robust in mechanical structure; and
- low cost in both materials and fabrication.

Under the various constraints for specific RFID applications, several aspects should be considered in RFID tag antenna design [12]:

Frequency band. The antenna type is dependent on the operating frequency. In LF and HF RFID applications, spiral coil antennas are most commonly used to capture the energy from the reader by coupling. At UHF and MWF frequencies, dipole antennas, meander line antennas, slot antennas, and patch antennas are widely employed.

Size. Tags are required to be small so that they can be embedded into or attached to a specific object such as a cardboard box, airline baggage strip, identification card or printed label. The size constraint is one of the challenges for RFID tag antenna design. Small size limits the coupling capability of the loop antenna especially at LF and HF, and results in low efficiency of the antenna at UHF and MWF. As a result, the reading distance of RFID systems will be reduced greatly.

Radiation patterns. Some applications require a specific directivity pattern of the tag antenna such as omnidirectional, directional or hemispherical coverage.

Sensitivity to objects. The performance of a tag will be changed when it is placed on different objects such as cardboard boxes with different contents, and degraded on lossy objects such as plastic bottles containing water and oil, or metallic cans. The tag antenna is required to be tuned for optimal performance on a particular object or designed to be less sensitive to the various types of object on which the tag is attached.

Cost. An RFID tag must be a low-cost device for massive applications. This imposes restrictions on both the antenna structure and the choice of materials for constructing the tag, including the microchip. The materials used for the construction of the tag antenna are conducting strip/wire and supporting dielectric. The dielectrics include flexible polyester for LF and HF and rigid printed circuit board substrates like FR4 for UHF and MWF. For further cost reduction, all-printed RFID tags have been reported that use screen printing or ink-jet printing techniques [13, 14].

Reliability. An RFID tag must be a reliable device that can cope with variations in temperature, humidity, and stress, and survive processes such as label insertion, printing, and lamination.

3.3.1 Near-field RFID Tag Antennas

3.3.1.1 Equivalent Circuit of Inductive Coupling Tag

Most near-field RFID systems employ inductive coupling for energy transfer and communications between readers and tags. A typical passive inductive coupling RFID system (as shown in Figure 3.3) consists of a reader, a reader coil antenna and a tag with a coil antenna. The big coil antenna generates a magnetic field to supply the tag with power and collect information from it. The tag usually comprises a small coil antenna and a microchip which contains an internal capacitance. The tag coil antenna and the on-chip capacitance form a resonant circuit at a resonant frequency which is the same as or close to the operating frequency of the reader. The resonant circuit provides a high voltage which is rectified to supply the power to the tag's circuit. The magnitude of the voltage is mainly determined by the properties of the tag coil antenna. Therefore, the coil antenna of the tag determines the performance of an RFID system which has specified EIRP and preselected microchip. By optimizing the design of the coil antenna, it is able to obtain maximum voltage to 'wake up' the microchip, thus achieving a maximum reading distance for the RFID system.

Figure 3.6 shows an inductive coupling tag and its equivalent circuit. The equivalent circuit of the coil antenna contains a voltage source U_i that represents the induced voltage. The inductance L of the coil antenna is determined by its geometry, the property of the material and the number of windings. The resistor R_A represents the ohmic loss which degrades the magnitude of the induced voltage. The capacitor C_p represents the parasitic capacitance of the coil antenna and/or the capacitance of additional capacitor which is used to form a resonant circuit. The microchip is represented by its internal capacitance C and the load R_L, which determines the power consumption of the circuit.

The voltage U_o at the load resistor R_L in the equivalent circuit shown in Figure 3.6 can be described by the following formula [2, 5]:

$$U_o = \frac{U_i}{j\omega(L/R_L + RC) + (1 - \omega^2 LC + R/R_L)}. \tag{3.1}$$

where the capacitors C_p and C_r are lumped as a common capacitance C, and $\omega = 2\pi f$, f being the operating frequency.

3.3.1.2 Resonance

To maximize the voltage on the tag circuit, the parallel capacitor and the coil antenna should form a resonant circuit with a resonant frequency which is the same as or close to the operating frequency of the reader. In a 13.56 MHz system with anti-collision procedures, the resonant frequency selected for the tag is often 1–5 MHz higher to minimize the effect

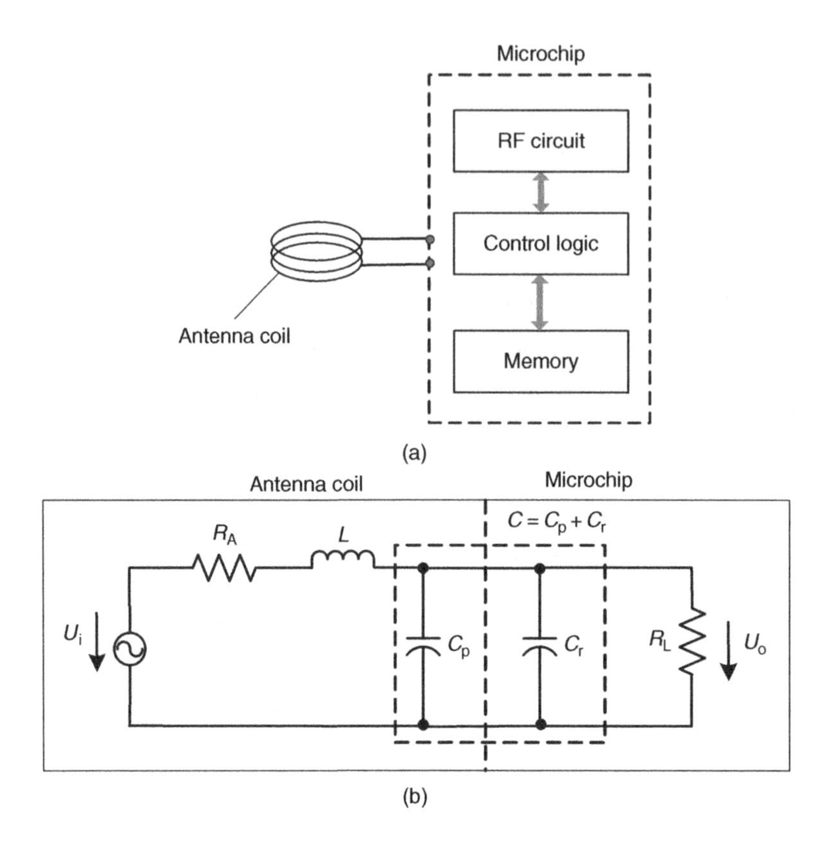

Figure 3.6 (a) Typical inductive coupling tag and (b) its equivalent circuit.

of the interaction between tags on the overall performance because the overall resonant frequency of the two tags directly adjacent to one another is always lower than the resonant frequency of a single tag [2]. The resonant frequency of the parallel resonant circuit can be calculated using the Thomson's equation:

$$f = \frac{1}{2\pi\sqrt{L \cdot C}}. \tag{3.2}$$

3.3.1.3 Inductance

For tag antenna design with a prior selected microchip with an internal capacitance C_r, the task is to configure a coil antenna to resonate at the operating frequency. If C_p is known (compared to C_r, C_p is normally very small in the HF band, so $C \approx C_r$), the required inductance of the coil antenna is given by

$$L = \frac{1}{(2\pi f)^2 C}. \tag{3.3}$$

The coil antenna is usually structured on a substrate, typically made of polyethylene terephthalate (PET), polyvinyl chloride (PVC), or polyamide and consists of wound wire or etched copper/aluminum strips. Conductive polymeric thick film pastes can also be used for the coil antenna by screen printing or dispensing for cost reduction. Etched or screen printed coil antennas are suitable for HF systems because low inductance is required. There are many types of inductors that can be used to realize the required inductance, and the spiral inductor is widely used in HF RFID systems.

A typical spiral inductor is illustrated in Figure 3.7. The conductive strip is wound either clockwise or counterclockwise. This configuration ensures that the current in adjacent tracks is in phase. The resulting mutual inductance yields a significant increase in the spiral inductor's self-inductance. Connecting the microchip to the open ends of the spiral antenna forms a tag. The microchip can be directly connected to the inner and outer ends of the spiral inductor. It is more convenient to use an underpass to connect the end of the outer turn to the centre for microchip assembly if more windings are required for a larger inductance.

The inductance of the spiral inductor is determined by its area ($L \times D$) and the number of windings [15]. The width of the tracks and the spacing between them are usually uniform, although they can be non-uniform. The spiral inductor can be any closed loop such as a square, rectangle, triangle, circle, semi-circle, or ellipse. The square spiral inductor has been widely used in practical applications because of its simple layout.

No analytical formula can be used to calculate the inductance of such spiral inductors. The calculation must instead be done by numerical methods. Many commercial software tools such as ADS, IE3D, and Microwave Studio can be used for this purpose. Figure 3.8 shows the calculated inductance of the spiral inductor shown in Figure 3.7(b) with length $L = 50$ mm, width $D = 40$ mm, strip width $W = 1$ mm, strip spacing $S = 0.5$ mm, number of windings $N = 6$, polyester substrate, $\varepsilon_r = 4$, and thickness $h = 50\,\mu$m. The calculation was done using IE3D software, based on the method of moments [16]. The inductance is about 2.3 μH from 10 MHz to 17 MHz.

3.3.1.4 Parallel Capacitance

If the coil antenna is predetermined, the required capacitance for the parallel capacitor, C, for a specific operating frequency is given by

$$C = \frac{1}{(2\pi f)^2 L}. \tag{3.4}$$

For tags operating in frequency range below 135 kHz, a chip capacitor ($C_p \approx$ 20–220 pF) is generally required to achieve resonance at the desired frequency. At high frequencies (13.56 MHz, 27.125 MHz), the required capacitance is usually so low that it can be provided by the internal capacitance of the microchip and the parasitic capacitance of the coil. In general, the internal capacitance of the microchip is fixed, thus the inductance of the coil antenna has to be modified for circuit resonance by varying its geometry.

3.3.1.5 *Q* Factor

To characterize the coil antenna, the quality factor Q is commonly used. The Q factor is a measure of the ability of a resonant circuit to retain its energy. A high Q means that a circuit leaks very little energy, while a low Q means that the circuit dissipates a lot of energy.

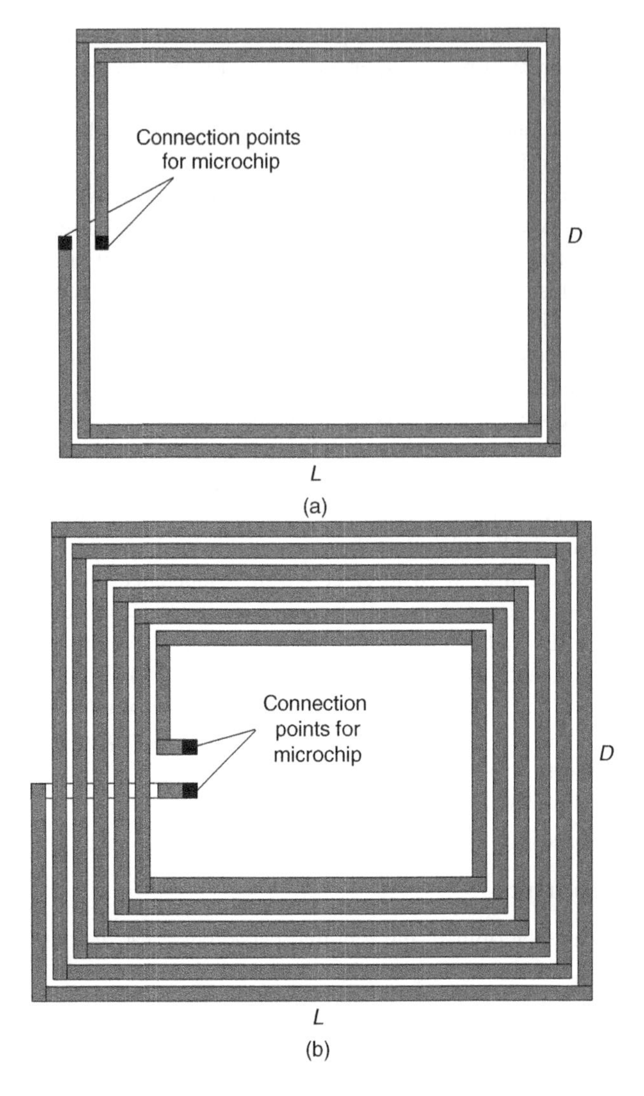

Figure 3.7 Spiral antenna: (a) same layer connection; (b) using two metal layers with an underpass.

In Figure 3.6, the entire tag circuit can be considered as a parallel RLC circuit, where R represents the entire ohmic losses of the tag, including the ohmic loss of the coil antenna and the series resistance of the microchip. In this case, Q can be defined as

$$Q = \frac{2\pi f L}{R}. \tag{3.5}$$

The induced voltage of the coil antenna is proportional to the Q factor. Usually, the Q factor is maximized for a long reading distance, but it has to be noted that a high Q factor limits the bandwidth of transmitted data. Therefore, the typical Q value for most tag coil antennas is about 30–80.

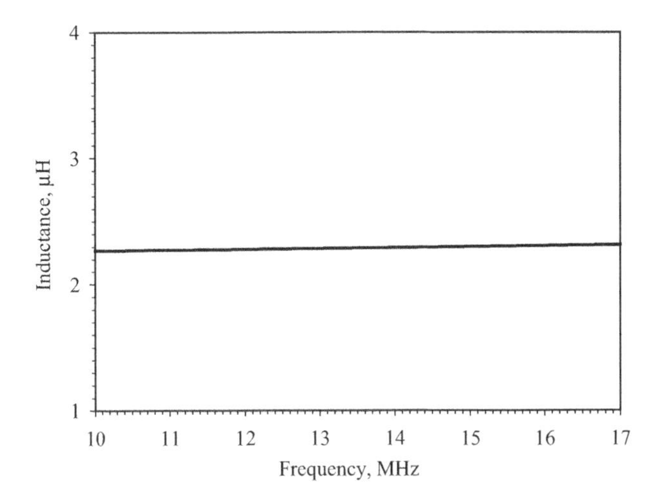

Figure 3.8 Simulated inductance of spiral coil antenna by IE3D.

3.3.1.6 Case Study

In this section, an example is presented to demonstrate the method for designing a coil antenna with a prior selected microchip. Important issues such as the essential procedures of getting necessary information from a microchip datasheet, determining required inductance, configuring and simulating the coil antenna, and calculating the Q factor will be addressed.

The EM4006 microchip [17] is a CMOS integrated circuit used in electronic read-only transponders and operating at 13.56 MHz. Generally, the characteristics of the microchip can be found in the datasheet provided by the manufacturer. The most important pieces of information for coil antenna design are the internal capacitance of the chip and the pad position configuration of the microchip. The electrical parameters and pad position of EM4006 are shown in Table 3.3 and Figure 3.9, respectively. Once we have the information, we can carry out the coil antenna design.

Calculating Required Inductance of Coil Antenna

The internal capacitance, C_{RES}, shown in Table 3.3, is required in coil antenna design. It is found that the typical value of C_{RES} is 94.5 pF at 13.56 MHz. Using (3.3), the required inductance of the coil antenna is:

$$L = \frac{1}{(2 \times 3.14 \times 13.56 \times 10^6)^2 \times 94.5 \times 10^{-12}} = 1.46\,\mu H \tag{3.6}$$

Coil Antenna Configuration and Simulation

Having calculated the required inductance of the coil antenna, the next step is to configure a coil antenna according to the specific design requirements such as size constraint and the properties of the substrate used. The shape of the coil can be square, rectangle, triangle, ellipse, or any other closed structures. Figure 3.10 shows the coil antenna layout in IE3D with the parameters: $L = 47$ mm, $D = 47$ mm, strip width $W = 1$ mm, strip spacing $S = 0.5$ mm. The substrate is polyester, 50 μm thick ($\varepsilon_r = 4.0$, $\tan\delta = 0.002$). With the input impedance

Table 3.3 Electrical characteristics of EM2006.

Parameter	Symbol	Test Conditions	Min	Typ.	Max.	Unit
Supply voltage	V_{DD}		1.9			V
Supply current	I_{DD}			60	150	μA
Rectifier Voltage drop	V_{REC}	$I_{C1C2} = 1$ mA, modulator switch on $V_{REC} = (V_{C1} - V_{C2}) - (V_{DD} - V_{SS})$			1.8	V
Modulator ON DC voltage drop	V_{ON1}	$I_{VDD\ VSS} = 1$ mA	1.9	2.3	2.8	V
	V_{ON2}	$I_{VDD\ VSS} = 10$ mA	2.4	2.8	3.3	V
Power on reset	V_R		1.2	1.4	1.7	V
		$V_R - V_{MIN}$	0.1	0.25	0.5	V
Coil 1 – Coil 2 capacitance	C_{RES}	$V_{coil} = 100$m V RMS $f = 10$ kHz	92.6	94.5	96.4	pF
Series resistance of CRES	R_S				3	Ω
Power supply capacitor	C_{sup}				140	pF

$V_{DD} = 2\,V$, $V_{SS} = 0\,V$, $f_{C1} = 13.56$ MHz sine wave, $V_{C1} = 1.0V$pp centered at $(V_{DD} - V_{SS})/2$, $T_a = 25°$C, unless otherwise specified.

$Z_A = R_A + X_A = 0.58 + j124.0\,\Omega$ obtained from IE3D, the inductance of the coil antenna is given by

$$L = \frac{X_A}{2\pi f} = \frac{124.0}{2 \times 3.14 \times 13.56 \times 10^6} = 1.46\,\mu H. \tag{3.7}$$

The Q factor of the tag is given by

$$Q = \frac{X_A}{R_A + R_S} = \frac{124.0}{0.58 + 3} = 34.6 \tag{3.8}$$

Antenna Pad Configuration

Much attention should be devoted to the antenna pad position configuration when the coil layout is made. The coil tracks at the input of the antenna (antenna pad) must be adequately configured to fit the pad position of the microchip for tag assembly. A proper antenna pad configuration enables the microchip to be easily affixed to the antenna, which reduces rejection rate and the cost of the tag.

Referring to Figure 3.9, the microchip has two pads C_1, C_2 which are fixed on the ends of the coil antenna. The distance between C_1 and C_2 is 0.74 mm. The antenna pad should be configured so that the microchip can be placed and aligned on it properly. The details of the antenna pad are shown in Figure 3.11. The width of the strips is tapered from 1.0 mm to 0.5 mm; the spacing changes from 1.0 mm to 0.2 mm. This arrangement ensures the proper affixing of the microchip to the coil antenna as long as the outline of the microchip is kept within the area of the antenna pads.

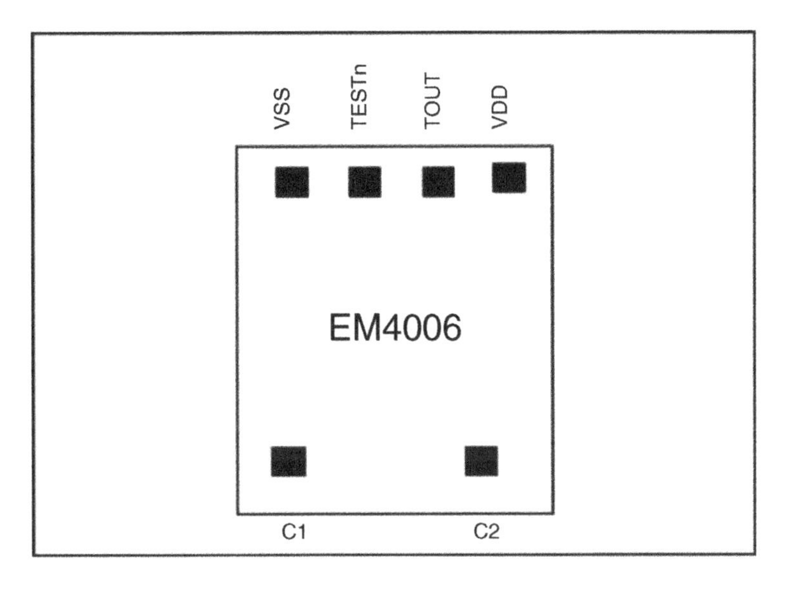

(a)

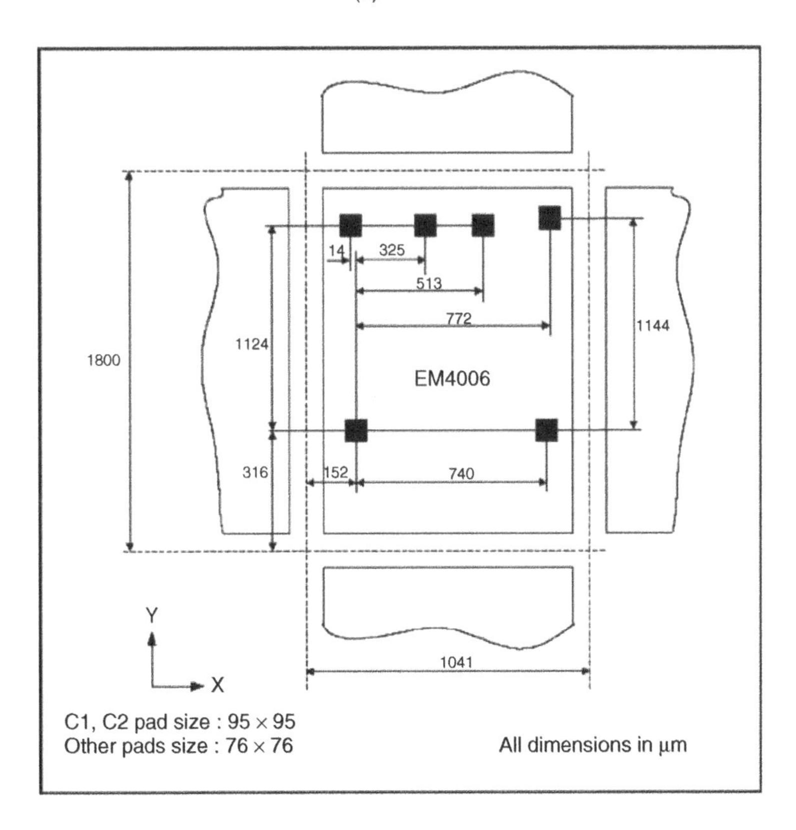

(b)

Figure 3.9 Microchip pad information: (a) pad assignment; (b) pad position.

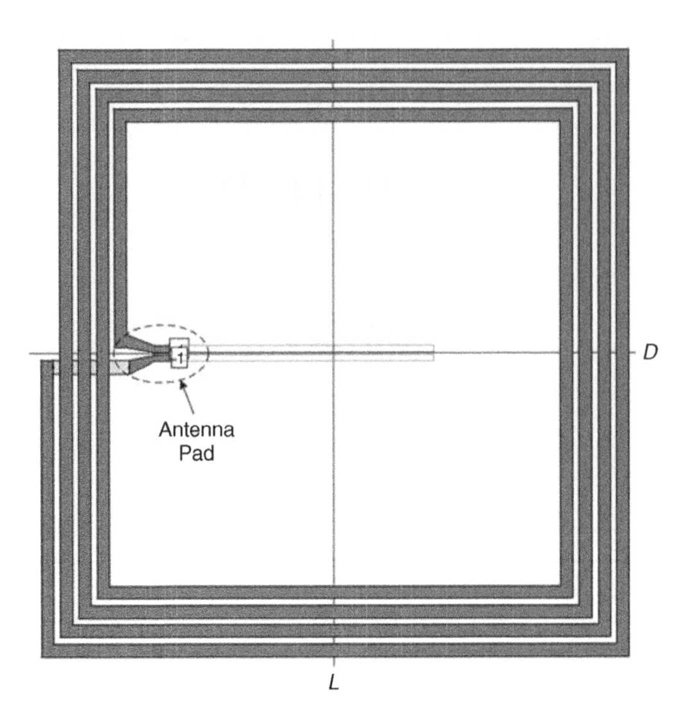

Figure 3.10 Tag coil antenna configuration.

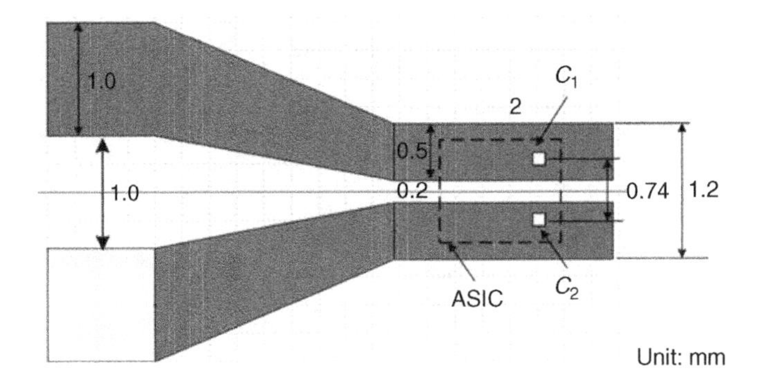

Figure 3.11 Details of the antenna pad configuration.

3.3.2 Far-field RFID Tag Antennas

For far-field RFID systems, the tag antenna design plays a vital role in system efficiency and reliability since the operation of passive RFID tags is based on the EM field they receive from the readers. Figure 3.3 illustrates the operating principles of a passive far-field RFID system. The reader sends out a continuous wave RF signal containing alternating current power and clock signal to the tag at the carrier frequency at which the reader operates. The RF voltage induced on the antenna terminals is converted to direct current which powers up the microchip. A voltage of about 1.2 V is necessary to energize the microchip for reading purposes. For writing, the microchip usually needs to draw about 2.2 V from the reader's signal. Then the microchip sends back the information by varying complex RF input impedance. The impedance typically toggles between two different states (conjugate matched and some other impedance) to modulate the backscattering signal. When receiving this modulated signal, the reader decodes the pattern and obtains the tag information.

3.3.2.1 Radio Link

In an RFID system, the reading distance is constrained by the maximum distance at which the tag can receive just enough power to turn on and scatter back, and the maximum distance at which the reader can detect this backscattered signal. The reading distance of an RFID system is the smaller of these two distances. Typically the reader sensitivity is high enough, therefore the reading distance is determined by the former distance. The reading distance is also sensitive to the tag orientation, the properties of the objects to which the tag is attached, and the propagation environment.

Power Link (Reader to Tags)

Consider the RFID system shown in Figure 3.4, where the output power of the reader is $P_{\text{reader-tx}}$, the gain of the reader antenna is $G_{\text{reader-ant}}$, the distance between the reader antenna and the tag is R, and the gain of the tag antenna is $G_{\text{tag-ant}}$. According to the Friis free-space transmission formula, the power received by the tag antenna is [18]:

$$P_{\text{tag-ant}} = \left(\frac{\lambda}{4\pi R}\right)^2 P_{\text{reader-ant}} G_{\text{reader-ant}} G_{\text{tag-ant}} \chi \tag{3.9}$$

where λ is the wavelength in free space at the operating frequency and χ is the polarization matching coefficient between the reader antenna and tag antenna. If the two antennas are perfectly matched in polarization, χ will be 1 or 0 dB. For most of far-field RFID systems, the reader antenna is circularly polarized while the tag antenna is linearly polarized, hence χ will be 0.5 or -3 dB.

Part of the power received by the tag antenna is delivered to the terminating microchip, and it can be expressed as:

$$P_{\text{tag-chip}} = \tau P_{\text{tag-ant}} \tag{3.10}$$

where τ is the power transmission coefficient determined by the impedance matching between the tag antenna and the microchip.

The maximum reading distance for a radio power link is obtained when $P_{\text{tag-chip}}$ is equal to the threshold power of the microchip, $P_{\text{tag-threshold}}$, which is the minimum threshold power to power up the microchip on the RFID tag:

$$R_{\text{power-link}} = \frac{\lambda}{4\pi}\sqrt{\frac{P_{\text{reader-tx}} G_{\text{reader-ant}} G_{\text{tag-ant}} \chi \tau}{P_{\text{tag-threshold}}}}. \tag{3.11}$$

For convenience, (3.11) can be modified as:

$$R_{\text{power-link}} = 10^{\alpha} (\text{m}) \tag{3.12}$$

where

$$\alpha = 27.6 - 20\log[f(\text{MHz})] + P_{\text{reader-tx}}(\text{dBm}) + G_{\text{reader-ant}}(\text{dBic}) + \frac{G_{\text{tag-ant}}(\text{dBi}) + \chi(\text{dB}) + \tau(\text{dB}) - P_{\text{tag-threshold}}(\text{dBm})}{20}.$$

Backscatter Communication Link

The backscatter communication link from the tags to the reader is largely dependent on the backscatter field strength of the tag. Based on a monostatic (backscattering) radar equation [19], the amount of modulated power received by the reader is given by:

$$P_{\text{reader-rx}} = \frac{\lambda^2}{(4\pi)^3 R^4} P_{\text{reader-tx}} G^2_{\text{reader-ant}} \chi \sigma, \tag{3.13}$$

where σ is the radar cross-section (RCS) of the RFID tag.

When the received power is equal to the reader's sensitivity, $P_{\text{reader-threshold}}$, the maximum distance for backscatter communication link can be obtained:

$$R_{\text{backscatter}} = \sqrt[4]{\frac{\lambda^2}{(4\pi)^3} \frac{P_{\text{reader-tx}} G^2_{\text{reader-ant}} \chi \sigma}{P_{\text{reader-threshold}}}} \tag{3.14}$$

Again (3.14) can be expressed in a modified form:

$$R_{\text{backscatter}} = 10^{\beta}(m) \tag{3.15}$$

where

$$\beta = 16.6 - 20\log[f\,(\text{MHz})] + P_{\text{reader-tx}}\,(\text{dBm}) + 2G_{\text{reader-ant}}(\text{dBic}) + \frac{\chi(\text{dB}) + \sigma(\text{dBsm}) - P_{\text{reader-threshold}}(\text{dBm})}{40}.$$

From (3.11) and (3.14) it is observed that the reading distance is determined by the output power of the reader, $P_{\text{reader-tx}}$ and the gain of the reader antenna, $G_{\text{reader-ant}}$, the gain of the tag antenna, $G_{\text{tag-ant}}$, the polarization matching coefficient, χ, the power transmission coefficient of the tag, τ, the RCS of the tag, σ, the threshold power of the microchip, $P_{\text{tag-threshold}}$, and receiver sensitivity of the reader, $P_{\text{reader-threshold}}$. The last two parameters are predetermined for a prior selected reader and microchip. The remaining parameters can be optimized to achieve a longer reading distance. The above-mentioned parameters will be addressed in the following sections.

3.3.2.2 EIRP and ERP

As mentioned in Section 3.3.2.1, the maximum reading distance is proportional to the output power of the reader and the gain of the reader antenna. Higher output power and gain of the reader antenna can offer a longer reading distance. However, the output power is always limited by national licensing regulations.

EIRP is the measure of the radiated power which an isotropic emitter (i.e. $G = 1$ or 0 dB) will need to supply in order to generate a defined radiation power at the reception location as at the device under test [2]:

$$P_{\text{EIRP}} = P_{\text{reader-tx}} G_{\text{reader-ant}} \tag{3.16}$$

In addition to the EIRP, ERP is frequently used in radio regulations and in the literature. The ERP relates to a dipole antenna rather than an isotropic emitter. It expresses the radiated power which dipole antenna (i.e. $G = 1.64$ or 2.15 dB) will need to supply in order to generate a defined radiation power at the reception location as at the device under test.

It is easy to convert between the two parameters:

$$P_{\text{EIRP}} = 1.64 P_{\text{ERP}} \tag{3.17}$$

Table 3.2 summarizes regulated EIRP or ERP in the UHF band for different countries/regions.

3.3.2.3 Tag Antenna Gain

The tag antenna gain, $G_{\text{tag-ant}}$, is the other important parameter for the reading distance. The range is largest in the direction of maximum gain which is fundamentally limited by the size, radiation patterns of the antenna, and the frequency of operation. For a small dipole-like omnidirectional antenna, the gain is about 0–2 dBi. For some directional antennas such as the patch antenna, the gain can be up to 6 dBi or more.

3.3.2.4 Polarization Matching Coefficient

The polarization of the tag antenna must be matched to that of the reader antenna in order to maximize the reading distance, which can be characterized by the polarization matching coefficient, χ. For far-field RFID systems, the reader antenna is always circularly polarized because the orientation of the tag is random. Using a linearly polarized tag antenna will result in a polarization mismatch loss, i.e. $\chi = 0.5$ or -3 dB. A circularly polarized tag antenna is preferable for some specific applications because the signal can be increased by 3 dB.

3.3.2.5 Power Transmission Coefficient

Referring to Figure 3.12, consider a tag antenna with a maximum effective aperture $A_{\text{e-max}}$ (in square meters), situated in the field of the reader antenna with the power density S (watts per square meter). It takes in power from the wave and delivers it to the termination, namely the microchip with load impedance Z_{T}. Part of the power received by the tag antenna is delivered to the termination while the rest of the power is reflected and re-radiated by the antenna. The amount of the power delivered to the microchip can be quantified by using the power transmission coefficient, τ. Let the power antenna received from the incident wave be $P_{\text{tag-ant}}$, and the power delivered to the chip $P_{\text{tag-chip}}$. Then

$$P_{\text{tag-ant}} = S A_{\text{e-max}}, \tag{3.18}$$

$$P_{\text{tag-chip}} = \tau P_{\text{tag-ant}}. \tag{3.19}$$

The power transmission coefficient, τ, is determined by the impedance matching between the tag antenna and the microchip. Proper impedance matching between antenna and microchip is of paramount importance in RFID since IC design and manufacturing are a big and costly venture. RFID tag antennas are normally designed for a specific microchip available on the market. Adding an external matching network with lumped elements is usually prohibitive in RFID tags due to cost and fabrication issues. To alleviate this situation, the tag antenna can be directly matched to the microchip which has complex impedance that varies with the frequency and the input power supplied to the microchip.

In the equivalent circuit shown in Figure 3.12(b), $Z_T = R_T + jX_T$ is the complex chip impedance and $Z_A = R_A + jX_A$ is the complex antenna impedance. The chip impedance includes the effects of chip package parasitics. Both Z_A and Z_T are frequency-dependent. In addition, the impedance Z_T may vary with the power delivered to the chip.

To describe the transmission of the power waves, we introduce a power wave reflection coefficient Γ [20]:

$$\Gamma = \frac{Z_T - Z_A^*}{Z_T + Z_A}, \quad 0 \le |\Gamma| \le 1 \tag{3.20}$$

The power delivered to the chip is

$$P_{\text{tag-chip}} = \left(1 - |\Gamma|^2\right) P_{\text{tag-ant}}. \tag{3.21}$$

The power transmission coefficient can be expressed as:

$$\tau = \frac{P_{\text{tag-chip}}}{P_{\text{tag-ant}}} = 1 - |\Gamma|^2 = \frac{4R_A R_T}{(R_A + R_T)^2 + (X_A + X_T)^2}, \quad 0 \le \tau \le 1. \tag{3.22}$$

When the antenna is conjugately matched to the chip, i.e. $R_T = R_A$ and $X_T = -X_A$, then $|\Gamma| = 0$, $\tau = 1.0$, and the corresponding maximum transferred power is

$$P_{\text{tag-chip-max}} = P_{\text{tag-ant}} = SA_{e-\max}. \tag{3.23}$$

When the antenna is shorted, the chip resistance $R_T = 0$ and the chip reactance $X_T = -X_A$, $|\Gamma|$ will be unity and τ zero. Thus, there is no power delivered to the chip.

It is convenient to relate the power transmission coefficient, τ, to another widely used parameter, return loss (RL), for describing the impedance matching characteristics. The return loss is defined as:

$$RL = 10\log_{10}(|\Gamma|). \tag{3.24}$$

It is convenient to obtain the return loss from simulation or/and measurement. With the return loss, the corresponding reflection coefficient and the power transmission coefficient can be easily calculated. Table 3.4 shows the corresponding reflection coefficient and power transmission coefficient for different return losses. Figure 3.13 illustrates the relationship between the power transmission coefficient and the return loss.

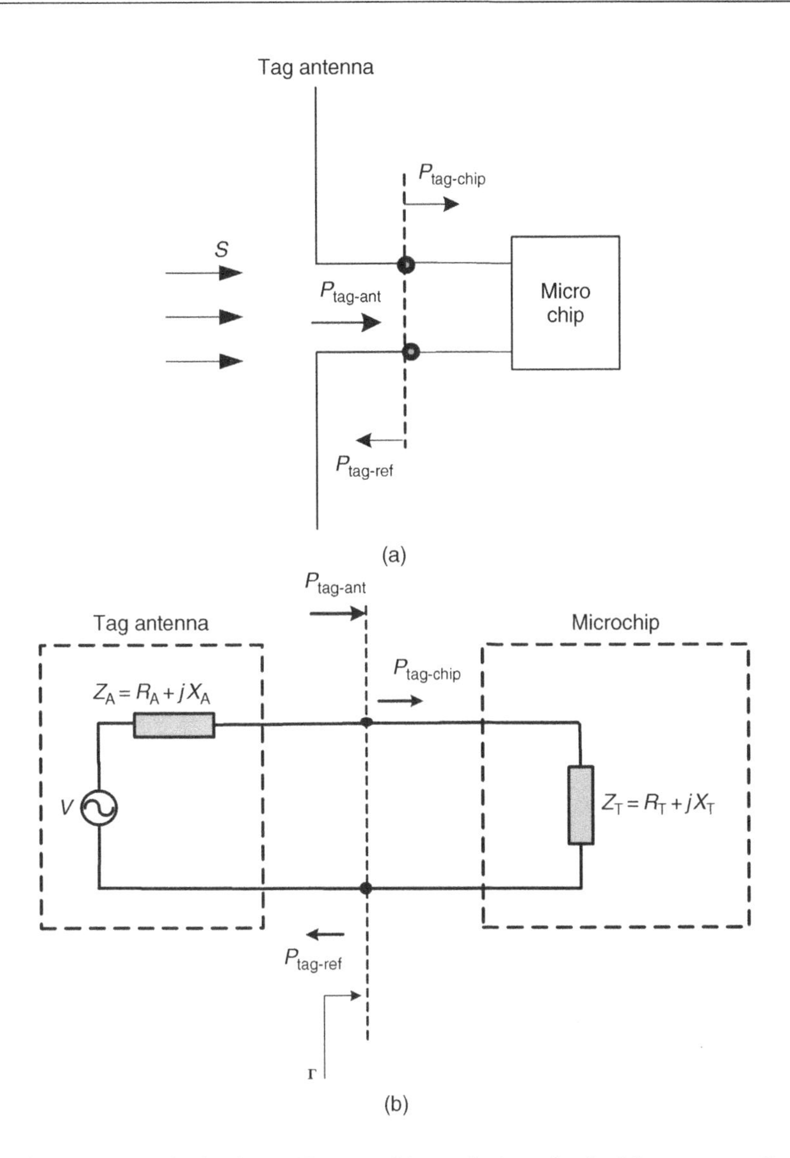

Figure 3.12 Power transfer in the RFID tag and its equivalent circuit: (a) power transfer in RFID tag configuration; (b) equivalent circuit.

3.3.2.6 Antenna RCS

As seen from (3.14), the maximum distance for the backscatter communication link is proportional to the RCS of the RFID tag antenna. In this section, the method of quantifying the RCS of a RFID tag antenna will be introduced.

Radar cross-section definition

The RCS is a measure of the amount of power scattered in a given direction when an object is illuminated by an incident wave. The IEEE defines RCS as 4π times the ratio of the power per unit solid angle scattered in a specified direction to the power per unit area in

Table 3.4 Reflection coefficient and transmission coefficient as function of return loss.

Return Loss(dB)	Reflection coefficient ($\|\Gamma\|$)	Transmission coefficient(τ)	Transmission coefficient(τ, dB)	Return loss (dB)	Reflection coefficient ($\|\Gamma\|$)	Transmission coefficient (τ)	Transmission coefficient (τ, dB)
0.0	1.0000	0.0000	$-\infty$	−10.0	0.3162	0.9000	−0.4576
−0.5	0.9441	0.1087	−9.6357	−11.0	0.2818	0.9206	−0.3594
−1.0	0.8913	0.2057	−6.8683	−12.0	0.2512	0.9369	−0.2830
−1.5	0.8414	0.2921	−5.3454	−13.0	0.2239	0.9499	−0.2233
−2.0	0.7943	0.3690	−4.3292	−14.0	0.1995	0.9602	−0.1764
−2.5	0.7499	0.4377	−3.5886	−15.0	0.1778	0.9684	−0.1396
−3.0	0.7079	0.4988	−3.0206	−16.0	0.1585	0.9749	−0.1105
−3.5	0.6683	0.5533	−2.5703	−17.0	0.1413	0.9800	−0.0875
−4.0	0.6310	0.6019	−2.2048	−18.0	0.1259	0.9842	−0.0694
−4.5	0.5957	0.6452	−1.9031	−19.0	0.1122	0.9874	−0.0550
−5.0	0.5623	0.6838	−1.6509	−20.0	0.1000	0.9900	−0.0436
−5.5	0.5309	0.7182	−1.4378	−22.0	0.0794	0.9937	−0.0275
−6.0	0.5012	0.7488	−1.2563	−24.0	0.0631	0.9960	−0.0173
−6.5	0.4732	0.7761	−1.1007	−26.0	0.0501	0.9975	−0.0109
−7.0	0.4467	0.8005	−0.9665	−28.0	0.0398	0.9984	−0.0068
−7.5	0.4217	0.8222	−0.8504	−30.0	0.0316	0.9990	−0.0043
−8.0	0.3981	0.8415	−0.7494	−35.0	0.0178	0.9997	−0.0013
−8.5	0.3758	0.8587	−0.6614	−40.0	0.0100	0.9999	−0.0004
−9.0	0.3548	0.8741	−0.5844	−45.0	0.0056	1.0000	−0.0000
−9.5	0.3350	0.8878	−0.5169	−50.0	0.0033	1.0000	−0.0000

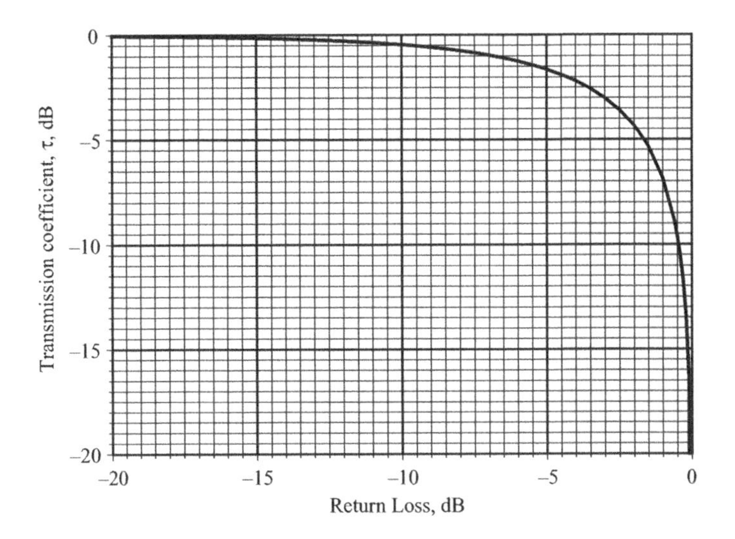

Figure 3.13 Transmission coefficient vs return loss.

a plane wave incident on the scatter from a specified direction. More precisely, it is the limit of that ratio as the distance from the scatter to the point where the scattered power is measured approaches infinity [19]:

$$\sigma = \lim_{R\to\infty} 4\pi R^2 \frac{|E_{\text{scat}}|^2}{|E_{\text{inc}}|^2}. \tag{3.25}$$

Where E_{scat} is the scattered electric field from the object and E_{inc} is the incident field on the object.

The RCS can also be given in another form:

$$\sigma = 4\pi \mathrm{R}^2 \frac{\mathrm{S}_{\text{scat}}}{\mathrm{S}_{\text{inc}}} \tag{3.26}$$

where S_{scat} indicates the scattered power density, S_{inc} is the incident power density at the scattering object, and$R$ denotes the distance from the object.

The RCS is given in units of square meters. However, this does not necessarily relate to the physical size of an object although it is generally true that larger physical objects have larger radar cross-sections. Typical values of RCS span from $10^{-5}\,\text{m}^2$ for insects to $10^{+6}\,\text{m}^2$ for a large ship. Due to the large dynamic range of the RCS, a logarithmic power scale is most often used with the reference value of $\sigma_{\text{ref}} = 1\,\text{m}^2$:

$$\sigma_{\text{dBsm}} = \sigma_{\text{dBm}^2} = 10\log_{10}\left(\frac{\sigma_{\text{m}^2}}{\sigma_{\text{ref}}}\right) = 10\log_{10}\left(\frac{\sigma_{\text{m}^2}}{1}\right). \tag{3.27}$$

There are three types of scattering from an object: *monostatic or backscattering*, where the incident and pertinent scattering directions are coincident but opposite in sense; *forward scattering*, where the incident and pertinent scattering directions are the same; and *bistatic scattering*, where the two directions are different. In far-field RFID systems, backscattering is used in the transmission of data from a tag to a reader.

The RCS of an object is dependent on a range of parameters, such as size, shape, material, surface structure, polarization, and the operating wavelength. The dependence of the RCS on the operating wavelength generally divides objects into three categories:

- *Rayleigh range.* The wavelength is much greater than the object dimensions, so there is little variation in phase over the length of the body. For objects smaller than around half the wavelength, the RCS exhibits a λ^{-4} dependency and therefore the reflective properties of objects smaller than 0.1λ can be disregarded in practice.
- *Resonance range.* The wavelength is comparable with the object dimensions – typically the object dimensions are taken to be between λ and 10λ. In this range, the electromagnetic energy shows a tendency to stay attached to the object's surface to create surface waves including traveling waves, creeping waves, and edge traveling waves. Objects with sharp resonance, such as sharp edges, slits and points, may at certain wavelengths exhibit resonance set-up of RCS. Under certain circumstances, this is particularly true for antennas that are being irradiated at their resonant wavelength.
- *Optical range.* The wavelength is much smaller than the dimensions of object. In this case, only the geometry and position (angle of incidence of the electromagnetic wave) of the object influence the RCS.

Antenna Scattering

Backscattering RFID systems employ antennas with different configurations as scatters. As antennas are always designed to transmit and receive EM waves, they are generally regarded as having two modes of scattering [19, 21]. The first is structural mode, the scattering that occurs because the antenna is of a given shape, size, and material and is independent of the fact that the antenna is specially designed to transmit or receive RF energy. The second is antenna mode, the scattering that has to do directly with the fact that the antenna is designed to radiate or receive RF energy and has a specific radiation pattern. The principles of antenna scattering modes are presented in Figure 3.14.

The antenna RCS, σ, can conceptually be defined as

$$\sigma = \sigma_{\text{struct}} + \sigma_{ant}. \tag{3.28}$$

Although the concept of dividing the antenna RCS into two components is simple and easily grasped, it should be noted that there is no formal definition of these scattering modes [19].

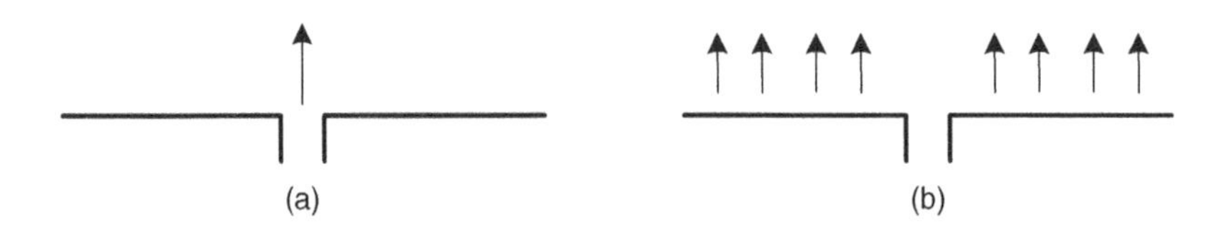

Figure 3.14 Antenna scattering: (a) antenna mode; (b) structural mode.

Antenna-mode RCS Equations

As discussed in Section 3.3.2.5, a tag antenna situated in the field radiated from the reader antenna collects the power from the incident wave and delivers part of it to the termination, namely the microchip with load impedance Z_T. The rest of the power is re-radiated into space by the tag antenna. The tag and its equivalent circuit are redrawn and shown in Figure 3.15 to illustrate the scattering mechanism of the antenna. The real part of the antenna impedance is split into two parts: the radiation resistance, R_r, and the ohmic loss resistance, R_L.

Referring to Figure 3.15, the voltage source represents an open circuit RF voltage applied on the terminals of the receiving antenna, and produces a current I through the antenna impedance Z_A and the terminating impedance Z_T. The current I is determined by the quotient

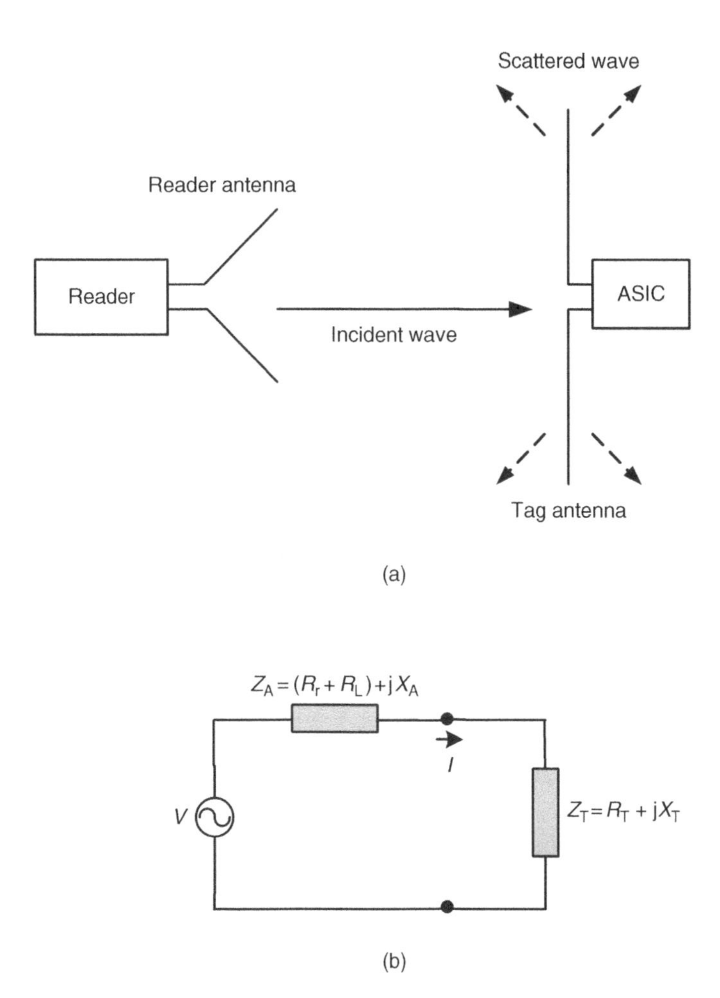

Figure 3.15 Schematic diagram of far field RFID tag scattering: (a) plane wave receiving and re-radiating of tag antenna; (b) equivalent circuit of the tag.

of the induced voltage V and the series connection of the individual impedances [18]:

$$I = \frac{V}{Z_A + Z_T} = \frac{V}{(R_r + R_L + R_T) + j(X_A + X_T)} \quad (3.29)$$

where I and V are the RMS or effective values.

The power delivered by the antenna to the microchip is

$$P_{\text{tag-chip}} = |I|^2 R_T = \frac{|V|^2 R_T}{(R_r + R_L + R_T)^2 + (X_A + X_T)^2}. \quad (3.30)$$

The effective aperture A_e of the antenna is the quotient of the received power $P_{\text{tag-chip}}$ and the incident wave density:

$$A_e = \frac{P_{\text{tag-chip}}}{S} = \frac{|V|^2 R_T}{S[(R_r + R_L + R_T)^2 + (X_A + X_T)^2]} \quad (3.31)$$

If the terminating impedance is the complex conjugate of the antenna impedance, i.e. $R_T = R_L + R_r$, and $X_A = -X_T$, the maximum effective aperture of the antenna is obtained:

$$A_{e-\max} = \frac{V^2}{4SR_T} \quad (3.32)$$

As can be seen from Figure 3.15, the current also flows through the antenna impedance Z_A. The real part of the impedance, R_A, has two parts: the ohmic loss resistance R_L, and the radiation resistance R_r, $R_A = R_L + R_r$. Therefore, some of the power will be dissipated as heat in the antenna as given by:

$$P_L = |I|^2 R_L. \quad (3.33)$$

The reminder is 'dissipated' in the radiation resistance. In other words, it is re-radiated into space by the antenna. This re-radiated power can be written as:

$$P_s = |I|^2 R_r = \frac{|V|^2 R_r}{(R_r + R_L + R_T)^2 + (X_A + X_T)^2}. \quad (3.34)$$

The scattering aperture of the antenna, A_S, or the antenna-mode RCS of the antenna can be defined as the ratio of the re-radiated power to the power density of the incident wave:

$$\sigma_{\text{ant}} = A_S = \frac{P_s}{S} = \frac{|V|^2 R_r}{S[(R_r + R_L + R_T)^2 + (X_A + X_T)^2]}. \quad (3.35)$$

If the antenna is operating under a maximum power transfer condition and lossless, that is $R_L = 0$, $R_r = R_T$, and $X_A = -X_T$, in which case

$$\sigma_{\text{ant}} = A_S = \frac{V^2}{4SR_r} \quad (3.36)$$

Therefore, in the case of conjugate impedance matching condition, $\sigma_{ant} = A_S = A_{e-max}$. This suggests that only half of the total power drawn from the incident wave is supplied to the terminating resistor R_T; the other half is re-radiated into the space by the antenna. When antenna is resonant short circuited [18], with R_T =0, and $X_T = -X_A$. The antenna-mode RCS of the antenna can be expressed as

$$\sigma_{\text{ant-max}} = A_{S-max} = \frac{V^2}{SR_r} = 4A_{e-max} \tag{3.37}$$

As a result, for the resonant short-circuit condition, the antenna-mode RCS is 4 times as great as its maximum effective aperture.

For the case when the antenna is open circuited, i.e. $Z_T \to \infty$; it is easy to get:

$$\sigma_{\text{ant-min}} = A_{\text{S-min}} = 0|_{Z_T \to \infty}. \tag{3.38}$$

The antenna-mode RCS can thus take any desired value in the range 0–$4A_{\text{e-max}}$ at varying values of the terminating impedance Z_T (as shown in Figure 3.16). In Particular, the antenna-mode RCS is ideally 4 times (or 6 dB) larger for the resonant short circuit relative to the conjugate matched case. This property is utilized for the data transmission from tag to reader in backscattering RFID systems.

It should be noted that the RCS can only be precisely calculated for simple structures – spheres, flat surfaces, and the like. Analytical derivation of the RCS of an antenna with

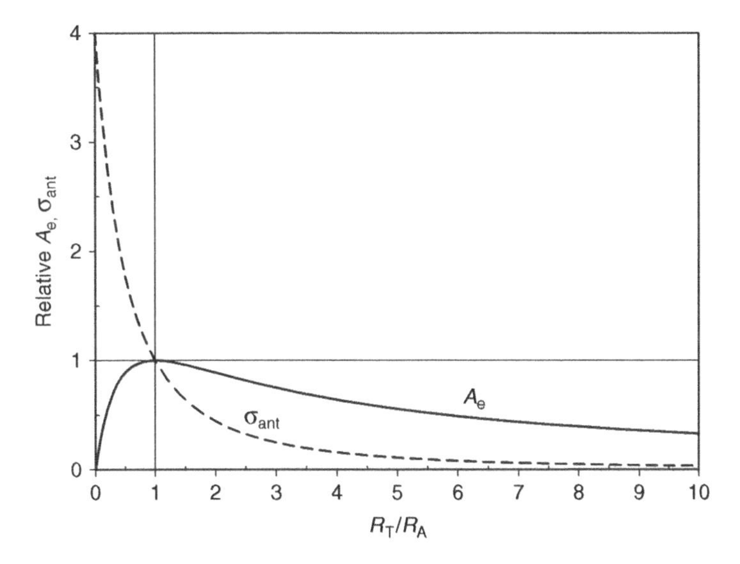

Figure 3.16 Variation of the effective aperture, antenna-mode RCS as a function of the ratio of the resistance R_T/R_A. (When $R_T/R_A = 1$ the antenna is operating under a conjugate matched condition. The case $R_T/R_A = 0$ represents a resonant short circuit at the terminals of the antenna. It is assumed that $R_L = 0$, and $X_A = -X_T$.)

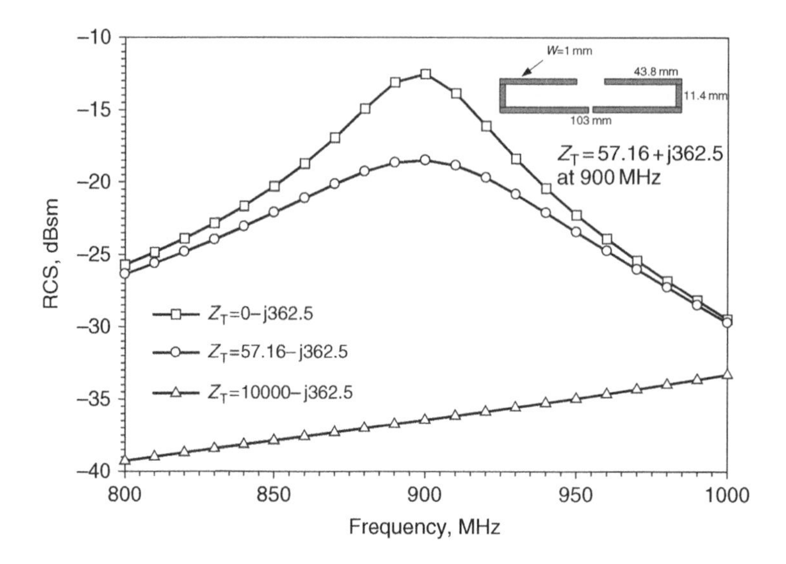

Figure 3.17 RCS of a folded dipole antenna with different terminating impedances, calculated by IE3D.

an arbitrary structure is difficult. Numerical methods such as the method of moments are generally used for antenna RCS calculation.

As an example to show the variation of antenna RCS with different terminating impedances, the RCS of a folded dipole antenna with three terminating conditions is shown in Figure 3.17. The impedance of the antenna is $Z_A = 57.16 + j362.5\,\Omega$ at 900 MHz. When the antenna is resonant short-circuited, i.e. $Z_T = 0 - j362.5\,\Omega$, the RCS of the antenna is maximum, $\sigma_{\text{ant-short}} = -12.55$ dBsm. When the antenna is conjugate matched, i.e. $Z_T = 53.16 - j362.5\,\Omega$, $\sigma_{\text{ant-matched}} = -18.51$dBsm. It is observed that $\sigma_{\text{ant-short}}$ is 6 dB larger than $\sigma_{\text{ant-matched}}$. For a higher terminating impedance of $Z_T = 1000 - j362.5\,\Omega$, as R_T/R_A is large, the RCS of the antenna decreases drastically ($\sigma = -36.42$ dBsm).

3.3.2.7 Case Study

Microchip Information

A chip with a plastic thin shrink small-outline package is used in this example. The electrical properties and outline of this chip are shown in Table 3.5 and Figure 3.18, respectively. It is found that the input impedance of the microchip is $Z_T = 11.5 - j422\,\Omega$ and its threshold operating power is -13 dBm at 915 MHz.

Tag antenna configuration and simulation

Several types of antennas have been reported to be used as passive far-field tag antennas, including meander line antennas, [22, 23], folded dipole antennas [24, 25], loop antennas [26, 27], slot antennas [28, 29], inverted-F antennas [30], planar inverted-F antennas [31, 32], slotted planar inverted-F antennas [33], patch antennas [34] and so on. Each type of antenna has its inherent characteristics for specific applications. For instance, a folded dipole antenna is used to demonstrate the design procedure. Its impedance can be adjusted easily by tuning

Table 3.5 Table 3.5 Electrical characteristics of the microchip.

Symbol	Parameter	Conditions	Min.	Typ.	Max.	Unit
Z_{867}	Input impedance	$T = 22^{\circ}C$, $f = 867$ MHz[a]		12.7−j457		Ω
$\mathbf{Z_{915}}$		**$T = 22^{\circ}C$, $f = 915$ MHz**[a]		**11.5−j422**		Ω
Z_{1450}		$T = 22^{\circ}C$, $f = 2450$ MHz[a]		3.7−j60.2		Ω
Z_{867}	Minimum operating power	$f = 867$MHz[b]		−14		dBm
$\mathbf{Z_{915}}$		**$f = 915$ MHz**[b]		**−13**		**dBm**
Z_{1450}		$f = 2450$ MHz[b]		−8		dBm

[a] Measured at typical 'Minimum operating power'.
[b] Values apply for operation with low modulation index (18%) and high return data rate (4 times forward link).

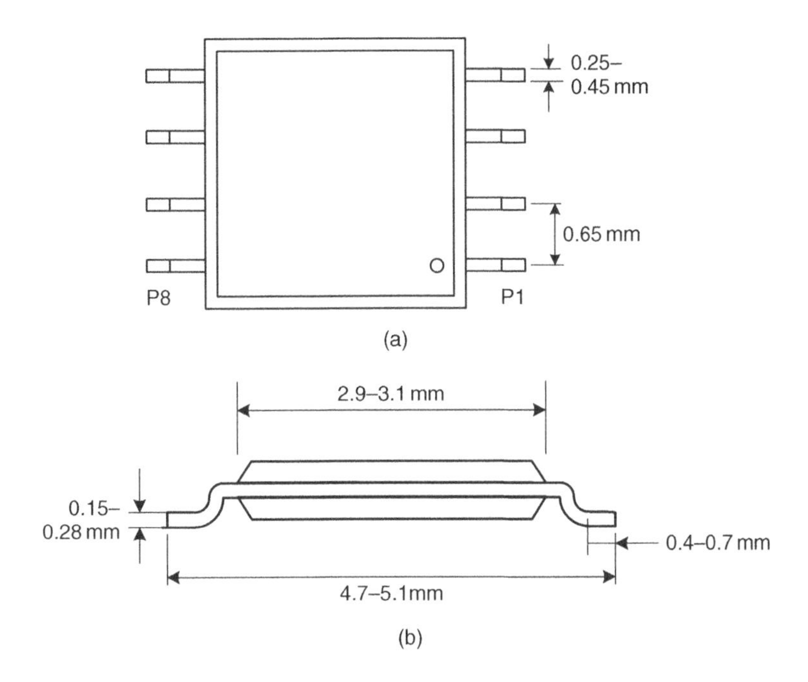

Figure 3.18 Package outline of the chip: (a) top view; (b) side view.

the length and width of the dipole. The antenna is structured on an FR4 substrate of 20 mils thickness, as shown in Figure 3.19.

According to (3.11) and (3.14), in order to determine the reading distance of the tag, the gain of the tag antenna, $G_{tag\text{-}ant}$, the power transmission, τ, and the RCS of the tag antenna, σ should be known. The gain of tag antenna cannot be directly obtained from simulation or measurement because the calculated or measured gain takes into account the mismatch loss of the antenna, and that mismatch loss is based on the fact that the terminating impedance is 50 Ω. However, the RFID tag antenna is always matched to the impedance of the terminating microchip, and the mismatch loss between the antenna and

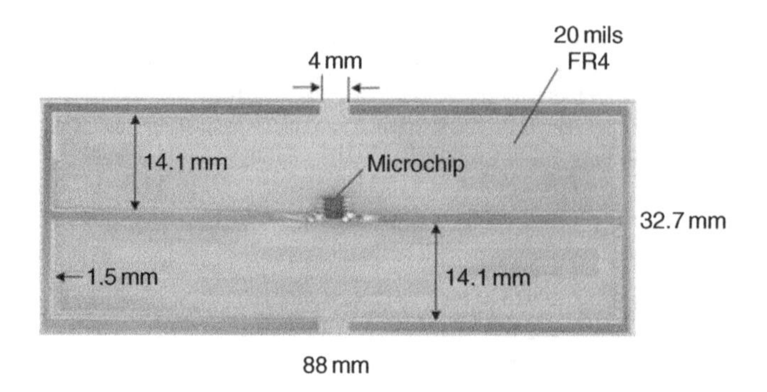

Figure 3.19 RFID tag with a folded dipole antenna.

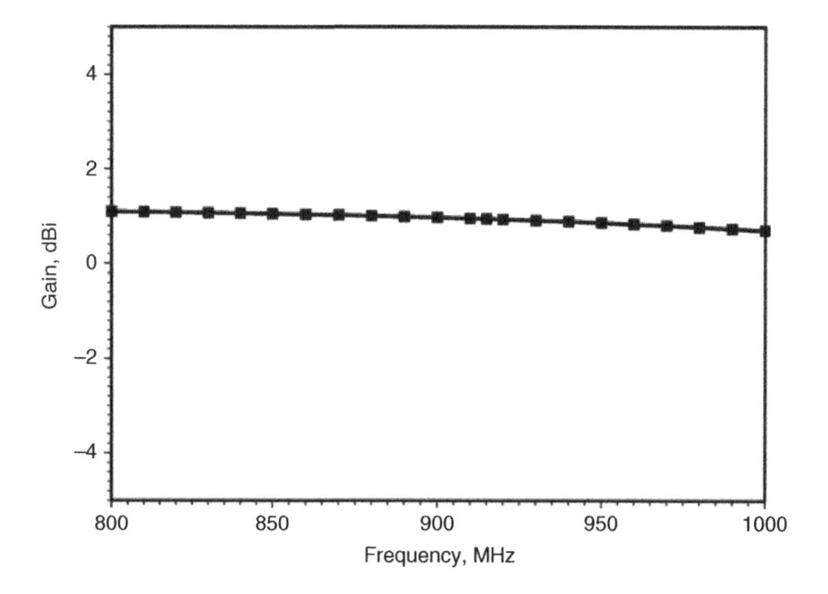

Figure 3.20 Gain of the folded dipole antenna, excluding the mismatch loss, calculated by IE3D.

the terminating microchip is quantified by the transmission coefficient, τ. The gain of the tag antenna can be calculated by multiplying the antenna directivity with the radiation efficiency. The calculated gain of the folded dipole antenna, shown in Figure 3.20, is 0.94 dB at 915 MHz.

The calculated input impedance of the antenna is shown in Figure 3.21. The real part varies from 23 to 55 Ω, while the imaginary part is from 250 to 650 Ω over 800–1000 MHz. The impedance is $Z_A = 34.0 + j428.8\,\Omega$ at 915 MHz. The return loss can be calculated through (3.22) and (3.24) by using Z_A and Z_T; it can also be obtained in IE3D by taking the terminating impedance to be $Z_C = 11.5 - j422\,\Omega$. The results are shown in Figure 3.22; the return loss is -5.84 dB at 915 MHz and the corresponding transmission coefficient τ is 0.74 or -1.32 dB.

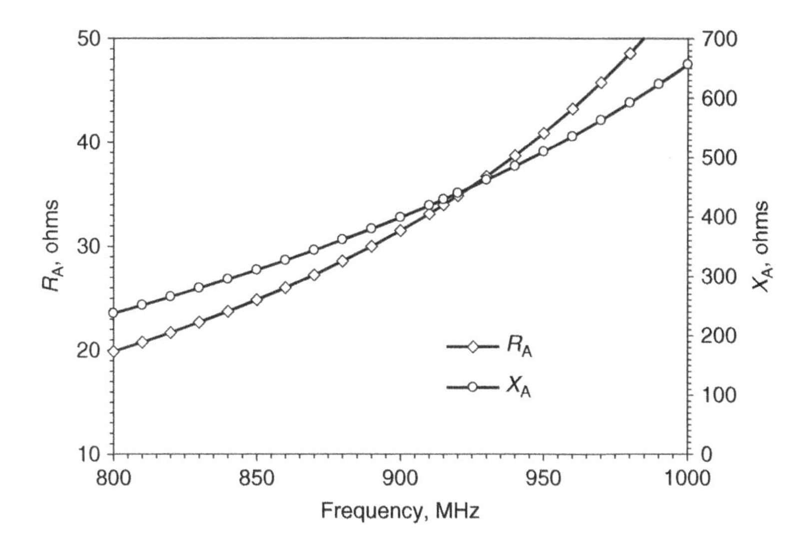

Figure 3.21 Impedance calculated by IE3D.

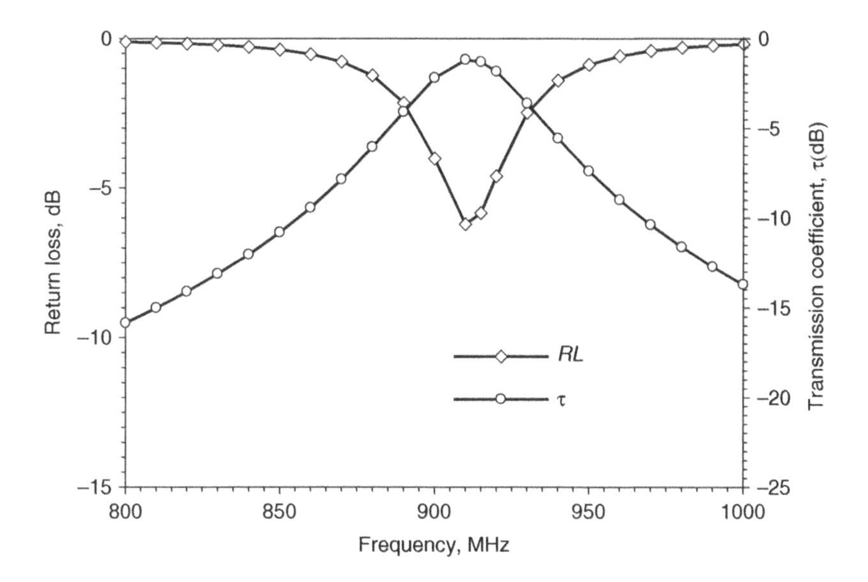

Figure 3.22 Calculated return loss and transmission coefficient.

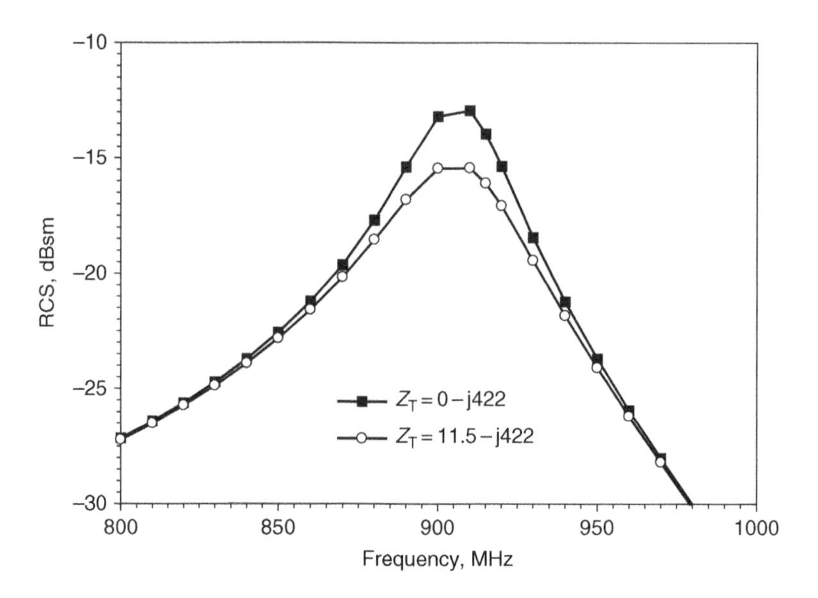

Figure 3.23 RCS of the folded dipole antenna, calculated by IE3D.

As discussed in Section 3.3.2.6, the RFID tag changes its chip impedance to modulate the backscattering signal. The microchip can be in either the short-circuited or other impedance condition. The simulated RCS of the antenna under the two impedance conditions is shown in Figure 3.23, where the value of σ is $-13.96\,\text{dBsm}$ for the short-circuited condition and $-16.11\,\text{dBsm}$ for $Z_T = 11.5 - j422\,\Omega$.

Reading Distance Calculation

Assume that the tag is used with a reader having the following parameters: operating frequency $f = 915\,\text{MHz}$; EIRP $= 4\,\text{W}$ ($P_{\text{reader-tx}} = 30\,\text{dBm}$, $G_{\text{reader-ant}} = 6$ dBic); sensitivity of the reader $P_{\text{reader-threshold}} = -65$ dBm. The parameters of the tag antenna are: antenna gain $G_{\text{tag-ant}} = 0.94\,\text{dB}$; power transmission coefficient of the tag $\tau = -1.32\,\text{dB}$; RCS of the tag $\sigma = -16.11\,\text{dBsm}$ (the smaller value); the polarization matching coefficient χ is -3 dB because the reader antenna is circularly polarized while the folded dipole is linearly polarized; the threshold operating power of the chip $P_{\text{tag-threshold}} = -13$ dBm. The reading distance can be calculated by (3.12) and (3.15):

$$R_{\text{power-link}} = 5.00\,\text{m}, \tag{3.39}$$

$$R_{\text{backscatter}} = 13.54\,\text{m}. \tag{3.40}$$

It is evident that the power-up distance, $R_{\text{power-link}}$, is smaller than the backscattering distance, $R_{\text{backscatter}}$. The reading distance of the system is thus determined by the smaller distance. This implies that the main concern in a tag antenna design should be the gain of the antenna and impedance matching between the antenna and the microchip.

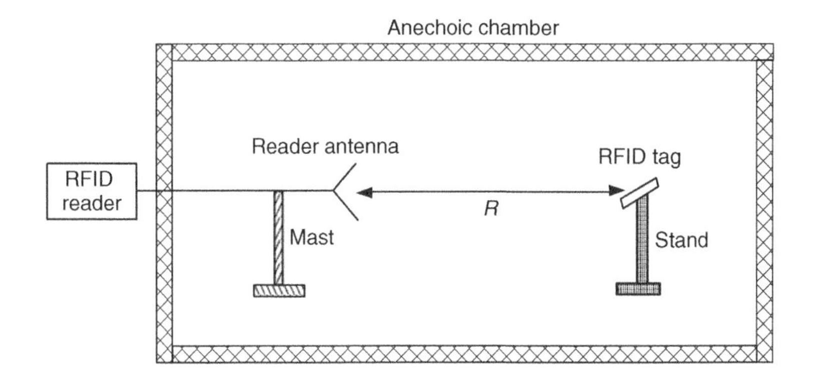

Figure 3.24 Reading distance measurement in an anechoic chamber.

Reading Distance Measurement

The reading distance can be measured by using a reader including a reader antenna with a known EIRP. To get more accurate results, the measurement should be conducted in an anechoic chamber to avoid multipath effects. The maximum distance at which the tag can communicate with the reader is recorded. The measurement set-up shown in Figure 3.24 uses a SAMSys MP9320 reader with 36 dBm EIRP, and the dimensions of the anechoic chamber are 10 m $\times$ 4 m $\times$ 4 m. The measured maximum reading distance is 4.85 m, which agrees with the calculated value (5.0 m) very well. The difference of 0.15 m may be caused by variation of the parameters of the microchip in terms of the load impedance and the minimum operating power level.

3.4 Effect of Environment on RFID Tag Antennas

RF tag performance can be affected by many factors, in particular the EM properties of objects near or in contact with the tag antenna. To date, there have been few published reports on this topic. Foster and Burberry [11] have performed basic measurements of an antenna near several objects. Raumonen *et al.* have studied a similar problem through simulations [35]. Dobkin and Weigand have also shown experimentally a decrease in the reading distance of several RF tags near a metal plate and a water-filled container, along with changes in the RF tag antenna input impedance near a metal sheet [36]. Griffin *et al.* have presented the power and backscatter communication radio link budgets that allowed the RF tag designer to quantify the effects of RF tag material attachment. The term 'gain penalty' has been introduced to describe the decrease in the RF tag antenna gain from its free-space value [37].

Among the objects the tags can attached to, metal and lossy materials such as metal cans and water are of great interest. The environment effect of these two objects is different for near-field coil antennas and far-field tag antennas because the EM field behavior differs dramatically in these two systems.

Water and other liquids have less effect on near-field RFID tags because the longer wavelengths of near-field systems are less susceptible to absorption. On the other hand, far-field RFID systems have a shorter wavelength and are more susceptible to absorption by liquids. In practical applications, near-field tags are more suited for tagging water- or liquid-bearing containers. Far-field tags can be made to work, but their effective reading distance would be drastically reduced.

While all frequencies cannot penetrate through a metal object, absorption by the object affects near-field and far-field tags differently. The reading distance of near-field tags may be reduced, whereas that of far-field tags may be increased if there is a certain air gap between the tag antennas and the metal surface where the object acts as a reflector of the tag antenna. This situation is unique for each particular application using far-field tags on the metal surfaces, and the same predictable results cannot be obtained in all cases. In situations where metallic materials are in part of the application, it is best to make use of the metal as part of the tag antenna by using the metal as the antenna ground. If this is not possible, shielding techniques are required.

3.4.1 Near-field Tags

3.4.1.1 Effects of Metal Material on Tag Antenna

Figure 3.25 shows the influence of the metal on the performance of a near-field tag. When the coil antenna is oriented parallel to a metal surface, the magnetic field generated by the coil

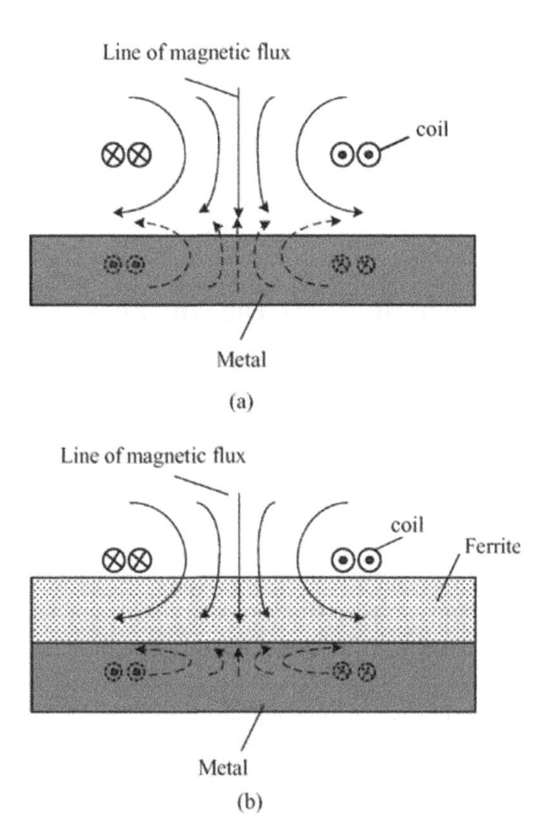

Figure 3.25 Coil antenna close to a metal surface: (a) magnetic field distribution of the coil antenna with a metal surface; (b) Using ferrite shielding to reduce the metal effect.

antenna reaches the surface of the metal and currents are induced within the metal to satisfy the boundary condition on the metal surface; that is, the magnetic field which is normal to the metal surface is zero. These currents, known as eddy currents, oppose the variation of the magnetic field that induced them and result in ohmic losses. The eddy currents can severely weaken the magnetic fields, and this leads to a reduction in the inductance of the coil antenna, thereby detuning the tag [2]. Systems with high Q factor are particularly sensitive to these changes.

Inserting highly permeable ferrite between the coil antenna and metal surface is an effective way to prevent the eddy current effect. Most of the magnetic field generated by the coil antenna will be concentrated in the ferrite material, therefore very little magnetic field can reach the metal surface and induce eddy currents.

When ferrite is used to shield the metal, it is necessary to take into account the fact that the inductance of the coil antenna may be significantly increased because of the high permeability of the ferrite material. As such, the coil antenna should be adjusted in order to maintain the resonant frequency.

Figure 3.26 shows the calculated inductance variation of the coil antenna over a metal plate. The coil antenna configuration is described in Section 3.3.1.6 and shown in Figure 3.10. It is placed at the centre of the metal plate 200 mm long and 200 mm wide. The inductance of the antenna is decreased when it is close to the metal surface. When the coil antenna is directly attached to the metal plate ($d = 0$ mm), the inductance drops to 0.8 μH at 13.56 MHz, which is about half of the inductance (1.52 μH) in free space, resulting in considerable detuning of the tag. As the coil antenna is moved away from the metal plate, the inductance of the coil antenna is gradually increased. At a larger distance of

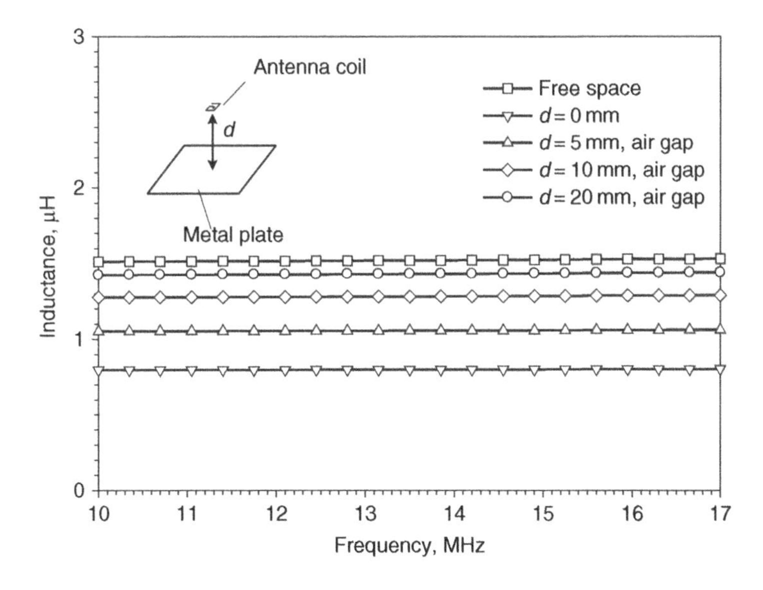

Figure 3.26 Inductance variations with metal environment, calculated by IE 3D; the dimension of the metal is 200 mm × 200 mm.

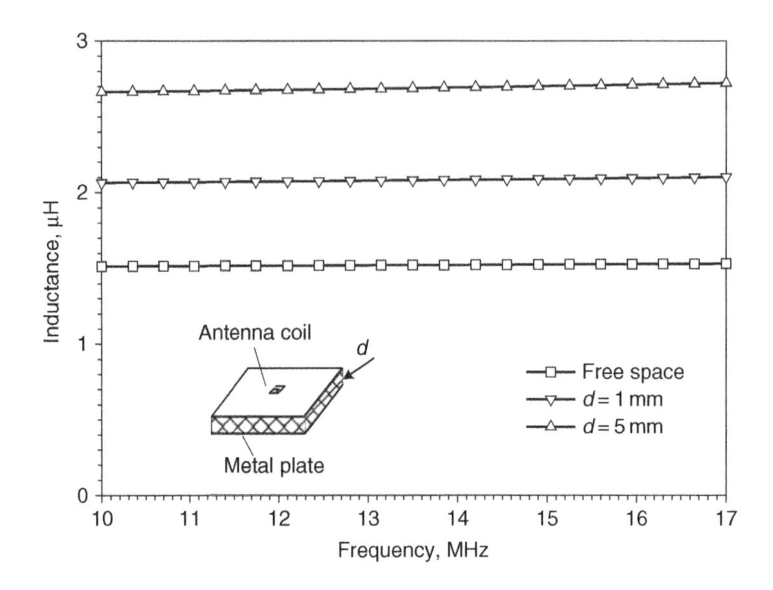

Figure 3.27 Inductance of the coil antenna on a metal plate with ferrite slab shielding. Dimension of the metal is 200 mm × 200 mm (by IE3D).

$d = 20$ mm, the effect of the metal is weakened, and the inductance is very close to that in free space.

Figure 3.27 shows the inductance of the coil antenna over a metal plate with the presence of ferrite material between them. The inductance of the coil antenna is 2.7 μH at 13.56 MHz when a 1 mm thick ferrite slab ($\varepsilon_r = 1.0$, $\mu_r = 20.0$) is used.

3.4.1.2 Effects of Water on Tag Antenna

Figure 3.28 illustrates the inductance of the coil antenna when it is close to or inside water. The size of the water model is 250 mm × 80 mm × 80 mm; and the electrical parameters are $\varepsilon_r = 77.3$, $\tan\delta = 0.048$ [37, 38]. It is observed that water results in very little reduction of the inductance of the coil antenna even when the coil antenna is placed inside the water. This suggests that the near-field RFID system is preferable for applications which require the tag to be attached to objects containing water or other liquids, such as soft drinks, milk, or oil.

3.4.2 Far-field Tags

In contrast with near-field RFID systems where the coil antenna is used for energy coupling, the effect of the environment effect on far-field RFID tags is more complicated. When a far-field tag is placed close to or directly affixed on an object, the performance of the tag will be affected by following factors:

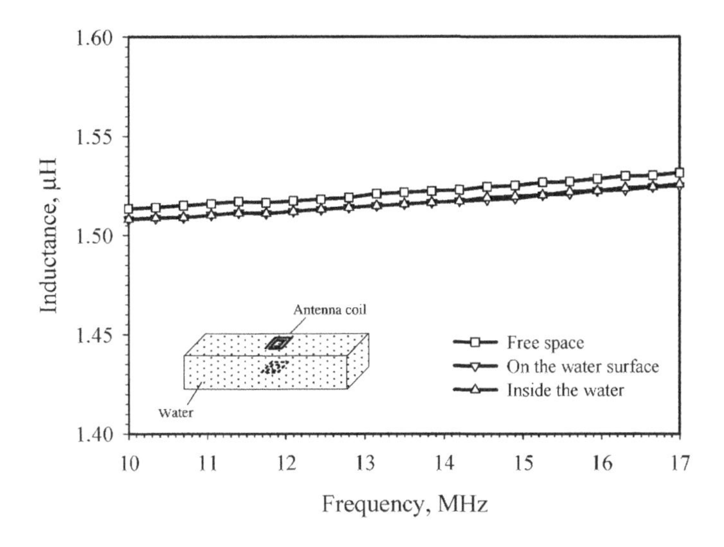

Figure 3.28 Inductance of coil antenna with water, calculated by IE3D.

- The properties of the object. The effect of a metal object is very different from that of a liquid object such as water.
- The radiation characteristics of the tag antenna. Generally, a directional antenna is less affected than an omnidirectional antenna.
- The positioning of the tag on the object, i.e. the distance of the tag from the object and the orientation of the tag.

From (3.14), it is known that for a far-field RFID system with a predetermined EIRP, the reading distance is determined by the characteristics of the tag including the gain of the tag antenna, $G_{\text{tag-ant}}$, power transmission coefficient, τ, and the threshold operating power of the chip, $P_{\text{threshold-chip}}$. Of these parameters, $G_{\text{tag-ant}}$ and τ will be dependent on the environment since the radiation characteristics and input impedance of the tag antenna will vary. In other words, the effect of the attached object on the performance of the tag can be assessed by the variation in the gain and input impedance of the tag antenna.

3.4.2.1 Effects of Metal Material on Tag Antenna

For a tag antenna which has its own ground plane (such as a patch antenna, planar inverted-F antenna (PIFA), or inverted-F antenna (IFA)), the metal effect is limited. The radiation patterns will only experience a slight distortion, while the input impedance will remain unchanged. For an antenna which does not have a ground plane such as a dipole, folded dipole, or meander line dipole, the radiation characteristics and input impedance of the antenna will be distorted when the tag is placed on or near a metal object.

The basic mechanism of the metal effect on an antenna is that the tangential electric field on the metal surface goes to zero. When an antenna such as dipole is in front of a conducting plane and oriented perpendicular to the plane, the effect of the conducting plane can be characterized by using an image source [39]. The direction of the image is the same as the original source to satisfy the boundary condition on the image plane, which results in an increase of the directivity of the antenna.

When the antenna is oriented parallel to the plane, the effect of the conducting plane can also be characterized by an image antenna which is equidistant below the image plane and oppositely directed. Generally, the directivity of the parallel positioned antenna is decreased, especially when the distance is very small because of the field cancellation of the two oppositely directed sources. When the antenna is far away, the directivity may be enhanced for some specific distance, depending on the antenna configuration, the distance, and the operating frequency.

Figures 3.29–3.32 show the effects of RFID tag antenna placed close to a metal plate. The antenna used is the folded dipole antenna discussed in Section 3.3.2.7; the antenna is positioned parallel to and above a metal plate measuring 250 mm × 250 mm. The directivity, radiation efficiency, gain, and input impedance are investigated for the different distances from the metal plate. When the antenna is placed very close to the metal plate ($d = 1$ mm), the directivity of the antenna increases considerably to about 8 dBi. However, the radiation efficiency drops significantly to 0.67 %, which results in very low gain of −13.97 dBi at 915 MHz. The input impedance, especially the real part, changes dramatically as well; the change of impedance causes poor impedance matching with the chip, resulting in a transmission loss of −19.2dB at 915 MHz. The large gain reduction and poor impedance matching result in the reading distance dropping to 0.15 m. When the antenna is moved further away from the metal plate, the directivity of the antenna remains almost unchanged. Meanwhile, the radiation efficiency shows an increasing trend, which results in antenna gain increasing. The real part of the input impedance shows little change, while the imaginary part tends to values close to those in free space. The reading distance of the tag is observed to have an increasing trend when it is far from the metal plate. The reading distance can be even more than that of the value in free space because the metal plate acts as a reflector

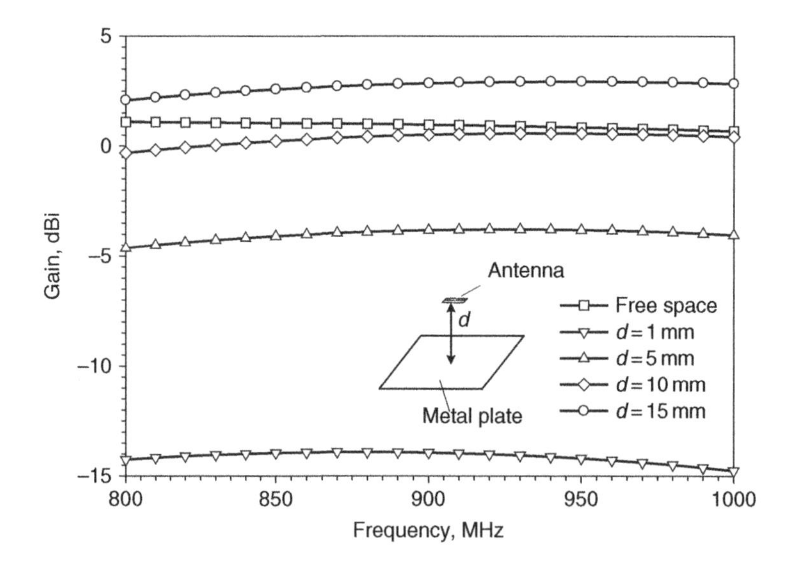

Figure 3.29 Gain of the antenna as a function of distance from metal plate (calculated by IE3D).

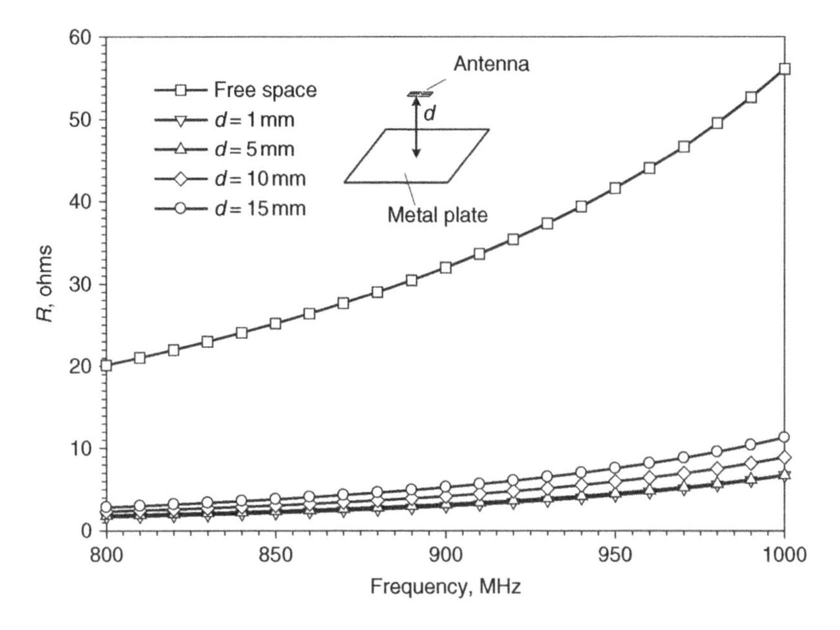

Figure 3.30 Real part of the input impedance of the antenna as a function of distance from metal plate (calculated by IE3D).

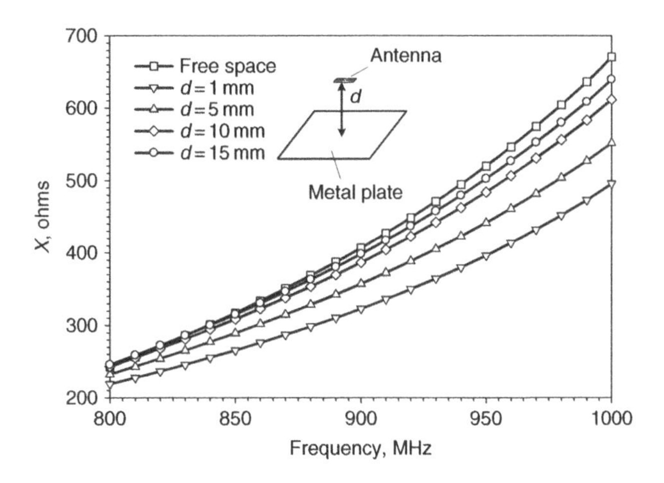

Figure 3.31 Imaginary part of the input impedance of the antenna as a function of distance from metal plate (calculated by IE3D).

which enhances the tag antenna gain. The detailed data on the metal effect on the tag antenna and reading distance at 915 MHz are summarized in Table 3.6. In conclusion, a metal object will degrade significantly the reading range of the tag when the tag is directly attached or very close to it. When the tag is allowed to be properly placed far from the metal object,

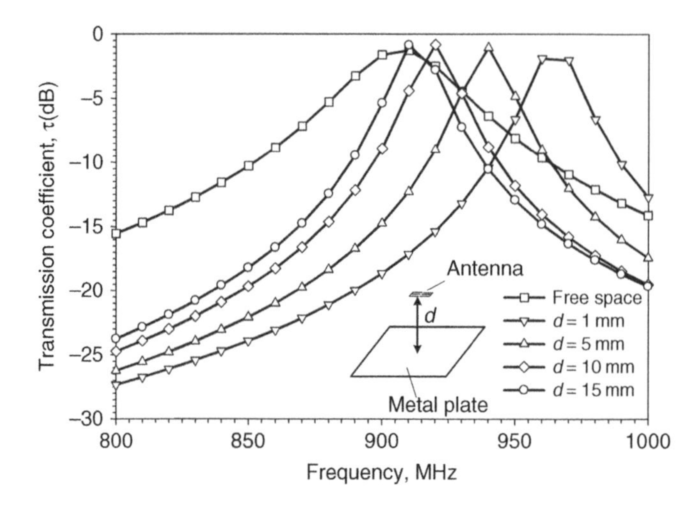

Figure 3.32 Transmission coefficient of the antenna as a function of distance from metal plate (calculated by IE3D).

Table 3.6 Effect of metal on the tag at 915 MHz.

	Directivity (dBi)	Radiation efficiency (%)	Gain (dBi)	Input impedance (Ω)	Transmission coefficient (dB)	Reading* range (m) $R_{power-link}$
Free space	1.81	81.84	0.94	33.65 + j427.00	−1.32	5.00
$d = 1$ mm	7.79	0.67	−13.97	3.20 + j336.00	−17.17	0.15
$d = 5$ mm	8.08	6.47	−3.80	3.45 + j372.50	−12.27	0.82
$d = 10$ mm	8.11	17.48	−0.53	4.49 + j404.30	−4.40	2.96
$d = 15$ mm	8.10	30.01	2.88	7.08 + j423.80	−0.86	6.60

[a] The system parameters were described in Section 3.3.1.6.

the reading distance of the tag may be enhanced because the metal object functions as a reflector.

3.4.2.2 Effects of Water on Tag Antenna

Figures 3.33–3.36 show the characteristics of an RFID tag antenna which is placed close to a water cuboid. As in the case of the metal plate, the antenna used is a folded dipole antenna and is positioned parallel to and above a water cuboid measuring 250 mm × 80 mm × 80 mm; $\varepsilon_r = 77.3$ and tan $\delta = 0.048$. The directivity, radiation efficiency, gain, and input impedance are investigated for the different distances away from the water cuboid. When the antenna is placed close to water ($d = 1$ mm), the directivity of the antenna increases while the radiation efficiency decreases significantly, which results in a reduction in the antenna

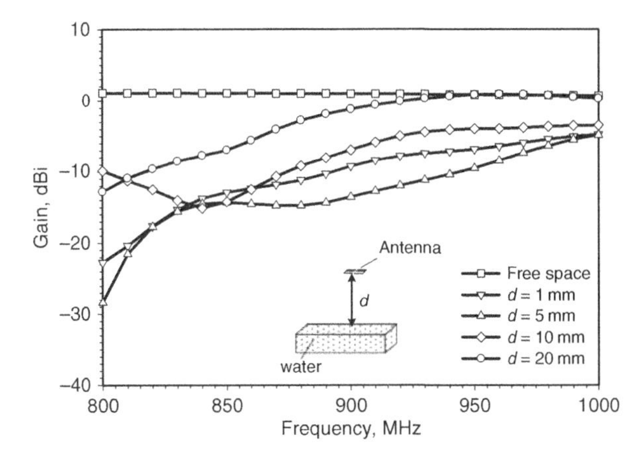

Figure 3.33 Gain of the tag antenna as a function of distance from water (calculated by IE3D).

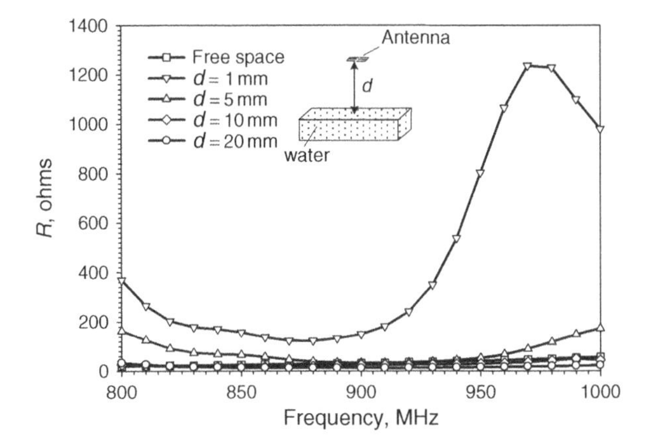

Figure 3.34 Real part of the input impedance of the antenna as a function of distance from water (calculated by IE3D).

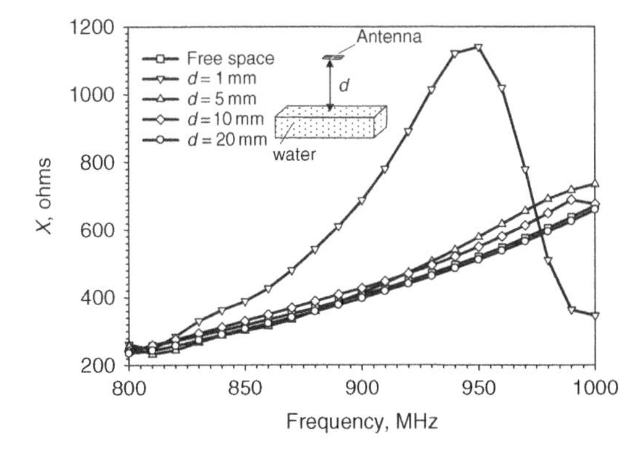

Figure 3.35 Imaginary part of the input impedance of the antenna as a function of distance from water (calculated by IE3D).

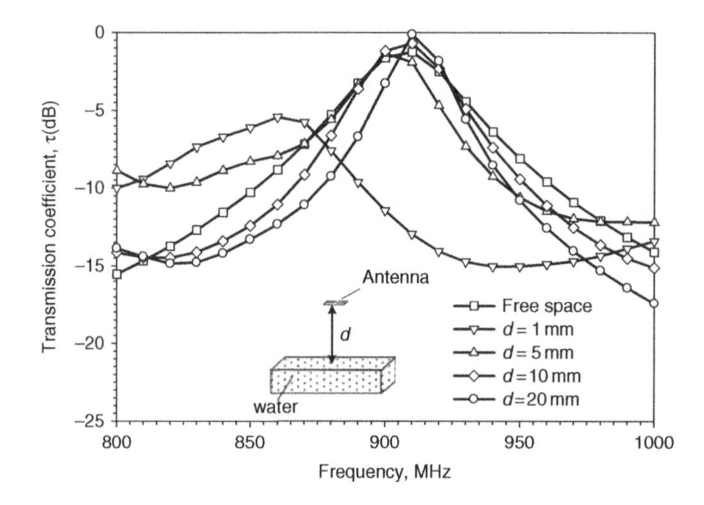

Figure 3.36 Transmission coefficient of the antenna as a function of distance from water (calculated by IE3D).

Table 3.7 Effect of water on the tag at 915 MHz.

Free Space	Directivity (dBi)	Radiation efficiency (%)	Gain(dBi)	Input impedance (ohms)	Transmission coefficient (dB)	Reading* distance(m) (power-link)
Free Space	1.81	81.84	0.94	33.65 + j427.00	−1.32	5.00
$d = 1$ mm	3.99	5.78	−8.39	181.30+j779.70	−12.96	0.45
$d = 5$ mm	2.44	3.03	−12.74	32.56 + j441.50	−1.90	0.97
$d = 10$ mm	2.64	14.17	−5.85	16.36 + j414.10	−0.47	2.52
$d = 20$ mm	4.61	30.28	−0.58	12.13 + j417.90	−0.14	4.81

[a] The system parameters were described in Section 3.3.2.7.

gain. In contrast with the metal plate, the water will always cause a reduction in the gain regardless of the distance between the water and the antenna. As antenna is moved further away, the antenna gain approaches the value obtained in free space. The input impedance shows a smooth variation except when the antenna is very close to the water ($d = 1$ mm).

The effect of the water on the tag antenna and reading distance at 915 MHz are summarized in Table 3.7. When the tag is very close to water, the reading distance drops significantly to 0.45 m. As the tag is moved further away, the effect of the water is decreased and the reading distance is enhanced.

3.4.3 Case Study

The results of measurements of the effect of various objects on a tag antenna are reported in this section. The measurement set-up is shown in Figure 3.37. The effect of the objects on

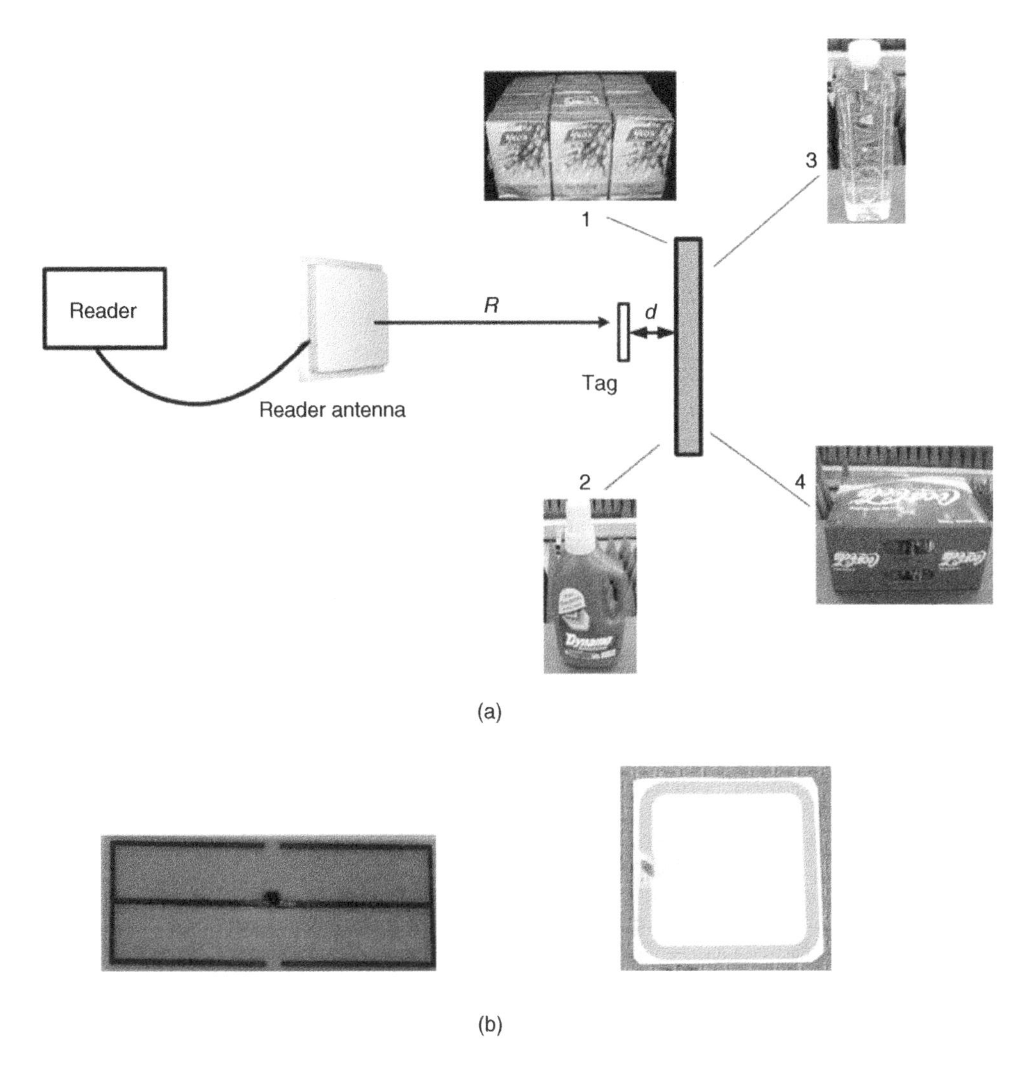

Figure 3.37 Measurement set-up for evaluating the effect of objects on an RFID tag: (a) measurement set-up and the selected items; (b) tags used in the evaluation.

the tag performance is evaluated by comparing the maximum reading distance. The tag is attached to four common household items. The items are packed soft drinks, liquid detergent, mineral water, and can-packed Coca-Cola. The first three items are categorized as lossy materials which have different water contents in them. The can-packed Coca-Cola is chosen to evaluate the effect of the metal object.

Two types of tag are selected for evaluating the effect of the four selected objects. One is the UHF tag discussed in Section 3.3.2.7, and the other is a tag developed by Philips at 13.56 MHz (I-code I) which is commercially available. The tags are mounted near or on the surfaces of the objects with different separations, d.

Table 3.8 Measured results for HF tag attached to different objects.

	Reading distance (R, cm)				
(d, mm)	Yeo's drink	Detergent	Mineral water	Canned Coca-Cola	Remark (free space)*
0	35	38	34	0	
1	36	38	38	0	
5	38	38	38	0	
10	38	38	36	23	41
15	39	39	38	25	
20	39	39.5	39	27	

*The reader used in the measurements is an Ormon V720S-BC5D4.

Table 3.9 Measured results for UHF tag attached to different objects.

	Reading distance (R, m)				
(d, mm)	Yeo's Drink	Detergent	Mineral water	Canned Coca-Cola	Remark (free space)*
0	0	0	0	0	
1	0.05	0.07	0.02	0.2	
5	0.34	0.75	0.28	0.5	
10	0.60	1.53	0.60	3.43	4.85
15	1.58	2.12	1.50	3.03	
20	2.25	2.90	2.10	3.01	

*The reader used in the measurements is SAMSys MP9320.

The results are tabulated in Tables 3.8 and Table 3.9. For an HF tag, the effect of water is minimal: only a few centimeters variation in the reading distance is observed. However, it is very sensitive to the metal object as it can be observed that the tag cannot be detected even when it is 5 mm away from the metal cans. For a UHF tag, both the lossy materials and metal object have a severe effect on tag performance. When the tag is affixed directly on the objects, the reading distance is zero. The reading distance of the tag is observed to increase when the tag is placed far away from the lossy objects because of their high dielectric loss.

On the other hand, the variation of the reading distance of the tag on the metal object shows a different trend. The tag cannot be detected when it is very close to the metal can. However, the reading distance is enhanced as the tag is moved away, and achieves the maximum at a specific separation ($d = 10$ mm). It should be noted that the results obtained here are specific to this particular scenario and will vary for different configurations.

3.5 Summary

This chapter has introduced RFID fundamentals and presented design considerations for RFID tag antennas as well as the method of evaluating the performance of the tags. From an antenna design point of view, RFID systems are preferably classified as near-field or far-field systems. Near-field RFID systems usually use inductive coupling for the energy transfer from readers to tags, and the load modulation technique for communication between reader and tags. In far-field RFID systems, the energy is transferred by capturing the electromagnetic waves, while the transmission of the information from tags to reader is achieved by using backscattered signals.

Generally, an RFID tag antenna is required to be small enough to be attached to or embedded into a specific object. It is often required to have specific radiation characteristics such as omnidirectional, directional, or hemispherical radiation patterns. The cost and reliability are the main considerations in mass production.

The antenna used in a near-field RFID tag is usually a coil. Such a coil is designed with a prior selected microchip. The coil antenna is configured to provide the inductance required for the circuit resonance at the operating frequency with a desired adequate Q factor. The spiral inductor is the most widely used and the inductance is determined by its geometrical parameters such as the length and width of the track, the separation between the tracks, and the number of windings.

Various types of far-field tag antenna have been reported. The antenna gain and impedance matching with the microchip are the main considerations in far-field tag antenna design. A high gain and good impedance matching will enable much power to be delivered to the microchip, providing a long reading distance.

In practical applications, RFID tags are always attached to specific objects. The varied characteristics of tag antennas suggest that tag performance is unavoidably affected by these object due to the EM coupling. For the near-field RFID tag antenna, the effect of the object is embodied by inductance reduction and field absorption, resulting in the detuning of the tag which weakens the signal and therefore causes a reduction in the reading distance. Generally, the near-field coil antenna is very sensitive to metallic objects but only slightly sensitive to lossy objects. Hence, a near-field RFID tag will be preferable for an application where the antenna is placed close to lossy materials.

Both lossy and metallic objects may considerably degrade the performance of far-field tag antennas. These objects mainly lower the radiation efficiency of the antenna, and also distort the impedance matching when the tag is placed very close to the objects. One way to minimize the effect of the object is to customize the RFID tag design by taking into account the property of the object during the antenna design. The other way is to adopt antennas which have their own ground plane. However, such antennas are usually bulky in size and their multilayer structures are not cost-effective for mass production.

References

[1] R. Want, An introduction to RFID technology. *Pervasive Computing*, 5(2006), 25–33.
[2] K. Finkenzeller, *RFID Handbook*, 2nd edn. Chichester: John Wiley & Sons, Ltd, 2004.
[3] T. Hassan and S. Chatterjee, A taxonomy for RFID. *IEEE System Sciences, 39th International conference*, Vol. 8, pp. 184b-184b, Jan. 2006.
[4] S. Cichos, J. Haberland, and H. Reichl, Performance analysis of polymer based antenna-coils for RFID. *IEEE Polymers and Adhesives in Microelectronics and Photonics International Conference*, pp. 120–124, June 2002.

[5] Item-level visibility in the pharmaceutical supply chain: A comparison of HF and UHF RFID technologies. http://www.tagsysrfid.com/modules/tagsys/upload/news/TAGSYS-TI-Philips-White-Paper.pdf.
[6] R.R. Fletcher, A low-cost electromagnetic tagging technology for wireless indentification, sensing and tracking of objects. Thesis, Massachusetts Institute of Technology, Cambridge, MA, 1993.
[7] G. Backhouse, RFID: Frequency, standards, adoption and innovation. *JISC Technology and Standards Watch*, May 2006. http://www.rfidconsultation.eu/docs/ficheiros/TSW0602.pdf.
[8] EPCglobal, Regulatory status for using RFID in the UHF spectrum. http://www.epcglobalcanada.org/docs/RFIDatUHFRegulations20060606.pdf.
[9] Anonymous, A summary of RFID Standards. *RFID Journal*. http://www.rfidjournal.com/article/articleview/1335/2/129/.
[10] Class 1 Generation 2 UHF Air Interface Protocol Standard Version 1.0.9. http://www.epcglobalinc.org/standards_technology/EPCglobalClass-1Generation-2UHFRFIDProtocolV109.pdf.
[11] P.R. Foster and R.A. Burberry, Antenna problems in RFID systems. *Proceeding of the IEE Colloquium on RFID Technology*, pp. 3/1–3/5, October 1999.
[12] K.V.S. Rao, P.V. Nikitin, and S.F. Lam, Antenna design for UHF RFID tags: a review and a practical application. *IEEE Transactions on Antennas* and Propagation, 53(2005), 462–469.
[13] V. Subramanian, J.M. J. Frechet, P.C. Chang, D.C. Huang, J.B. Lee, S.E. Molesa, A.R. Murphy, D.R. Redinger, and S.K. Volkman, Progress toward development of all-printed RFID tags- materials, processes, and devices. *Proceedings of the IEEE*, 93(2005), 1330-1338.
[14] D.R. Redinger, S.E. Molesa, S. Yin, R. Farschi and V. Subramanian, An ink-jet-deposited passive component process for RFID. *IEEE Transactions on Electron Devices*, 51(2004), 1978– 1983.
[15] I.D. Robertson (ed.), *MMIC Design*. London: Institution of Electrical Engineers, 1995.
[16] IE3D version 11, Zeland Software, Inc., Fremont, CA.
[17] http://www.emmicroelectronic.com.
[18] J. Kraus, *Antennas*. New york: McGraw-Hill, 1988.
[19] E. Knott, J. Shaeffer, and M. Tuley, *Radar Cross Section*, 2nd edn. Boston: Artech House, 1993.
[20] K. Kurokawa, Power waves and the scattering matrix. *IEEE Transaction Microwave Theory and Techniques*, 13(1965), 194–202.
[21] K. Penttilä, M. Keskilammi, L. Sydänheimo and M. Kivikoski, Radar cross-section analysis for passive RFID systems. *IEE Proceedings: Microwaves, Antennas and Propagation.*, 153(2006), 103–109.
[22] G. Marrocco, A. Fonte, and F. Bardati, Evolutionary design of miniaturized meander-line antennas for RFID applications. *Proceedings of the IEEE Antennas and Propagation Society International Symposium*, Vol. 2, pp. 362–365, June 2002.
[23] G. Marrocco, Gain-optimized self-resonant meander line antennas for RFID applications. *IEEE Antennas and Wireless Propagation Letters*, 2(2003), 302–305.
[24] X.M. Qing and N.Yang, A folded dipole antenna for RFID. *Proceedings of the IEEE Antennas and Propagation Society International Symposium*, Vol. 1, pp. 97–100, June 2004.
[25] R.L. Li, G. DeJean, M.M. Tentzeris, and J. Laskar, Integrable miniaturized folded antennas for RFID applications. *Proceedings of the IEEE Antennas and Propagation Society International Symposium*, Vol. 2, pp. 1431–1434, June 2004.
[26] A.S. Andrenko, Conformal fractal loop antennas for RFID tag applications. *Proceedings of the IEEE Applied Electromagnetics and Communications International conference*, pp. 1–6, October 2005.
[27] P.H. Cole and D.C. Ranasinghe, Extending coupling volume theory to analyze small loop antennas for UHF RFID applications. *Proceedings of the IEEE International Workshop on Antenna Technology Small Antennas and Novel Metamaterials*, pp. 164–167, March 2006.
[28] S.Y. Chen and P. Hsu, CPW-fed folded-slot antenna for 5.8 GHz RFID tags. *IEE Electronics Letters*, 40 (2004), 1516–1517.
[29] S.K. Padhi, G.F. Swiegers, and M. E. Bialkowski, A miniaturized slot ring antenna for RFID applications. *Proceedings of the IEEE Microwaves, Radar and Wireless Communications International conference*, Vol. 1, pp. 318– 321, May 2004.
[30] L. Ukkonen, L. Sydänheimo, and M. Kivikoski, A novel tag design using inverted-F antenna for radio frequency identification of metallic objects. *Proceedings of the IEEE Advances in Wired and Wireless Communication International Symposium*. on, pp. 91–94, 2004.
[31] M. Hirvonen, P. Pursula, K. Jaakkola, and K. Laukkanen, Planar inverted-F antenna for radio frequency identification. *IEE Electronics Letters*, 40 (2004), 848–850.

[32] W. Choi, N.S. Seong, J.M. Kim, C. Pyo and J. Chae, A planar inverted-F antenna (PIFA) to be attached to metal containers for an active RFID tag. *Proceedings of the IEEE Antennas and Propagation Society International Symposium*, Vol. 1B, pp. 3–8, July 2005.

[33] H. Kwon, and B. Lee, Compact slotted planar inverted-F RFID tag mountable on metallic objects. *IEE Electronics Letters*, 41(2005), 1308–1310.

[34] L. Ukkonen, L. Sydänheimo, and M. Kivikoski, Patch antenna with EBG ground plane and two-layer substrate for passive RFID of metallic objects. *Proceedings of the IEEE Antennas and Propagation Society International Sysmposium*, Vol. 1, pp. 93–96, June 2004.

[35] P. Raumonen, L. Sydänneimo, L. Ukkonen, M. Keskilammi, and M. Kivikoski, folded dipole antenna near metal plate. *Proceedings of the IEEE Antennas and Propagation Society International Symposium*, Vol. 1, pp. 848–851, June 2003.

[36] D. M. Dobkin and S.M. Weigand, Environmental effects on RFID tag antennas. *Proceedings of the IEEE International Microwave Symposium*, pp. 135–138, June 2005.

[37] J.D. Griffin, G.D. Durgin, A. Haldi, and B. Kippelen, RFID tag antenna performance on various materials using radio link budgets. *IEEE Antennas and Wireless Propagation Letters*, 5 (2006), 247–250.

[38] A.R. Von Hippel(ed.), *Dielectric Materials and Applications*. New York: John Wiley & Sons, Inc.,1954.

[39] W.L. Stutzman and G.A. Thiele, *Antenna Theory and Design*, 2nd edn. New York: John Wiley & Sons, Inc., 1998.

4

Laptop Antenna Design and Evaluation

Duixian Liu and Brian Gaucher
Thomas J. Watson Research Center / IBM

4.1 Introduction

Wireless local area network (WLAN) use has increased tremendously over the past several years [1–7]. According to a new report [7], the WLAN market will grow at an annual rate of 30 % per year, and will hit $5 billion in 2006, The report also found that WLAN sales have increased 60 % compared to 2004. As a result, the unlicensed 2.4 GHz Industrial, Scientific and Medical (ISM) band has become very popular and is now widely used for several wireless communication standards. Examples include laptop computers with built-in 802.11 b/g WLAN capability and the newly developed Bluetooth™ technology for cable replacement to connect portable and/or fixed electronic devices. 802.11g devices can provide date rate up to 54 Mbps. For even higher data rate, 802.11a devices in the 5 GHz Unlicensed National Information Infrastructure (UNII) band with channel bonding techniques [8] or multiple input, multiple output (MIMO) technologies can be used [9, 10].

The initial implementations integrated these systems into portable platforms such as laptops using PC cards inserted into the PC card slot. However, laptop manufacturers have moved away from PC cards in favor of integrated implementations since wireless technologies have become more prevalent and lower cost. Integrated wireless solutions avoid the problematic issues of breakage and physical design constraints associated with external antennas. As a result, nearly all laptops on the market today have integrated WLAN devices. Until recently, system designers did not consider the wireless subsystem and the design did not include an antenna, when in reality integrated antennas can be a significant differentiator [11]. There are a plethora of articles [12–17] regarding all these systems, but few fully integrate the antenna as part of the system and platform, nor do they achieve the potential performance such integration can offer. The goal of this chapter is to highlight the specific design

Antennas for Portable Devices Zhi Ning Chen

challenges associated with antenna integration into laptops. The achievement of these goals will be illustrated through practical design examples including suggested test and integration methodologies to address the challenges outlined below.

There are three major challenges for antenna design associated with wireless integration into laptops. First, laptops are very densely packed electronic devices and there is little room for additional functions. Second, Federal Communications Commission (FCC) emission requirements have forced laptop manufacturers to make extensive use of conducting materials in the covers of the laptops or conducting shields just inside the laptop covers to minimize radiation from today's very high speed processors. Thus, it is difficult to place an antenna in an environment free enough of other conductors to create an efficient radiator. Third, the size, shape, and location of the antenna may be affected by other design constraints such as the mechanical and industrial design. It is therefore necessary to make engineering tradeoffs between the design, performance, and placement of the antenna on the one hand, and industrial and mechanical design, and the size of the laptop on the other. As an example, early results based strictly on analytical modeling, blind cut-and-try, or use of 'integratable' vendor solutions, yielded an integrated Bluetooth™ antenna solution incapable of reliable connectivity much beyond 1–3 meters [6], not even close to the Bluetooth™ advertised 'spec' of 10 meters. Surprisingly, vendor solutions that touted fully integrated design capability for Bluetooth™ were clearly advertising measurements of freestanding antennas. Once integrated with odd ground planes and cabling of a real system, the antennas fell far short of advertised performance. Selling an integrated system solution that falls short of user expectations creates disappointment, dissatisfaction, and will discourage wide acceptance of wireless technology. This is could be a disaster in the PC industry. Clearly, a better solution to this problem was required.

4.2 Laptop-Related Antenna Issues

4.2.1 Typical Laptop Display Construction

For most laptops with integrated wireless, the antennas are typically placed in the laptop display to ensure wireless connection performance. So it is necessary to have a basic understanding of the laptop display construction. Figure 4.1 shows a sketch of a basic display. It consists of a liquid crystal display (LCD) panel, two metal hinge bars (one on the left and one on the right of the display), a display cover, an optional thin metal foil, and a plastic bezel (not shown). The thin metal foil is used to prevent electromagnetic interference in the case where a plastic cover is used. If the display cover is made from metal, typically aluminum or magnesium, or carbon fiber reinforced plastic (CFRP), the thin metal foil is not required. In old laptops or new low end laptops, the LCD panel is much smaller than the display cover, so an antenna can be placed almost anywhere around the gap between the LCD panel and the cover and achieve reasonable performance. However, for the newer laptops, especially the high end laptops, the gap is very small, typically between 3–7 mm, and the display is very thin as well. As a result, the space for integrated antennas is very limited. We will discuss the antenna locations in laptop displays in more detail in later sections.

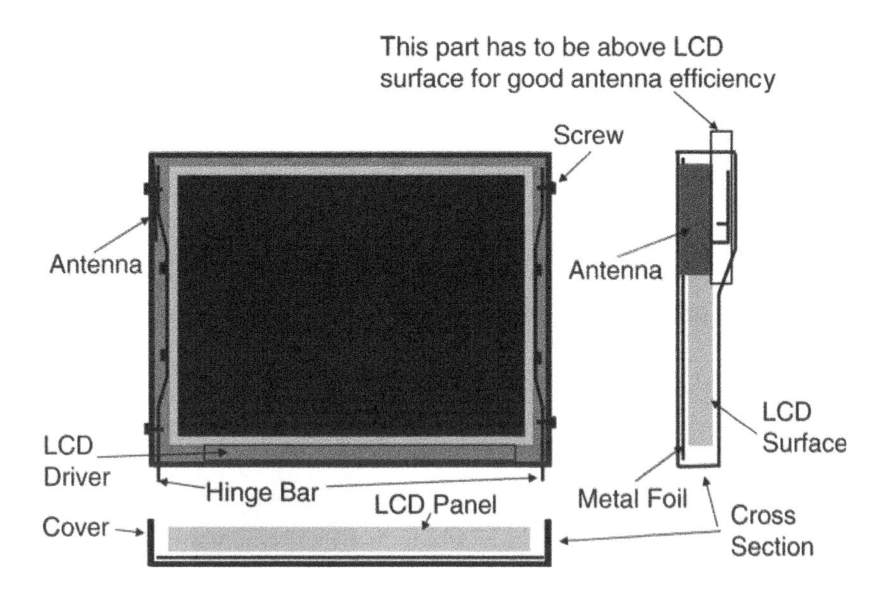

Figure 4.1 Basic laptop display construction. (Reproduced by permission of IBM.)

4.2.2 Possible Antennas for Laptop Applications

Figure 4.2 shows several possible antennas for laptop applications. Dipole and sleeve dipole antennas are basically the same, except that one is center fed and the other is end fed. Dipole antennas have wider bandwidth than sleeve dipoles, but sleeve dipoles are easier to use. In fact, sleeve dipoles were the first integrated antennas used in Apple iBook laptops. These antennas perform best if they are mounted on the top of the display. Helical and monopole antennas should also be placed on the top of the display to achieve their best performance. The helical antenna is physically small, but its bandwidth is narrower than that of the monopole antenna, making it problematic to match over the fairly broad ISM bands. In principle, traditional slot and patch antennas could be placed on the surface of the display, given their large size. However, these antennas have not been used due to mechanical and industrial design reasons. Ceramic chip antennas are typically helical or inverted-F (INF) antennas or their variations, with high dielectric loading to reduce the antenna size. They are small, but their bandwidth is too narrow. Slot and INF antennas belong to the same antenna category, and are good candidates for laptop applications because of their broader bandwidth characteristics. They are also very popular for laptop applications due to their overall performance, ease of integration, simple design and low cost.

For the traditional slot antenna [18], a slot, usually a half-wavelength long, is cut from a large (relative to the slot length) metal plate (see Figure 4.2). The center conductor of the coaxial cable is connected to one side of the slot. The outside conductor of the cable is connected to the other side of the slot. The slot antenna has a very large impedance at the center of the slot, and nearly zero impedance at the end of the slot. The feeding point is off-center to provide 50-ohm impedance and can be easily tuned by sliding it one way or the other. The slot antenna used in laptops is quite different from the traditional slot antenna. It is more like a loop antenna on an edge of a large metal plate.

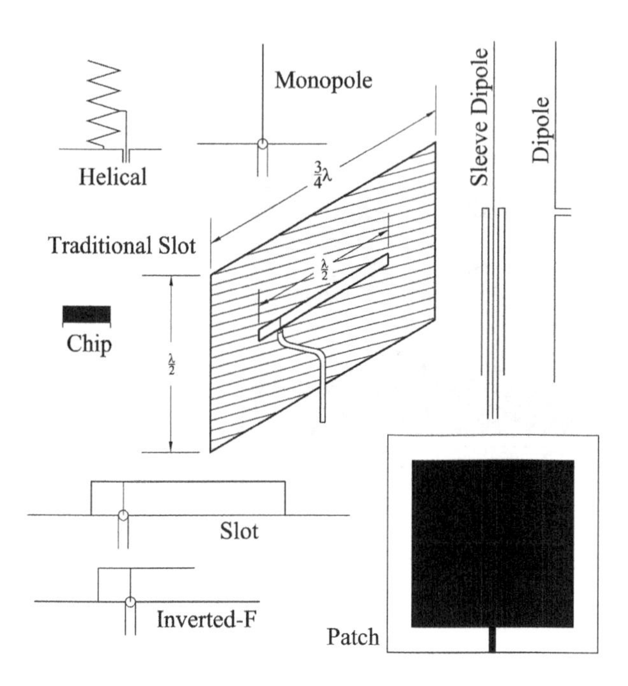

Figure 4.2 Possible antennas for laptop applications. (From [6]. Reproduced by permission of IBM.)

Slot and INF antennas have similar impedance characteristics [6, 19]. That is, moving the feed point to the slot end to decrease impedance (short end for the INF antenna) and moving the feed point to the slot center (open end for the INF antenna) to increase impedance. The slot length is a half-wavelength long for the slot antenna and a quarter-wavelength long for the INF antenna. Therefore, the length of the INF antenna is half the length of the slot antenna. This is an advantage for the INF antenna, since, in many applications, the space allocated for an antenna is very limited.

The slot antenna can be considered as a loaded version of the INF antenna. The load is a quarter-wavelength stub. Since the quarter-wavelength stub itself is a narrow band system, the slot antenna has narrower bandwidth than that of the INF antenna. This is another advantage the INF antenna has over the slot antenna.

The slot and INF antennas also have different radiation characteristics. For most implementations, the INF antenna has two polarizations and the radiation pattern is relatively omnidirectional. This is the third advantage it has over the slot antenna. The slot antenna primarily has one polarization and the radiation pattern is less omnidirectional than that of the INF antenna. However, the slot antenna tends to radiate more energy in the horizontal direction, and therefore has more useful energy for wireless LAN applications than the INF antenna does.

4.2.3 Mechanical and Industrial Design Restrictions

For laptop applications, the laptop itself is an integral part of the overall antenna system. Most antenna systems used for laptops can be considered as 'dipole-like' antennas. The antenna itself is one part (or monopole) of the dipole, and the other part is provided by the laptop. Antenna designers also view the laptop as the basic antenna element and the antenna itself as a tuning element. Since the laptop itself plays such a crucial role for the integrated antenna design, it is very important to study the antenna placement on laptops.

Figure 4.3 shows some typical antenna locations and antenna types for laptops. Even though sleeve dipole and monopole antennas have very good performance, they are mechanically weak, expensive to make, and are unattractive.[1] Industrial design trends discourage putting anything visible on the surface of the laptop display in order to maintain a thin and sleek appearance. Consequently, patch and chip antennas placed on the surface of the display are avoided. Chip and INF antennas have unacceptable performance if they are placed on the side of the laptop base. Base mounted antennas suffer not only from effects due to the blockage of the laptop system, especially the laptop display, but also from external environmental influences such as metal desks and the effects of users' hands or laps. A metal desk may significantly shift the tuning of base mounted antennas and create unwanted reflections that alter the omnidirectionality of the antenna. Absorption of the RF signal by a laptop user's hands and lap can have a dramatic effect on the effective antenna gain when the antenna is placed in the base of a laptop. Overall, an antenna should be placed on the top or close to the top of the display to achieve best coverage. The performance analysis below also supports placing antennas on displays.

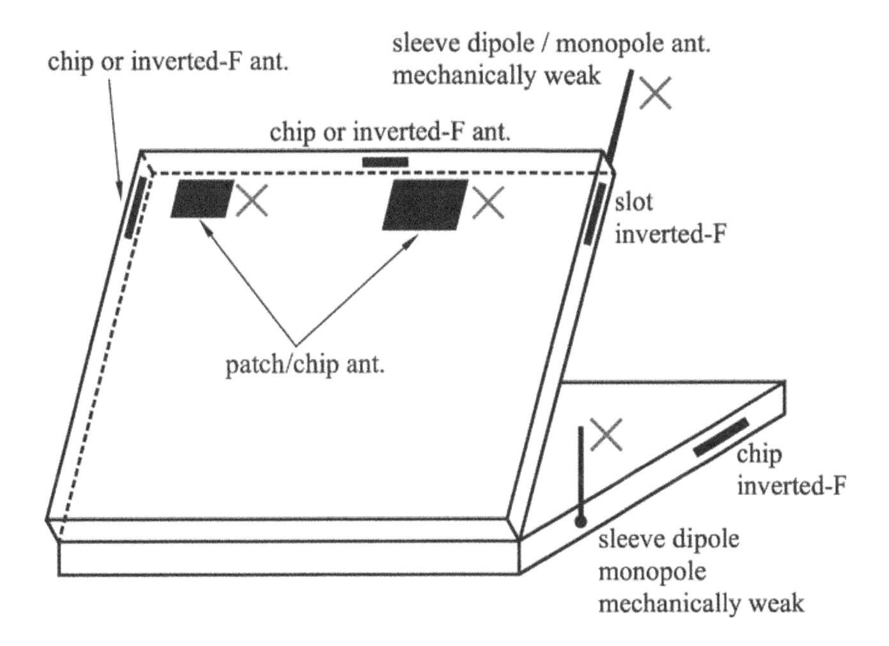

Figure 4.3 Possible antenna locations for several types of antennas. (From [6]. Reproduced by permission of IBM.)

[1] Note that the sleeve dipole used in the Apple iBook is embedded in the display, but the arrangement cannot be used in recent laptops due to the limited space available.

4.2.4 LCD Surface Treatment in Simulations

Since antennas are radiating devices, the antenna type, mounting location, mounting method and antenna environments such as antenna cover shape and material, as well as laptop structure and materials, all affect antenna performance. The integrated antenna locations inside laptops are not unique. They can be on the vertical or horizontal edges of the display. The locations can be near the end or the middle of a display edge. Depending on laptop platforms, the antenna can be mounted on a display hinge bar, an edge of a metal cover or on a separated ground plate. Display cover materials are a major issue. Currently four major cover materials are used: acrylonitrile butadiene styrene (ABS), CFRP, aluminum and magnesium. The electrical parameters such as dielectric constant and loss tangent (or conductivity) are unknown for CFRP. How to treat LCD in simulations is also an issue. Several researchers have treated the LCD as a metal box for simplicity [20–22]. However, their studies were concentrated on general antenna location evaluations. This assumption is not quite accurate for detailed antenna design and location study, since the LCD is part of the antenna system. Simulations indicate that if an INF antenna is placed on an LCD rim-like structure, one can observe hot spots periodically along the rim. This implies the rim itself is not good enough to be a reliable ground for the antenna. So the thinking is this: if the LCD surface behaves like a metal surface, the back metal plate with size GL × GW (see Figure 4.4) will have minimal effect on the antenna performance since the LCD is large and adequate to be a reliable ground plane for the antenna. On other the hand, if the LCD surface behaves like a plastic, the back metal plate will be required for reliable antenna performance.

Figure 4.4 shows the relevant parameter used for the LCD treatment study. The metal rim around the LCD panel is typically 5 mm wide. The LCD panel is about 5 mm thick. Figure 4.5 shows the simulated and measured standing wave ratio (SWR) of an INF antenna on an LCD panel with and without the metal plate. The solid and dashed lines are for the simulated results assuming the LCD surface to be metal with and without the metal plate,

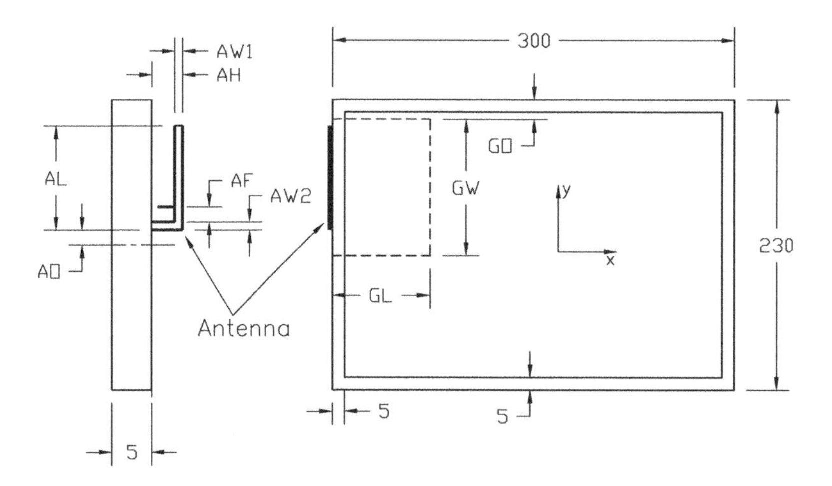

Figure 4.4 An INF antenna on a laptop LCD panel (GL = 70 mm, GW = 90 mm, GO = 25 mm, AO = 25 mm, AL = 28.5 mm, AH = 3 mm, AF = 2 mm, AW1 = AW2 = 2 mm. (Reproduced by permission of Lenovo.)

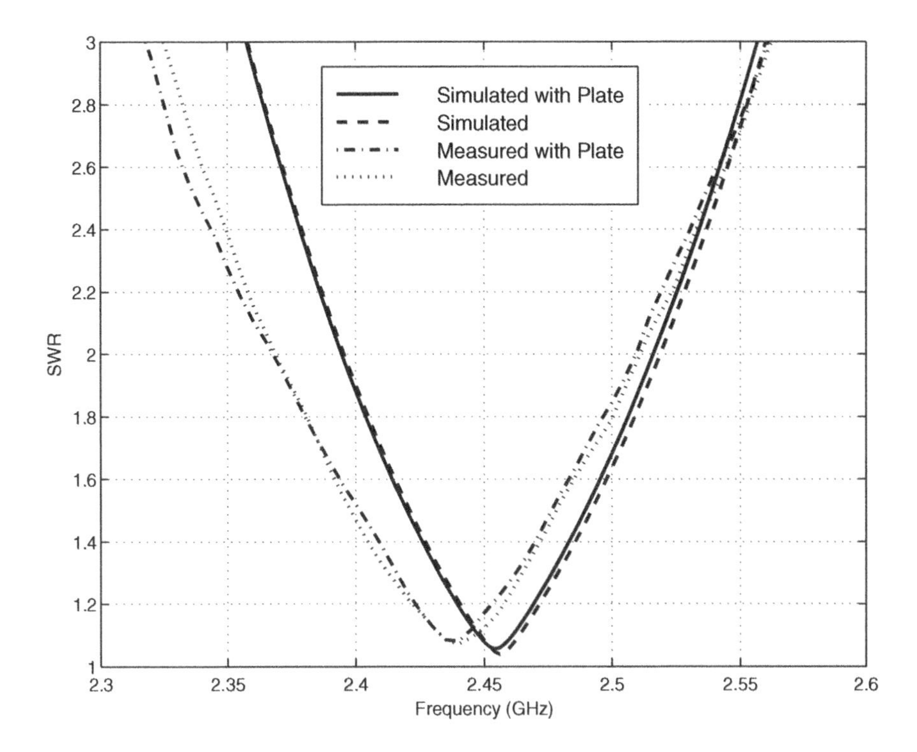

Figure 4.5 Measured and simulated SWR of an INF antenna on the edge of an LCD panel with and without a metal plate on the back. (Reproduced by permission of Lenovo.)

respectively. It is clear the metal plate has negligible effect on the SWR. The dash-dotted and dotted lines indicate the measured results with and without the metal plate, respectively. Again, the metal plate has negligible effect on SWR. However, there is a frequency shift between the measured and simulated results, and the measured SWR has wider bandwidths. This implies the LCD surface behaves more like a lossy metal.

4.2.5 Antenna Orientation in Display

When a planar antenna is placed in a laptop display, there are two major orientations used as shown in Figure 4.6 for slot antennas. Slot antennas are still used for the 2.4 GHz single band applications in laptops. In the first case, the antenna plane is parallel to the LCD plane. Since most LCD panels are almost as large as the LCD covers (or displays), there is no space to place the antenna inside the cover with this orientation, implying the antenna is external to the LCD cover. In the second case, the antenna plane is perpendicular to the LCD panel. This is a very compact way to integrate an antenna into a laptop display. The antenna performance is very similar in both cases when the laptop is open. However, when the laptop is closed, the first case still has good performance, since the laptop base has almost no interference with the antenna. For the second case, the antenna performance deteriorates tremendously, since the antenna is almost caged in by a metal box when the laptop is closed. In Bluetooth™ applications, the antenna has to work properly even when a laptop is closed.

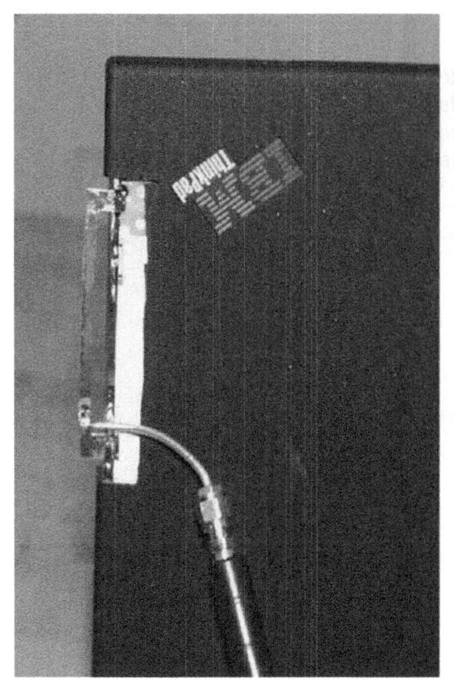

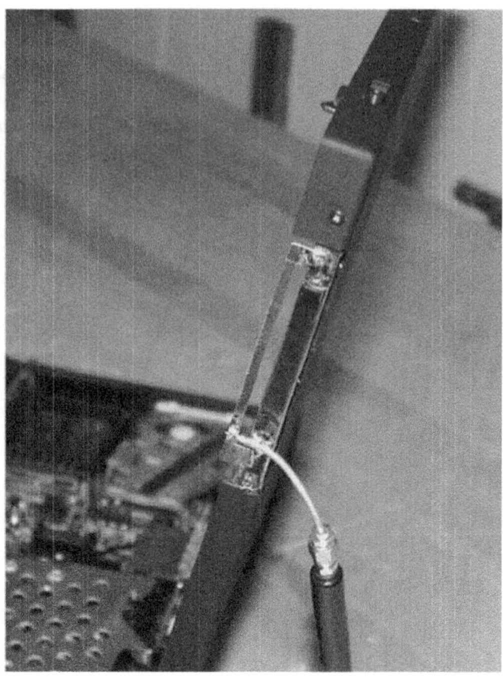

Figure 4.6 A slot antenna in parallel and perpendicular orientation on a laptop display. (Reproduced by permission of Lenovo.)

When an antenna is designed for one orientation, it usually does not work well for the other orientation. Figure 4.7 shows an INF antenna mounted parallel (left) and perpendicular (right) to an LCD panel-size metal plate. The antenna was optimized for the left case, so the antenna was centered around 2.45 GHz and well matched to the 50 Ω coaxial cable as shown in Figure 4.8 (solid curve). However, when the same antenna was mounted perpendicular to the metal plate, the center frequency shifted higher and the match became worse. It is expected the antenna performance will change much more than those shown in Figure 4.8 in a real laptop display environment. This indicates that it is difficult to design one antenna that fits in all laptops.

4.2.6 The Difference between Laptop and Cellphone Antennas

Many people would think laptop antenna design is easier than cellphone antenna design since laptops are much larger than cellphones, but this is a misconception. For cellphones being a radiating device, the system and industrial designers know that the antenna is a very important part of the cellphone design, so antenna space is allocated at the beginning of the design process. Usually the antenna is on the top of a cellphone. Cellphone antennas are usually close to a square shape and much thicker, more like a 3 D structure. Laptops started as non-radiating devices. As a result, the laptop system and industrial designers did not consider antennas in the early development, so the space allocated for them was usually

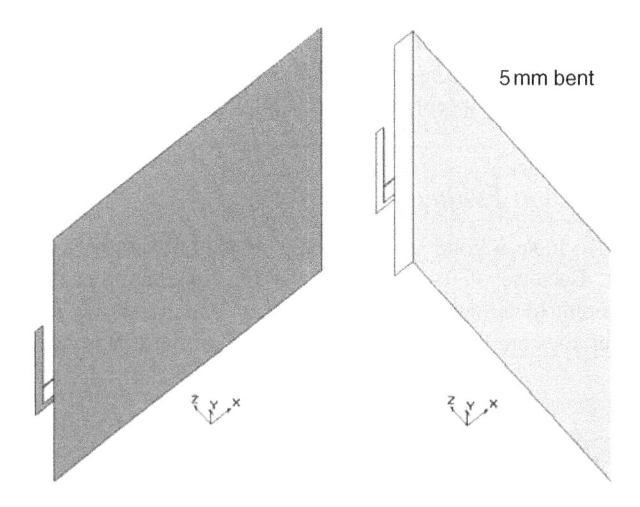

Figure 4.7 INF antenna mounted on an LCD-sized metal plate with parallel and perpendicular orientations. (Reproduced by permission of Lenovo.)

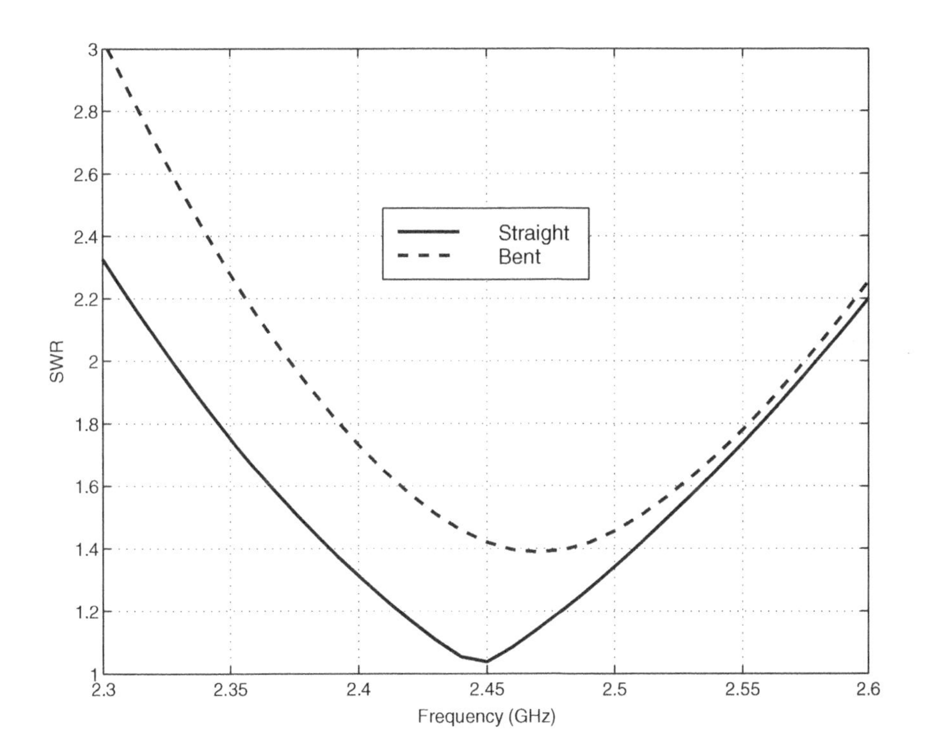

Figure 4.8 Measured and simulated SWR of an INF antenna on the edge of an LCD panel with/without a metal plate on back. (Reproduced by permission of Lenovo.)

very small and the location was often very bad. When an antenna is placed in the laptop display, the antenna is usually rectangular with a large length to width ratio. It is usually very thin, less than 0.5 mm. So a laptop antenna is typically a 2D structure. Therefore, a laptop antenna is much smaller in volume than a cellphone antenna.

4.2.7 Antenna Location Evaluations

It is very important to have a good understanding of the antenna performance effects due to antenna location. Because of its popularity, an INF antenna was used to examine its performance at different locations on a laptop. This antenna is shown in Figure 4.9. Since the antenna characteristics are dependent on its location on the laptop, an antenna tuned for

Figure 4.9 The INF antenna used for antenna location evaluations. (From [6]. Reproduced by permission of IBM.)

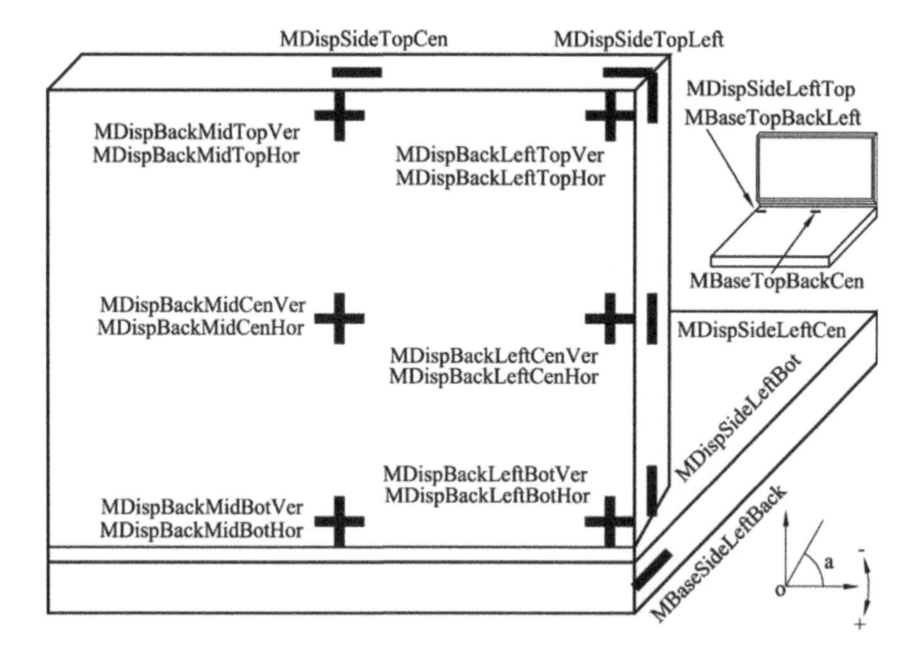

Figure 4.10 The INF antenna location and orientation on a laptop. (From [6]. Reproduced by permission of IBM.)

a particular location probably will not work as well for other locations. Therefore, some minor modifications to the antenna are necessary to ensure acceptable SWR in each case. The center frequency shift is left alone during the evaluations under the assumption that this is easily adjusted. Figure 4.10 shows the antenna locations and orientations. Due to laptop symmetry, only antenna locations on the left side of the laptop are considered. The laptop used has ABS plastic for its display and base covers. Other laptop models use CFRP (a very lossy material at radio frequencies) or metal covers, and the results shown here would not be applicable. Table 4.1 lists the peak and average gain values at these locations. In the table, 0 degrees represents the horizontal plane; negative and positive angles represent above and below the horizontal plane, respectively. The measurements have an azimuth scan from −180 to 180 degrees and elevation scan from −40 degrees (above horizontal plane) to 35 degrees (below the horizontal plane), both in 5-degree increments. The laptop was open with an angle between the cover and the base of 90 degrees. The frequencies listed in the table are those frequencies corresponding to the maximum average and peak gain values. The average gain is defined over an azimuth scan (360 degrees) for a given elevation angle. The table indicates that, except for the MBaseSideLeftBack location, placing the antenna high (center and top) or in vertical orientation tends to yield maximum radiation on or close to the horizontal plane. This is another indication that one should place the antenna as high as possible. Table 4.2 lists the center and resonating frequencies and 2:1 SWR of the

Table 4.1 Average and peak gain values (Freq in GHz, Ang in degrees, peak and average gains in dBi) vs. location.

Antenna location on laptop	Horizontal plane				Maximum average			Maximum peak		
	Freq	Ave	Freq	Peak	Ang	Freq	Ave	Ang	Freq	Peak
MdispSideTopCen	2.462	0.1	2.489	4.0	30	2.498	0.4	35	2.492	4.7
MdispSideTopLeft	2.468	−0.4	2.468	4.1	−40	2.489	1.2	−20	2.507	5.1
MdispSideLeftTop	2.600	−1.3	2.519	3.2	10	2.462	1.7	15	2.483	4.8
MdispSideLeftCen	2.480	0.2	2.462	4.1	35	2.447	1.1	35	2.440	5.2
MdispSideLeftBot	2.561	−1.9	2.450	3.5	35	2.444	2.2	35	2.426	5.4
MdispBackLeftTopVer	2.477	−0.9	2.498	3.5	15	2.417	0.7	20	2.417	5.0
MdispBackLeftCenVer	2.423	−0.2	2.426	3.5	−35	2.441	1.3	−40	2.423	4.3
MdispBackLeftBotVer	2.405	−1.6	2.396	3.2	−30	2.417	0.9	−30	2.408	5.5
MdispBackLeftTopHor	2.432	−1.0	2.432	2.5	−10	2.432	0.7	10	2.423	4.9
MdispBackLeftCenHor	2.408	0.4	2.417	5.3	5	2.423	0.9	−25	2.414	6.1
MdispBackLeftBotHor	2.408	−0.4	2.405	6.7	−35	2.453	2.0	10	2.435	6.9
MdispBackMidTopVer	2.441	−0.0	2.444	3.7	10	2.432	0.5	10	2.414	5.4
MdispBackMidCenVer	2.414	0.4	2.402	5.6	−30	2.432	1.5	5	2.417	5.7
MdispBackMidBotVer	2.423	−1.7	2.411	3.0	−35	2.423	−0.6	−35	2.432	4.1
MdispBackMidTopHor	2.354	−0.4	2.342	4.0	10	2.405	0.9	10	2.405	5.9
MdispBackMidCenHor	2.420	0.3	2.426	5.9	−30	2.417	1.0	5	2.420	6.4
MdispBackMidBotHor	2.411	−0.4	2.414	5.0	−40	2.456	2.8	−40	2.450	7.8
MbaseSideLeftBack	2.468	1.7	2.492	6.8	0	2.468	1.7	0	2.492	6.8
MbaseTopBackLeft	2.444	−1.4	2.438	3.4	−30	2.402	1.5	−35	2.405	7.0
MbaseTopBackCen	2.438	−2.6	2.402	1.3	−40	2.429	2.0	−35	2.426	7.1

Note: negative angles for above the horizontal plane. (From [6]. Reproduced by permission of IBM.)

Table 4.2 Center frequency and SWR values vs. location.

Antenna location on laptop	Freq and bandwidth (MHz)			SWR	
	Fcen	Fmin	Bandwidth	SWRcen	SWRmin
MdispSideTopCen	2500	2500	164	1.11	1.11
MdispSideTopLeft	2501	2495	154	1.10	1.08
MdispSideLeftTop	2483	2475	166	1.10	1.08
MdispSideLeftCen	2470	2475	164	1.18	1.17
MdispSideLeftBot	2490	2480	147	1.20	1.17
MdispBackLeftTopVer	2437	2435	137	1.19	1.18
MdispBackLeftCenVer	2425	2425	121	1.24	1.24
MdispBackLeftBotVer	2452	2445	142	1.25	1.24
MdispBackLeftTopHor	2445	2440	120	1.05	1.04
MdispBackLeftCenHor	2426	2425	96	1.19	1.18
MdispBackLeftBotHor	2429	2425	119	1.10	1.08
MdispBackMidTopVer	2428	2425	100	1.17	1.16
MdispBackMidCenVer	2427	2425	99	1.17	1.16
MdispBackMidBotVer	2429	2425	97	1.19	1.19
MdispBackMidTopHor	2427	2425	116	1.06	1.06
MdispBackMidCenHor	2422	2420	102	1.17	1.17
MdispBackMidBotHor	2442	2435	117	1.15	1.07
MbaseSideLeftBack	2460	2450	169	1.09	1.03
MbaseTopBackLeft	2416	2410	97	1.16	1.12
MbaseTopBackCen	2441	2440	88	1.09	1.09

(From [6]. Reproduced by permission of IBM.)

antenna at different locations. Note that the center frequency Fcen is slightly different from resonating frequency Fmin. The resonating frequency corresponds to the minimum SWR values. Table 4.2 indicates that the 2:1 SWR bandwidth will be wider if the antenna is placed on a small ground plane (side of display) or the edge of a large ground plane (backside of display). One has to remember that even though the laptop uses a plastic cover, metal foil and shields exist inside the cover to reduce emissions from laptops to meet FCC regulations.

4.3 Antenna Design Methodology

There is always an engineering tradeoff between technical rigor and time to market. There are myriad 3D electromagnetic tools for modeling antennas and devices, but even state of the art tools are incapable of timely accurate simulation of this problem. At the other end of the spectrum is the empirical approach of cut, try, and 'field test' in the laboratory. By themselves, neither solution is acceptable. A careful balance of the two, with a new test methodology and evaluation criteria, can provide an acceptable solution. The following describes a methodology that has been successfully used to design integrated antennas into laptop computers. There are three parts to the method: modeling, 'cut-and-try', and controlled measurement for comparison to specific metrics. While the method is not rigorous in the sense of producing a fully optimized antenna design that is completely characterized, it has proven to be a reasonably efficient technique for finding antenna designs with superior performance in the laptop environment: the highest data throughput, best range, and smallest number of dead zones.

4.3.1 Modeling

Depending on antenna types and implementation locations on a laptop, 3D antenna-modeling tools can be used. Modeling tools are very important for antenna structures such as patch antennas placed on a laptop display cover. However, simulation results from modeling tools can only be used as a guide for mobile antenna design. Since an antenna radiates, its performance is closely related to its environment. In most cases, modeling tools cannot treat these environments in detail due to geometries and computer compositions. Another big problem for mobile antennas is the small ground plane. Since the ground plane is small, the mobile device, in this case the laptop, is itself part of the antenna. Therefore, an antenna designed for freestanding operation will generally not work well when the antenna is installed on a laptop. For INF, slot, monopole, and dipole antennas placed on a laptop, cut-and-try design methods together with antenna measurements are more practical and productive approaches.

4.3.2 Cut-and-Try

Given the difficulty of modeling the antenna with all of the effects produced by the laptop, it is generally best to develop an antenna design which meets the size constraints imposed by the laptop mechanical and industrial designers. For example, an INF antenna might be modeled and built for use as a freestanding antenna. The next step would be to mount it in the laptop, observe the shift in its performance, and tune it for operation in the laptop environment. Clearly, some metric of the antenna performance is required.

4.3.3 Measurements

4.3.3.1 Standing Wave Ratio

Perhaps the most obvious metric is the antenna's center (or resonant) frequency and bandwidth. These parameters are easy to measure with a network analyzer and provide quick feedback on the effects of the laptop environment on antenna performance and on the tuning process itself. For these applications, the bandwidth is frequently defined as the frequency range over which SWR $< 2:1$. The goal of the design process is to have a bandwidth of the antenna covering the range of the radio frequency band, plus some margin for antenna manufacture. Since coax cables are lossy at 2.4 GHz and worse at 5 GHz, one needs to be careful to understand how this influences SWR measurements on long cables to achieve an accurate assessment of performance.

4.3.3.2 Average Gain

The SWR is a necessary but not a sufficient condition for antenna performance. To obtain another measure of the antenna performance it is necessary to consider, in more detail, the applications: WLAN and Bluetooth™. Both are indoor applications with operating ranges between 1 and 100 meters. Under these conditions, the signal at the receiving antenna is the sum of many scattered rays and, in the fringe areas (maximum range) of operation, there may be no dominant ray. In this case, one can assume the RF propagation environment may be described by Rayleigh statistics, which will be used later in the link budget description.

From the user's perspective, a good system is one that maintains a reliable high rate connection throughout the operating range and for any orientation of the laptop computer or position of the user. This requirement alone would argue for an omnidirectional antenna. However, as mentioned above, the received signal at the antenna comes from many different directions and the details of the antenna pattern are therefore 'blurred' by the RF scattering characteristics of the indoor environment. In fact, it can be argued that the most important metric of the antenna, after its center frequency and bandwidth, is its efficiency. That is, if the energy is radiated and not lost, it is a good antenna. Unfortunately, it is difficult to measure antenna efficiency in such conditions.

Another approach and methodology advocated here is to measure the 'average' gain of the antenna installed on the laptop in an anechoic chamber. There are numerous ways to define and measure the average gain. The most comprehensive would be to measure the antenna pattern over 4π steradians (all of the radiated energy), average the results over all angles and normalize the average with respect to an ideal isotropic radiator. This is, in principle, straightforward, but in practice is too tedious. The method used here determines the average gain from pattern measurements made in the horizontal (azimuth) plane for both polarizations of the electric field. The results are averaged over azimuth angles and normalized with respect to an ideal isotropic radiator. This is to say that the average gain at each frequency is calculated by

$$G_{\text{ave}} = 10\log_{10}\frac{\sum_{i=1}^{N}[G_h(\phi_i)+G_v(\phi_i)]}{N}, \tag{4.1}$$

where $G_h(\phi_i)$ and $G_v(\phi_i)$ correspond respectively to gain values of the vertical and horizontal polarized waves at the azimuth angle ϕ_i, and $\phi_1, \ldots, \phi_N$ are uniformly distributed over $[0, 360°]$.

This average gain is used in the link budget model to determine if the system performance is adequate. Clearly, this definition of average gain is not a comprehensive or rigorous characterization of the antenna's performance. It is, however, a measurement which can be done in a reasonable amount of time and is reproducible, since it is done in an anechoic chamber. In addition, results can be reproduced in different laboratories. It has proved to be a reasonable tradeoff between detailed measurements and the time constraints of developing antenna designs for products. Note that the 'average gain' definition does not need to be restricted to the horizontal plane, although this may be preferable; it can be defined over any elevation angle within $\pm 45°$ of the horizontal plane. For high performance applications, the average gain values from three or more different elevation angles (still within $\pm 45°$ of the horizontal plane) separated at least 15° can be further averaged. Of course, this average gain definition will place more restrictions on the antenna design.

Note that both antenna efficiency and radiation pattern are important for laptop applications. Some antennas have high efficiency but most radiation points to 'the sky'. As a result, the antenna still has weak performance. Ideally most of the radiated energy should stay within $\pm 45°$ of the horizontal pale when the laptop is open 90°. This is especially important for outdoor usage or indoor applications without much scattering and reflections. Integrated antennas should have omnidirectional radiation patterns on the horizontal plane. Major nulls on the radiation patterns should be avoided.

4.4 PC Card Antenna Performance and Evaluation

Nearly all laptop computers are equipped with one or two PC card slots for extended applications. Communications related PC cards, such as the Cisco Aironet wireless card, use the slot for the WLAN. The performance of these cards is laptop dependent. The antenna is typically placed at the outer end of a card to reduce the laptop effects, especially metal and carbon-filled plastic laptop cases, on the communication performance. Therefore, it is very useful to study signal strength versus the spacing between the laptop and the antenna.

The test setup shown in Figure 4.11 is for a 2.4 GHz Bluetooth™ radio subsystem. This experiment is intended to illustrate the effects of antenna placement and laptop materials. One laptop, an IBM 770 ThinkPad, has a popular vendor radio installed in the PC bay using an extender card so that the radio and its antenna are well removed from the conducting surfaces of the laptop. A simulated PC card slot opening was made from copper clad PC board material and was placed over the radio. This conducting surface represents shields or conducting plastics used in modern laptops. The card position could then be adjusted so that the antenna was outside the slot (positive displacement, *d*), flush with the opening of the slot ($d = 0$) or inside the slot (negative *d*). A second radio, installed in another laptop (not shown), was used to form a link and keep the radio under test transmitting. It was located so that its signal at the probe position was much weaker than the signal from the radio under test. The output of the log amplifier, which is proportional to the log of the power received at the probe antenna, was low pass filtered and displayed on an oscilloscope.

The experiment proceeded by setting the distance *d*, and measuring the output power of the radio under test. Since the probe antenna was not calibrated, only relative power levels were determined in the measurement. The position of the slot, *d*, was varied between −10 mm and 15 mm.

The results are shown in Figure 4.12 for each of the three possible carrier frequencies used by the radio (2.404 GHz, 2.441 GHz and 2.459 GHz). The relative output power of the transmitting radio is a function of its antenna position, d, relative to the conducting aperture

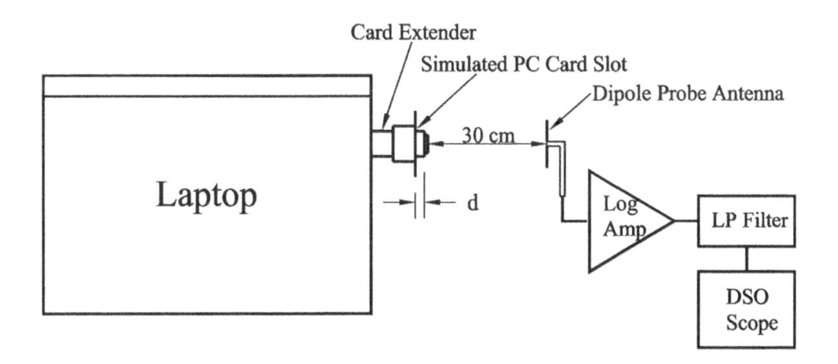

Figure 4.11 Top view of the PC card test setup (LP stands for low pass, DSO for digital storage oscilloscope). (From [6]. Reproduced by permission of IBM.)

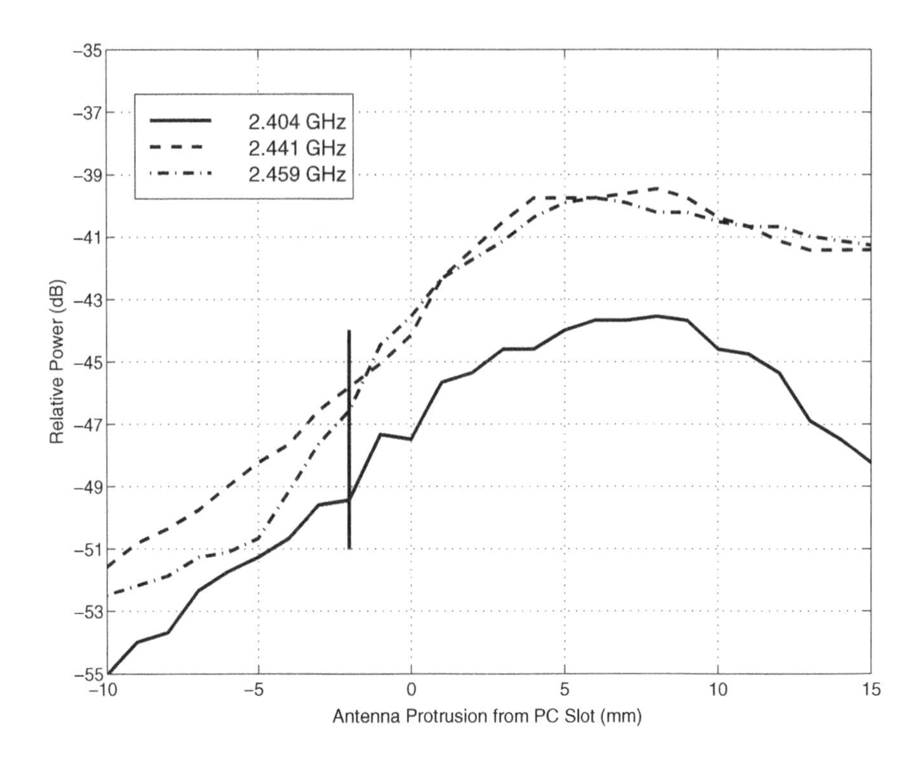

Figure 4.12 PC card result. (From [6]. Reproduced by permission of IBM.)

of the simulated PC card slot. Between -10 mm and $+4$ mm the sensitivity of the output power to this dimension is almost 0.8 dB/mm. The effect saturates at $d = 4$ to 5 mm. That is, once the antenna is 4 or 5 mm outside the conducting surface of the laptop there is little additional benefit to increasing the protrusion of the antenna. It should be noted that the transmitting power was measured only in one direction, which is nearly that of peak gain. Increasing the protrusion further might improve the omnidirectionality of the antenna.

The design of a high end laptop with a CFRP cover is indicated on the plot at $d = -2$ mm. The potential for improvement in antenna sensitivity is almost 6 dB if the antenna is moved to a position where it protrudes 4 mm from the laptop case.

To understand the impact the additional 6 dB has on the link range, consider the following: The radio vendor has reported a range with this radio of 5 m. We have developed a link budget model (see the next section) for the Bluetooth™ radio from which it is possible to determine the range as a function of antenna gain or output. Included in the model is a path loss exponent ($1/r^n$). For indoor environments and short range applications like Bluetooth™, $n = 2.5$ appears to be a reasonable value, while $n = 3.5$ is used for WLANs. With this model the 6 dB loss of power when the PC card radio is installed in the high end laptop reduces the range by over 40 %. If both ends of the link were to have these radios installed in this particular laptop model, the range would be expected to be about one third of the case where the output of the radios is unaffected by the laptops. The reduction in range for this laptop, based on the current measurements, is consistent with results reported by the vendor.

4.5 Link Budget Model

One metric of a radio system's performance, that receives a great deal of attention, is its range. Range is often quoted in the advertising of a product. Whether it is a cellphone, a cordless phone, or a WLAN, the user wants the longest possible range, while maintaining 'good' connectivity. When engineering a single subsystem of a radio, such as the antenna, it is necessary to know how 'good' it has to be. An approach used frequently is to develop a link budget model for the entire radio that can be used to understand the impact on system performance of each of its subsystems. A link budget model calculates the margin in the received signal to noise ratio (SNR) relative to the required SNR for acceptable error rate performance. It is an accounting of the signal power launched by the transmitter, the propagation and antenna losses, and the characteristics of the receiver such as noise figure and noise bandwidth. In dB, the link margin, LM, can be expressed as

$$LM = (S/N)_{\text{demod}} - (E_b/N_o) = S_{\text{demod}} - N_{\text{demod}} - (E_b/N_o), \tag{4.2}$$

where $(S/N)_{\text{demod}}$ is the effective SNR at the demodulator in dB and E_b/N_o is the SNR required for a low error rate also in dB.

S_{demod} can be divided into two parts: propagation losses and antenna characteristics; and receiver implementation. The basis, for the first part of S_{demod}, is the Friis transmission formula [18]. It gives the power, P_r, available at the terminal of the receiving antenna given the power, P_t, at the transmitting antenna. It assumes ideal free space propagation of an RF signal with matched transmitting and receiving antennas and polarization. P_r can be written as

$$P_r = P_t + 10\log_{10}(\lambda/4\pi r)^2 + G_t + G_r. \tag{4.3}$$

The separation distance of the antennas, r, is the range. λ is the wavelength in free space at the operating frequency. G_t and G_r are the gain values of the two antennas. The gain of an antenna is given by its radiation pattern and losses due to antenna construction materials. It is a parameter that is relatively independent of frequency in the band of interest so that the received power is proportional to the inverse square of the RF frequency. The transmitting antenna is taken to be an access point (AP) with $G_t = 0\,\text{dBi}$.

The term $(\lambda/4\pi r)^2$ is the free space path loss. It includes a term proportional to $1/r^n$. In free space $n = 2$. However, for indoor office environments there are walls and partitions that attenuate and reflect the signal, which can be accounted for with an effective path loss exponent $n > 2$. This work uses $n = 3.5$ for WLAN applications, based upon measured impulse channel sounding statistics across a variety of office structures and conditions.

In addition, there are numerous scatters. These have two effects on the propagating RF signal:

- Each signal reflection tends to randomize the polarization of the E field with respect to the polarization of the receiving antenna. This effect is included in the model with a polarization loss, PL, of 3 dB.
- The signal arriving at the receiving antenna is the vector sum of many signals. Since the environment is rarely static, the received signal depends on both position and time. This vector sum, commonly called fading, reduces the average value of the signal at the receiving antenna. The loss of signal is accounted for in the model with a Rayleigh fading parameter, R_f, approximately 7 to 8 dB, also based upon impulse channel soundings.

The second part of S_{demod} includes the cable loss, CL, of the cable connecting the antenna to the receiver's low noise amplifier, and any signal enhancements in the receiver such as equalizer gain, EQ, or diversity gain, DG. S_{demod} can be written:

$$S_{\text{demod}} = P_r - PL - R_f - CL + EQ + DG. \tag{4.4}$$

The noise at the demodulator, N_{demod}, may be written as

$$N_{demod} = -174\,\text{dBm} + NF + NBW, \tag{4.5}$$

where NF is the receiver noise figure and NBW is the receiver noise bandwidth in dB · Hz.

Through the use of a link budget model and antenna gain measurements made in a controlled environment (e.g. anechoic chamber) it is possible to calculate an 'average' range for the system and to understand how different antenna designs impact the range of the system. This method produces reproducible results; however, these should not be treated as equivalent to range measurements in any single environment, but as a statistical representation of what one might expect across a variety of environments. Such measurements are generally characterized by a wide spread in values that stem from the details of the scattering and losses in the propagating environment.

Table 4.3 Example of a link budget spreadsheet for the 802.11b WLAN.

Parameters	to AP 11 Mbps	Peer to Peer at Different Data Rate			
		11 Mbps	5.5 Mbps	2 Mbps	1 Mbps
Frequency (GHz)	2.45	2.45	2.45	2.45	2.45
Transmit power (W)	0.032	0.020	0.020	0.020	0.020
Transmit power (dBW)	−15.0	−16.9	−16.9	−16.9	−16.9
Tra. Antenna gain (dBi)	0.0	−2.0	−2.0	−2.0	−2.0
Polarization loss (dB)	3.0	3.0	3.0	3.0	3.0
EIRP (dBW)	−18.0	−21.9	−21.9	−21.9	−21.9
Range (m)	32.4	25.1	37.3	60.6	90.1
Path loss exponent (dB)	3.5	3.5	3.5	3.5	3.5
Free space path loss (dB)	88.6	84.7	90.7	98.1	104.1
Rec. antenna gain (dBi)	−2.0	−2.0	−2.0	−2.0	−2.0
Cable loss (dB)	1.9	1.9	1.9	1.9	1.9
Rake equalizer gain (dB)	0.5	0.5	0.5	0.5	0.5
Diversity Gain (dB)	5.5	5.5	5.5	5.5	5.5
Receiver noise figure (dB)	13.6	13.6	13.6	13.6	13.6
Data rate (Mbps)	11000	11000	5500	2000	1000
Required E_b/N_o (dB)	8.0	8.0	5.0	2.0	−1.0
Rayleigh fading (dB)	7.5	7.5	7.5	7.5	7.5
Receiver sensitivity (dBm)	−80.1	−80.1	−86.1	−93.5	−99.5
SNR (dB)	8.0	8.0	5.0	2.0	−1.0
Link margin (dB)	0.0	0.0	0.0	0.0	0.0

(From [6]. Reproduced by permission of IBM.)

Table 4.3 shows a link budget tool in spreadsheet form for the 802.11 b WLAN. In the range calculations, 7.5 dB of extra path loss due to the effects of multipath Rayleigh fading was assumed. Since data are typically sent in packets or blocks, a common system metric used to characterize performance is block error rate. This path loss represents an average power loss due to the effect of destructive multipath summation at the antenna while operating at the 10 % block-error rate system sensitivity point. The results are the ranges for a laptop to an AP (0 dBi transmitting antenna gain) at 11 Mbps data rate and to a peer laptop (−2 dBi transmitting gain) with different data rates at 11, 5.5, 2, and 1 Mbps, respectively.

4.6 An INF Antenna Implementation

Most laptop antennas used in laptops are related to INF antennas, whether they are single band or double band antennas. Figure 4.13 shows an INF antenna integrated in a laptop prototype for 2.4 GHz applications. The antenna is stamped from a brass sheet and mounted

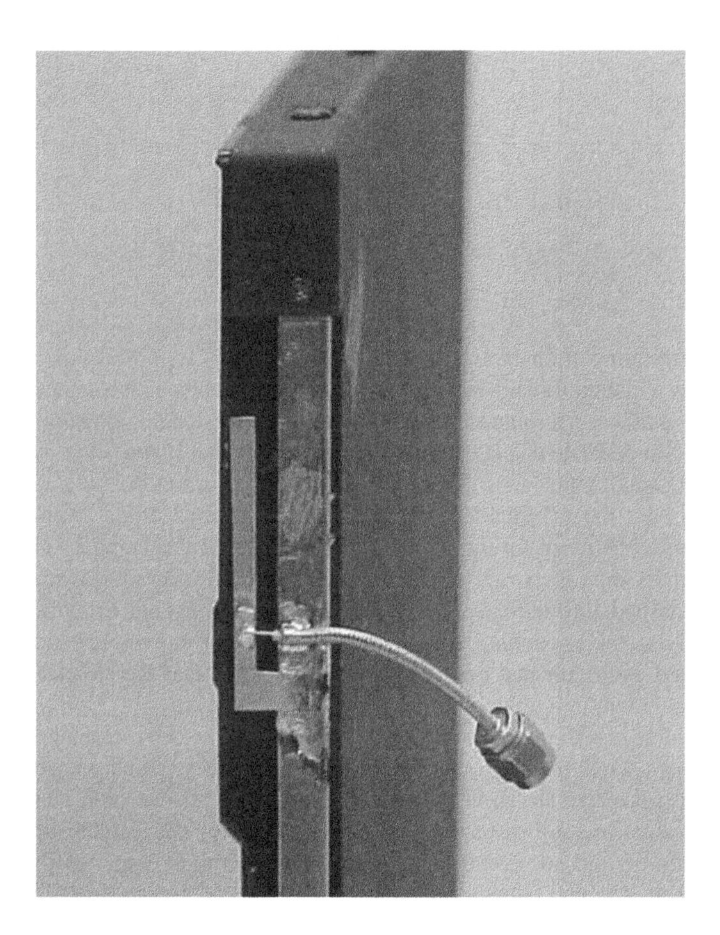

Figure 4.13 INF antenna integrated in a laptop prototype. (From [6]. Reproduced by permission of IBM.)

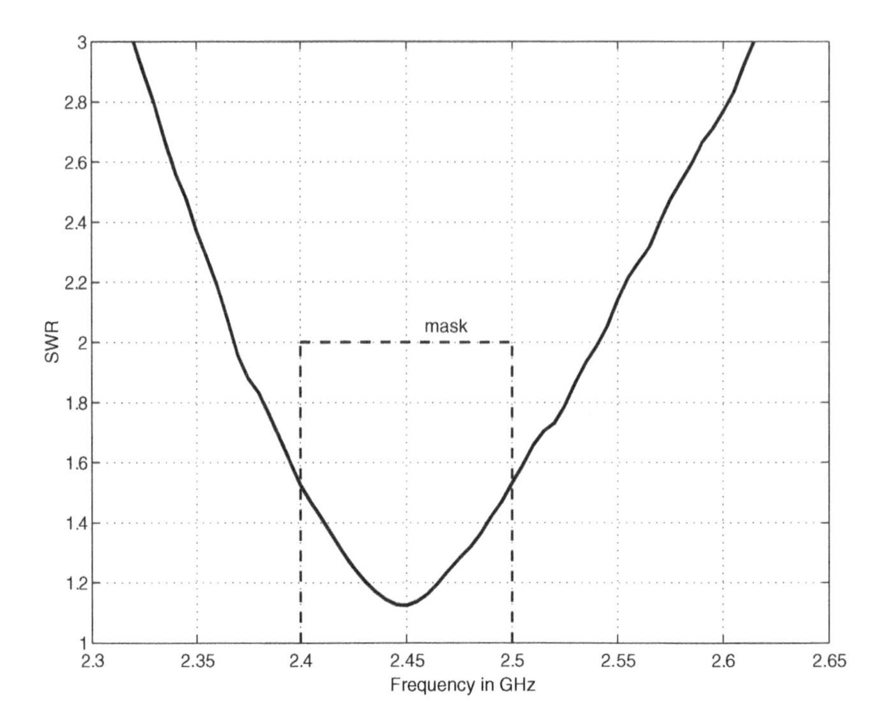

Figure 4.14 Measured SWR of the integrated laptop antenna at the 2.4 GHz band. (From [6]. Reproduced by permission of IBM.)

on a metal support frame of the laptop display. Since the metal support frame is connected to the laptop display and therefore provides a very large ground plane to the antenna, the antenna system has very stable performance. Thus even moving the feeding coaxial cable around, the antenna input impedance changes very little, a problem that plagues the measurement of even freestanding small antennas with long cables and poor grounds.

Figure 4.14 shows the measured SWR of the antenna. The vertical dashed line shows the recommended SWR mask for the 2.4 GHz band. Note that the 2.4 to 2.5 GHz frequency range, slightly wider than the strict US band, is used here to cover worldwide applications. The horizontal dashed line indicates 2:1 SWR. It is clear that the antenna has adequate SWR bandwidth and the maximum SWR is less than 1.6 over the whole band, allowing for manufacturing and environmental margin. It should be noted that the effective cable length in this test is zero, minimizing interpretation and use of this data.

Figure 4.15 shows the measured radiation patterns of the antenna in the horizontal plane when the laptop is open at 90°. The solid and dashed lines are for the horizontal and vertical polarizations, respectively. The dash-dotted line is for the total radiation pattern. The gain (average/peak) values are shown on the legend of the figure. The vertical polarization has larger gain value than the horizontal polarization. The overall average gain value is about 0 dBi, similar to an isotropic radiator. However, the peak gain value (2.6 dBi) is larger than that of a half-wavelength dipole antenna (2.14 dBi). This is due to the effect of the laptop display surface. In most countries, the peak gain values are also important and are tracked by the regulatory bodies, creating the need to balance and optimize the average and peak

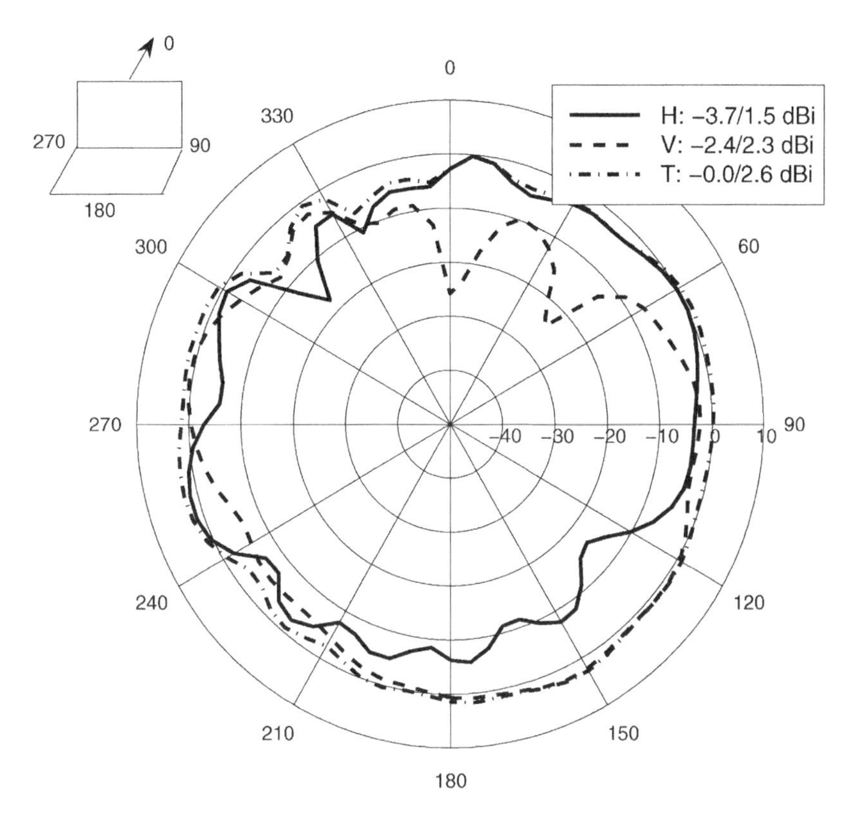

Figure 4.15 Radiation patterns of the integrated laptop antenna at 2.45 GHz. (From [6]. Reproduced by permission of IBM.)

values of each design. For the 2.4 GHz WLAN case in the USA, the FCC sets a 36 dBm EIRP maximum. With a one-watt transmitter output, the maximum peak gain for an antenna is therefore 6 dBi.

As mentioned earlier, the INF antenna is the simplest integrated antenna structure for laptop applications. The antenna can be stamped from a metal sheet, fabricated on a printed circuit board (PCB), or cut directly on the metal support structure or the metal foil used for RF shielding. This solution was arrived at through electromagnetic simulation, basic antenna experience, and materials analysis of the laptop plastics and nearby conductors as well as predefined gain and SWR metrics. The SWR and average gain defined above were arrived at through link budget models that attempt to statistically predict range/coverage in meters at specified throughput or data rates at a given reliability for that connection. The methodology ties the system to the antenna through simulation, empirical testing, and link performance.

4.7 Integrated and PC Card Solutions Comparison

Antenna and wireless system theories indicate that integrated wireless subsystems should outperform the PC card wireless system. Actual measurements for the two wireless systems confirm this conclusion. The IBM iSeries ThinkPad (a laptop model) with integrated wireless

Table 4.4 Integrated and PC Card Wireless Solutions SNR Comparison.

Distance (m)	0 (deg)		90 (deg)		180 (deg)		270 (deg)	
	Int	Card	Int	Card	Int	Card	Int	Card
0	59	54	52	47	54	45	49	49
5	53	50	49	43	49	49	53	50
10	45	37	47	35	45	38	43	36
15	42	31	51	34	46	35	45	26
20	40	22	52	32	45	35	45	26
25	41	33	49	29	43	30	48	22
30	37	21	46	30	43	30	46	24
35	42	22	43	31	42	29	45	20
40	34	23	46	23	42	23	46	27
45	37	29	46	25	42	25	46	28

(From [6]. Reproduced by permission of IBM.)

was used for this study. Two slot antennas were implemented in the ThinkPad, one on the upper left side and another on the top right edge of the display. An IBM High Rate Wireless LAN PC card was used for the comparison study. Table 4.4 lists the SNR values for distances from 0 to 45 meters with laptop orientation angles 0°, 90°, 180°, and 270°. The SNR values were obtained through the IBM WLAN Client Configuration Utility gain test program. Distances were measured from AP to laptop. Angle 0° is the laptop rear cover toward the north, 90° is toward the west (AP direction), 180° is toward the south, and 270° is toward the east. These actual tests indicate that integrated wireless is 47% better on average than the PC card version. When the laptop is far from the AP, the integrated antenna has much higher gain values than the PC card antenna, resulting in much higher SNR. Above 25 meters, the SNR for the integrated wireless system is more than 10 dB larger than that of the PC card system. The higher SNR values imply longer distance for the same data rate or higher data rate for the same distance.

As a practical example, an iSeries ThinkPad with the integrated antenna was tested against a PC card version and shown to have superior performance. The test was conducted on the fifth floor of an IBM building in Yamato Japan. This floor has three APs. When the RF signal was weak, the PC card switched to another AP, while the iSeries integrated antenna performance was still good and maintaining a connection to the same AP.

4.8 Dualband Examples

The 2.4 GHz ISM band has become extremely popular and is now widely used for several wireless communication standards. As a result, system interference and capacity are of concern. IEEE 802.11 a devices at the 5 GHz band do not have these concerns. For world-wide applications, an antenna covering the 5.15–5.85 GHz range is currently needed. Dualband antennas with one feed point have been proposed by many authors [23–53]. Most antennas proposed either provide inadequate coverage at the 5 GHz band or are not suitable for integration in portable devices. In this section we will present three designs that have been used in laptop computers.

4.8.1 An Inverted-F Antenna with Coupled Elements

This antenna structure [47] as shown in Figure 4.16 is a bent version of the closely coupled triband antenna proposed by Liu [51]. This antenna inherits many properties of the closely coupled antenna. Therefore, most conclusions drawn in [51] apply to the antenna here. For the low band (the 2.4 GHz band), the antenna behaves as an INF antenna. Much of the current flows in the INF section. The current in the L-shaped and tab sections is very weak, so it has negligible effect on the low band. At the middle and high bands, much of the current is concentrated either in the L-shaped section or on the tab section. The dominant effect is on the middle/high band resonance and the radiation pattern. However, since the INF section is fed directly, it has a relatively strong influence on the middle and high bands. The antenna behaves in a complicated way at the middle and high bands. Depending on the applications and available volume for antenna implementations, the middle and high bands can be exchanged. As referenced in Figure 4.17, R2 provides the middle band, while R3 provides the high band. Figure 4.17 also shows the evolution from the original triband antenna to the low profile triband antenna. For the WLAN applications, the middle and the high bands are combined to cover the 5 GHz band. As a result, the triband antenna is used as a dualband antenna in this case.

The resonant frequency of the low frequency band is determined primarily by L1+H1−W1 as shown in Figure 4.16. Increasing H1 and the width of the metal strips will widen the bandwidth of the antenna at the lower band. Moving the feed point FP horizontally will change the antenna impedance. Moving FP to the left (open) side will increase the impedance and to the right (grounded) side will reduce the impedance. Changing the feed point will have some effect on the resonant frequency as well. The middle and high band elements have negligible effects on the lower band. The middle band frequency is primarily determined by H2+L2. The impedance in this band is primarily determined by D12 and S2,

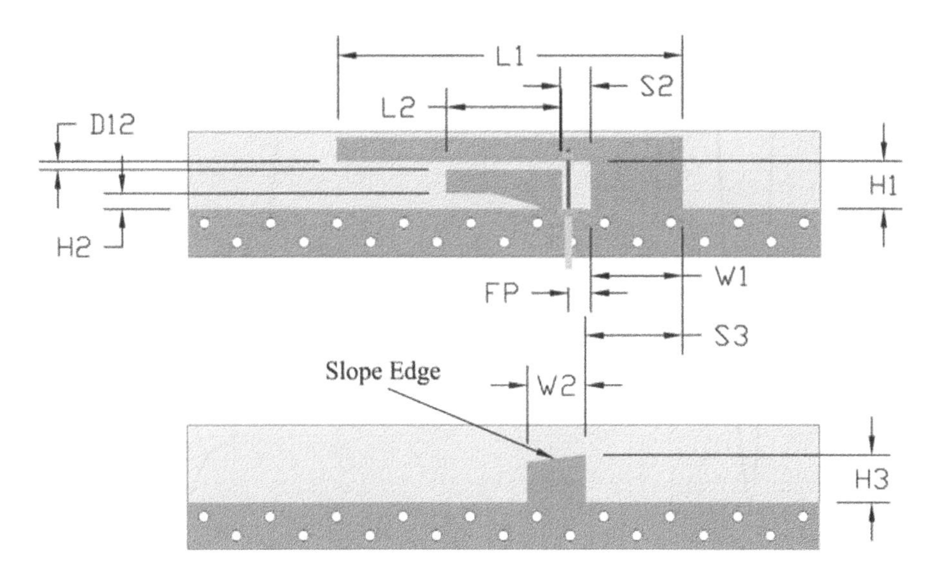

Figure 4.16 INF antenna with coupled elements implemented on PCB. (From [47]. Reproduced by permission of © IEEE.)

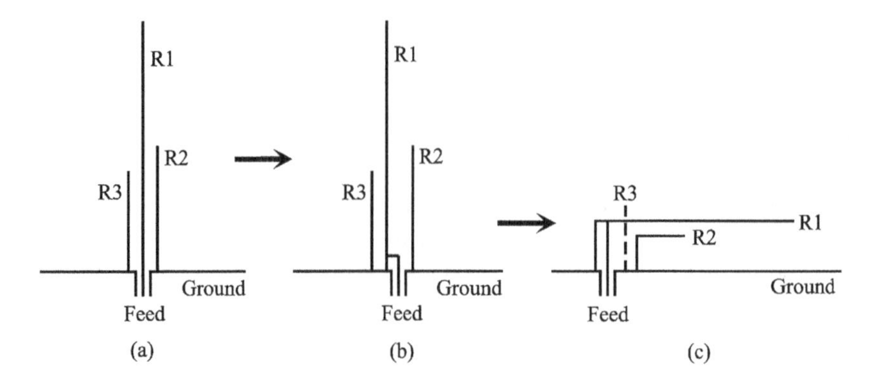

Figure 4.17 Triband antenna evolution.

the coupling distances. Generally speaking, reducing D12 and S2 will increase the coupling and consequently the impedance at this band. Widening the L2 width will broaden the impedance bandwidth. Tapering the corner near H2 seems to improve the bandwidth as well. The high band is primarily determined by H3, S3 and W2. H3 is the major controlling factor for adjusting the resonant frequency. S3 changes the coupling between this band and the lower band. The substrate thickness and the substrate dielectric constant will also affect the

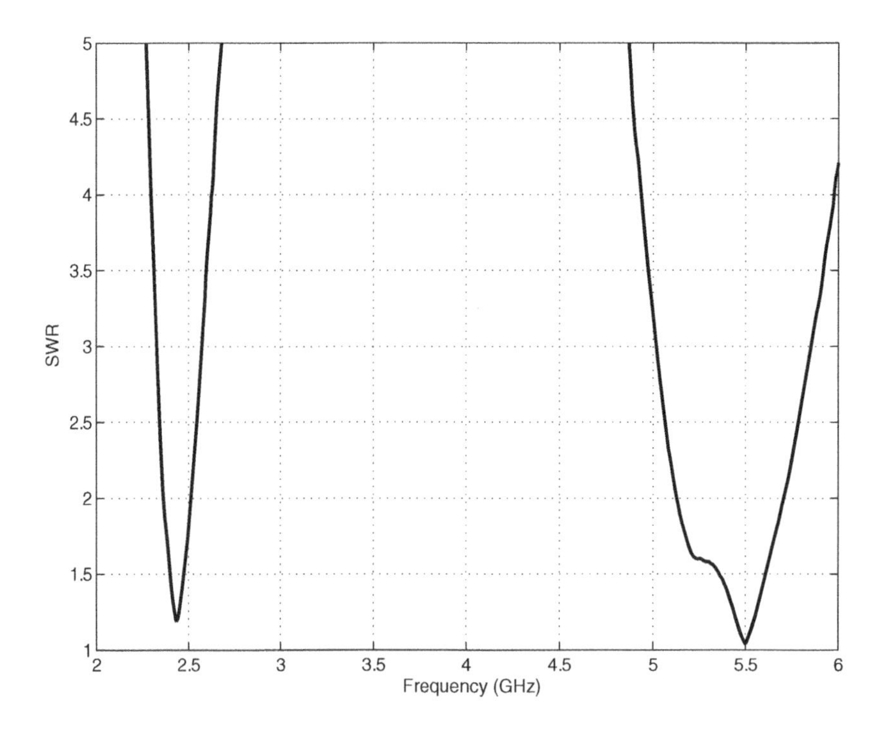

Figure 4.18 Measured SWR of the dualband prototype antenna.

coupling. Experiments indicate a sloped top edge of W2 will improve matching and widen the bandwidth.

A PCB version antenna prototype was made as shown in Figure 4.16 and mounted on the top right vertical edge of an IBM ThinkPad display. The display has a metal rim and physical supports, which provide additional ground plane to the antenna. In fact, the display itself is part of the overall antenna system. Figure 4.18 shows the measured SWR of the prototype antenna in the display with a metal cover. The feeding coaxial cable was very short and low loss. There are two resonances in the 5 GHz band. By adjusting the separation of the two resonant frequencies, the SWR bandwidth can be further improved. Figure 4.19 shows the measured radiation patterns of the antenna in the 2.4 GHz band with an elevation angle 20° above the horizontal plane. The overall radiation pattern is close to omnidirectional. Figure 4.20 shows the radiation patterns of the antenna in the 5 GHz band with an elevation angle 10° above the horizontal plane. The antenna average gain is nominally 0 dBi in both bands. Note that the radiation patterns with the best average gain values are typically on different elevation angles for dualband laptop antennas.

Figure 4.21 shows the final antenna design used in a commercial laptop product with metal covers based on the design shown in Figure 4.16.

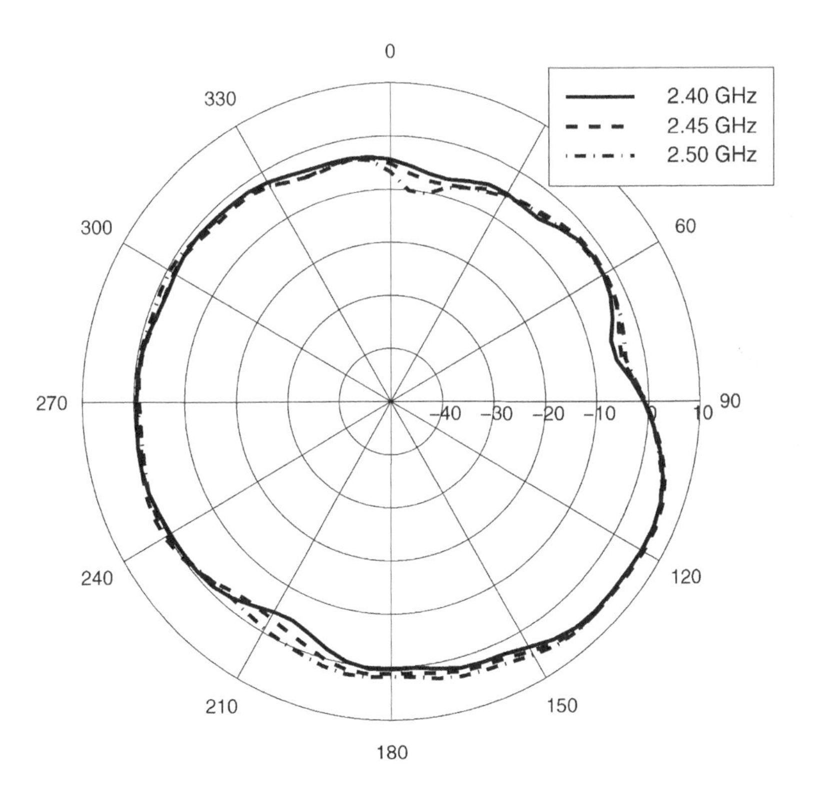

Figure 4.19 Measured radiation patterns of the dualband prototype antenna through the 2.4 GHz band in display. (From [47]. Reproduced by permission of © IEEE.)

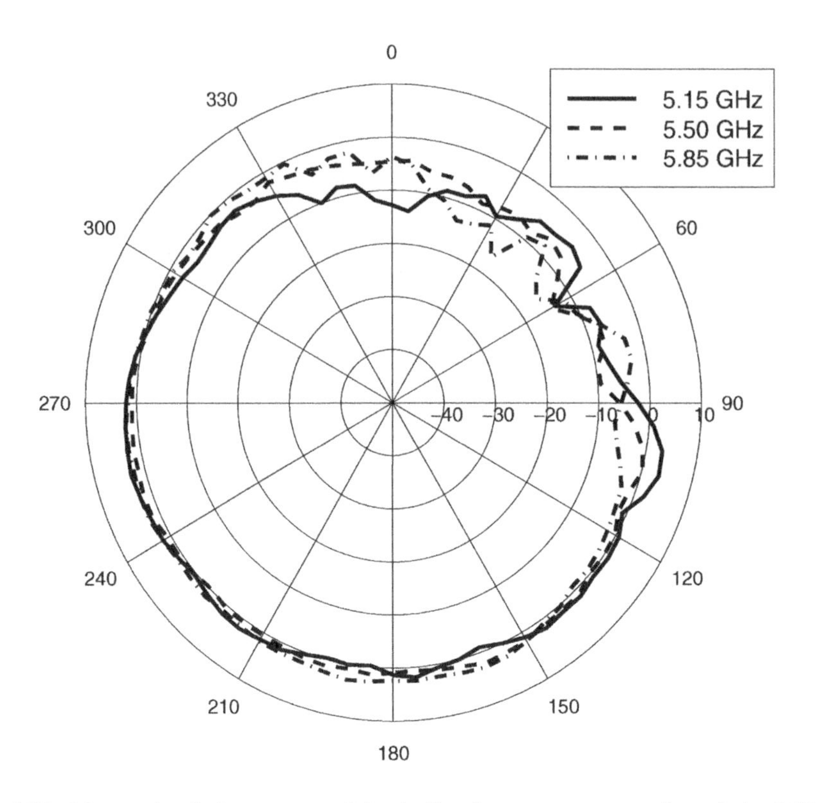

Figure 4.20 Measured radiation patterns of the dualband prototype antenna through the 5 GHz band. (From [47]. Reproduced by permission of © IEEE.)

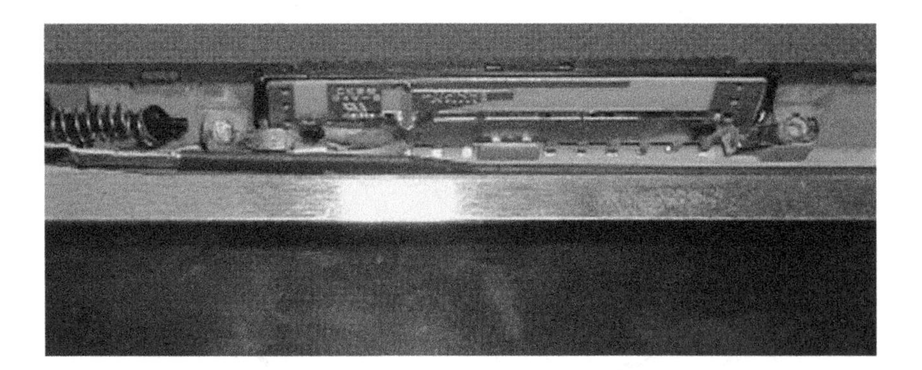

Figure 4.21 The final antenna design used in commercial laptops with metal display covers.

4.8.2 A Dualband PCB Antenna with Coupled Floating Elements

This antenna [48] is based on printed half-wavelength dipole antennas with some floating and closely coupled elements for dualband applications [46, 49]. In all the antenna structures in [44, 49], the feeding dipole covers the low band, and the coupled elements cover the high

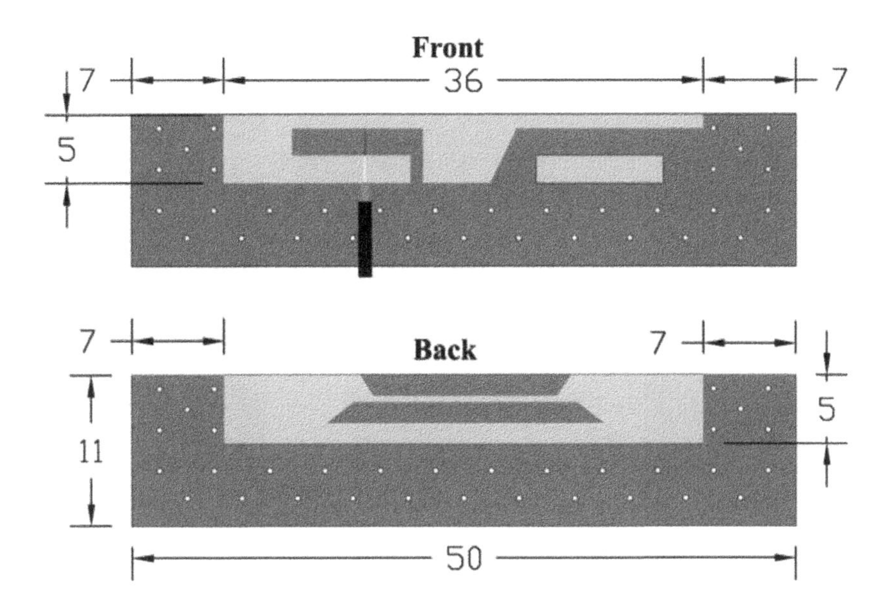

Figure 4.22 Major dimensions of the antenna in mm. (From [48]. Reproduced by permission of © IEEE.)

band/bands. Since the antenna is a half-wavelength dipole at the low band, the antenna size tends to be large. In those antenna designs, the antenna performs very well and has a wide bandwidth at the low band. If only one coupled element is used to cover the high band, the bandwidth, especially the gain bandwidth, is very narrow at the high band. So multi-coupled elements are used to cover the high band. However, the antenna presented here, shown in Figure 4.22, works in an opposite way. This antenna consists of an INF antenna and two coupled dipole elements (on the back side of the PCB) to cover the high frequency band and a coupled loop structure (on the front side) to cover the low frequency band. To improve the antenna performance, especially in the high frequency band, a thin (0.3 mm thickness), low-loss and low-*k* FR 4 PCB material (Megtron-5 from Matsushita Electric Works, Ltd) with a 3.5 dielectric constant at 1 GHz and 0.004 loss tangent is used. The antenna structure is small since it is a half-wavelength structure at the high band not at the low band.

Figure 4.22 shows the antenna structure in detail with the major antenna dimensions. Note that the antenna itself needs only an area of about $36 \times 5 = 180\,\text{mm}^2$; the remaining space is used to mount the antenna to a laptop display and to reduce the effects from the antenna environment. The additional ground is provided by the laptop display. A full wave method of moments tool [52] was used for the analysis of the antenna. Investigations had been carried out on the adjustment of the antenna characteristics with regard to the following parameters:

- Length, width and the end shape of the sub-resonators
- Shape of loop resonant enhancer
- Inverted-F type feeding element and its feeding point.

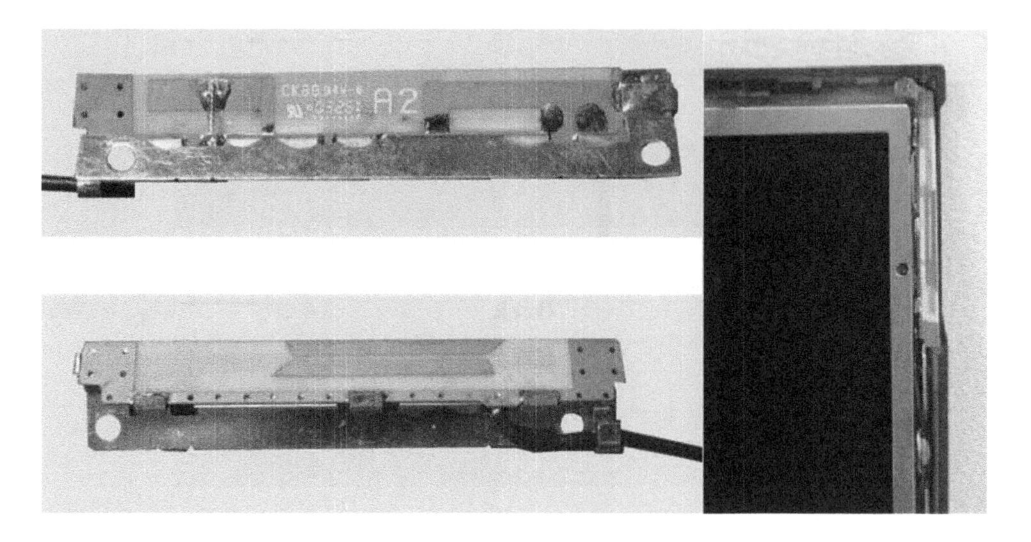

Figure 4.23 The final antenna used in commercial laptops. (From [48]. Reproduced by permission of © IEEE.)

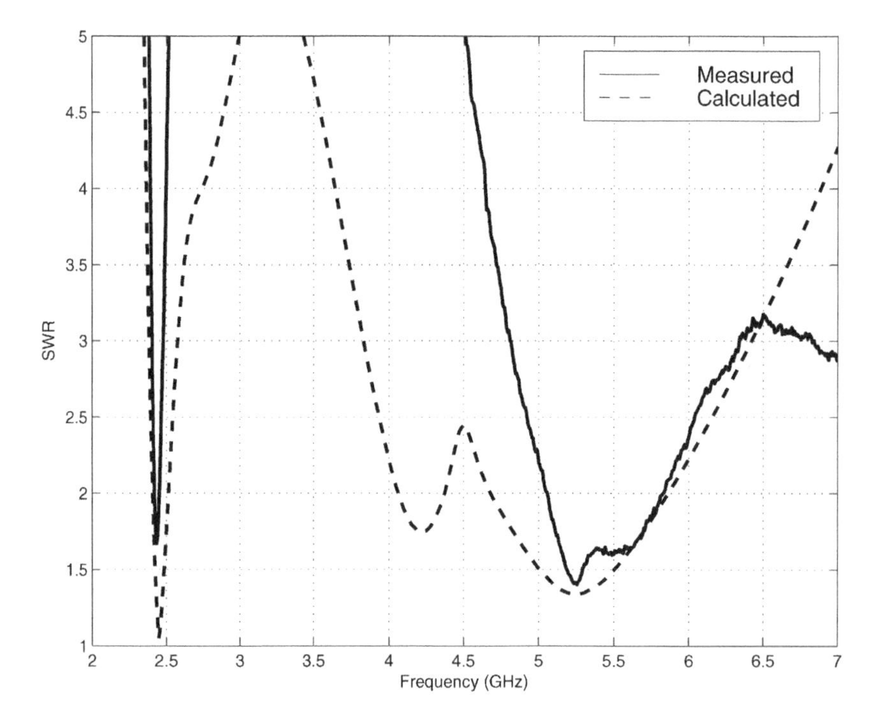

Figure 4.24 Measured and calculated SWR in free standing with 30 cm long coaxial cable. (From [48]. Reproduced by permission of © IEEE.)

Although RF power is fed by a micro coaxial cable for the prototype antenna, a simplified source was used for the calculation model.

An actual implementation of this antenna into a laptop with a magnesium cover is shown in Figure 4.23. In order to mount the antenna into the limited space at the top right side of the LCD panel cover and maintain stable grounding, a metal bracket was attached to the antenna assembly.

Both calculated and measured SWRs in free standing with 30 cm long feed cable are shown in Figure 4.24. During the calculation, an equivalent dielectric constant ($\epsilon_r = 3.2$) was used instead of the one ($\epsilon_r = 3.6$) specified by the PCB manufacture in order to reduce the number of unknowns, in other words, the size of the execution job. Both calculated and measured SWR show good impedance matches and resonant frequencies in the 2.4 and 5 GHz bands, although a slight difference is observed in the non-resonant frequency range. The wide bandwidth in the calculated SWR is due to the small ground plane. Note that the antenna has a large ground plane when it is mounted in the display.

The measured SWR in the actual laptop is shown in Figure 4.25. This shows a better SWR than the freestanding one. This is partly due to the longer (860 mm) coaxial cable and partly due to the lossy antenna environment. Figures 4.26 and 4.27 show the measured radiation patterns of the antenna in the display at 2.45 GHz and 5.25 GHz, respectively. There are no major nulls in the radiation patterns. The antenna has both strong horizontal and vertical polarizations at both bands.

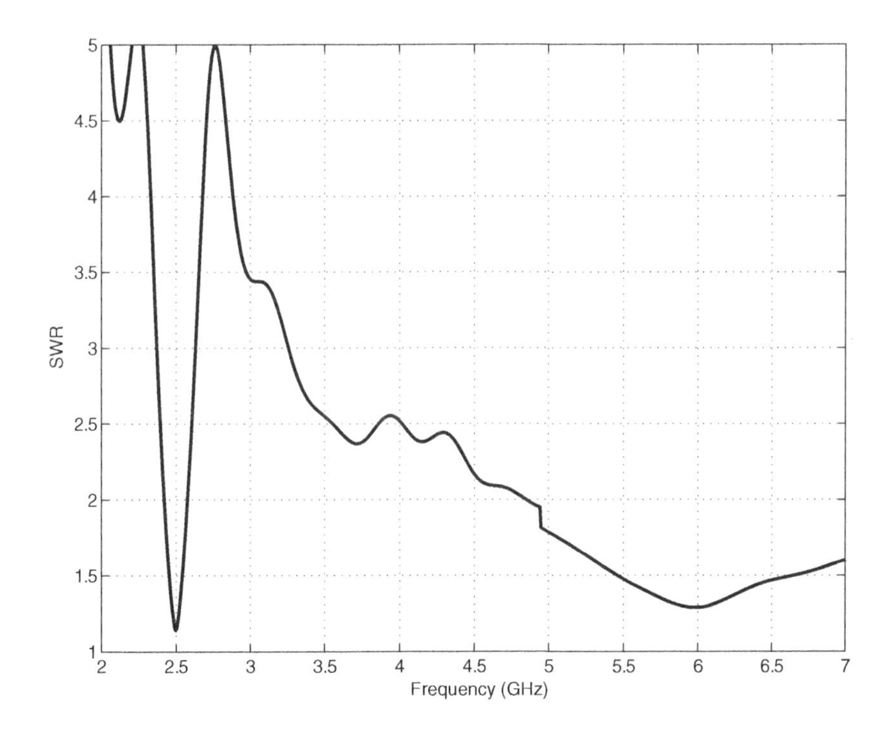

Figure 4.25 Measured SWR in laptop display with 85 cm long coaxial cable. (From [48]. Reproduced by permission of © IEEE.)

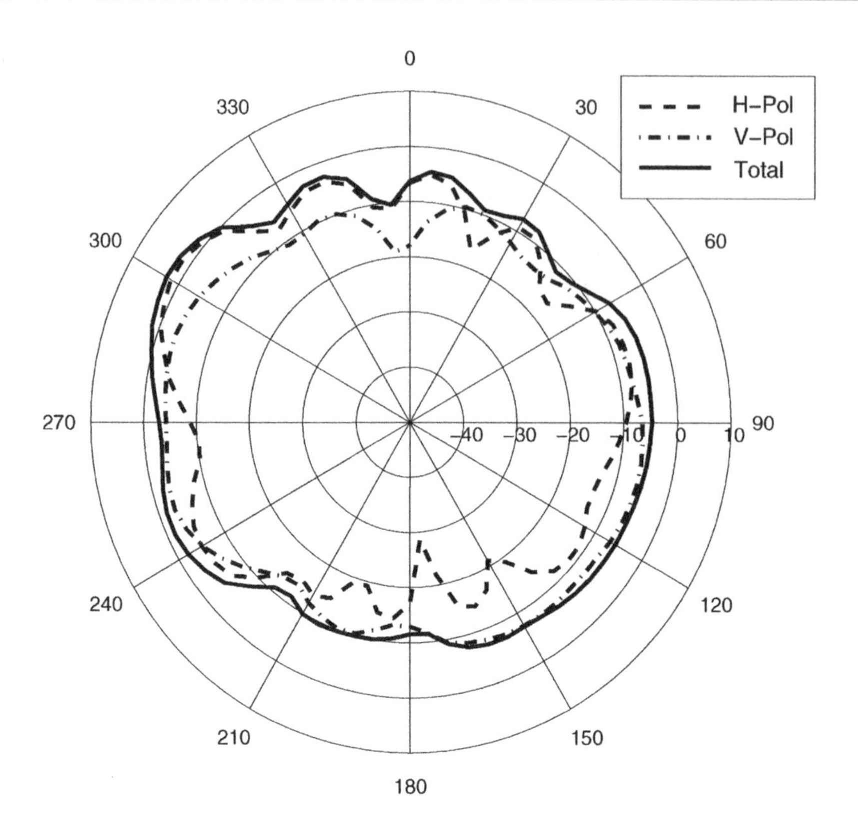

Figure 4.26 Measured radiation patterns at 2.45 GHz and 45° elevation angle with 86 cm long cable.

Figure 4.28 shows the measured average gain of the antenna through the 2.4 and 5 GHz bands with 30 and 86 cm long coaxial cables, respectively. The gain reduction due to cable loss is obvious. Considering the system has 860 mm of coaxial cable attached to the antenna which causes 2.7–5 dB loss, the estimated average gain is higher than −4 dBi in the frequency range between 3.5–7.7 GHz and 2.4–2.5 GHz.

4.8.3 A Loop Related Dualband Antenna

Hitachi Cable in Japan developed a very simple dualband antenna design using its proprietary thin film technology termed Multi-Frame Joiner (MTF) [53–55]. The 0.1 mm thick planar antenna structure including the ground plane is sandwiched with polyamide (with a 3.0 dielectric constant) film, giving the antenna stability and insulation from other metal devices. Since the antenna structure is only 0.2 mm thick, it can be bent into desired shapes without damaging the antenna structure. Unlike many flexible PCBs, antennas made from the Hitachi Cable thin film technology will stay in the bent position. Due to the thin structure, this antenna has its own large and reliable ground plane. In laptop applications, the ground plane is placed between the back side of the LCD panel and the display cover. So the ground plane does not take extra space.

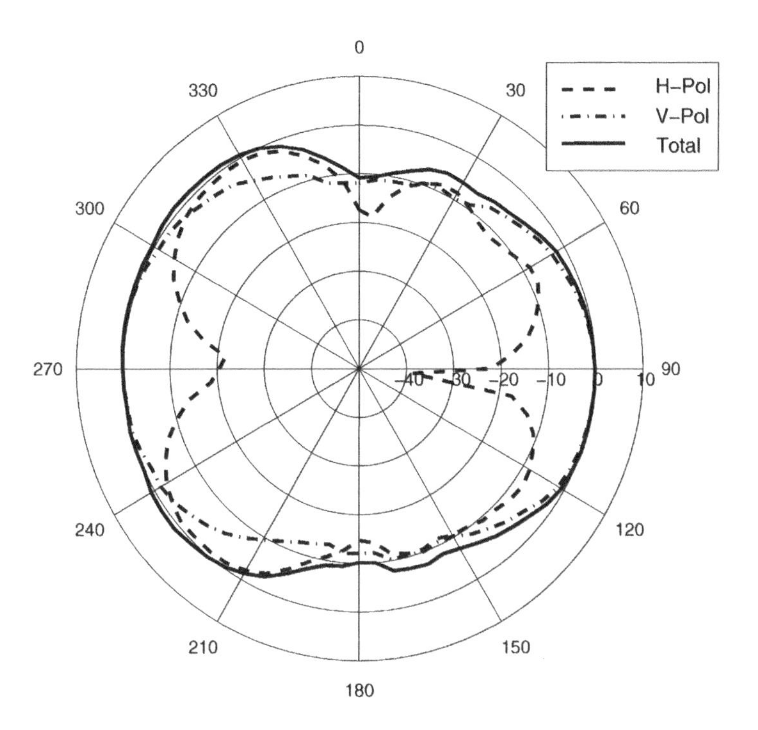

Figure 4.27 Measured radiation patterns at 5.25 GHz and 45° elevation angle with 86 cm long cable.

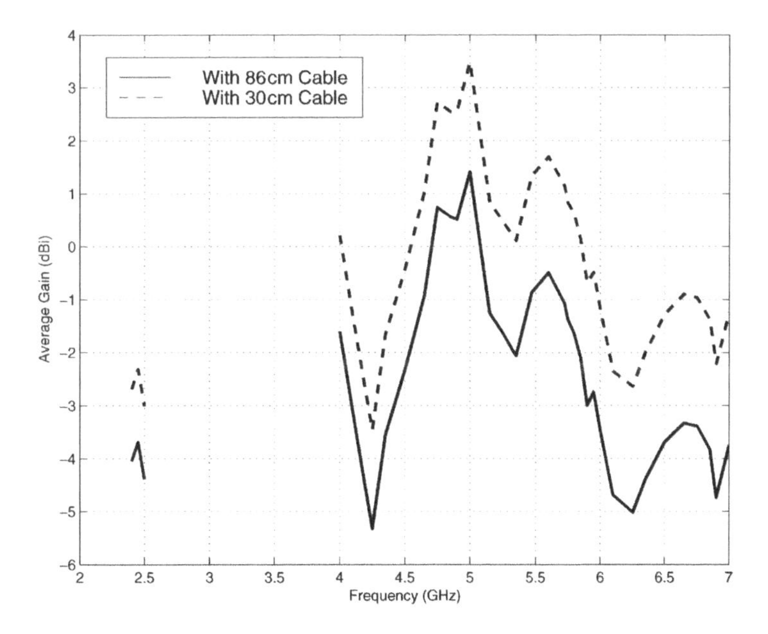

Figure 4.28 Measured average gain of the antenna in a metal cover laptop with different coaxial cable lengths. (From [48]. Reproduced by permission of © IEEE.)

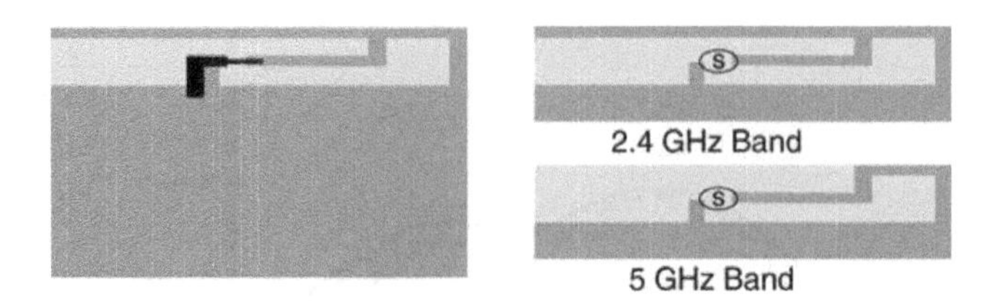

Figure 4.29 Hitachi Cable antenna and its equivalent structures.

Figure 4.29 shows a sketch of the original dualband antenna and its operating principles. For the 2.4 GHz band, the antenna can be considered as a variation of the INF antenna. So its performance is similar to that of the INF antenna. A half-wavelength loop is used to cover the 5 GHz band. Note that half of the loop is provided by the ground plane. Figure 4.30 shows the measured SWR of the antenna in a laptop (dashed curve). It is clear the design can cover both 2.4 and 5.15–5.825 GHz bands.

An improved design is shown in Figure 4.31 [50, 56, 57]. Compared to the original design, this new version has two changes. First, the 5 GHz loop is laid out differently (so is the feed location). But the major change is the use of a second radiator for the 5 GHz band so that the antenna has much wider bandwidth at 5 GHz. In fact, it can cover all the possible WLAN bands throughout the world in the 4.9–6.1 GHz frequency range. From the solid curve of

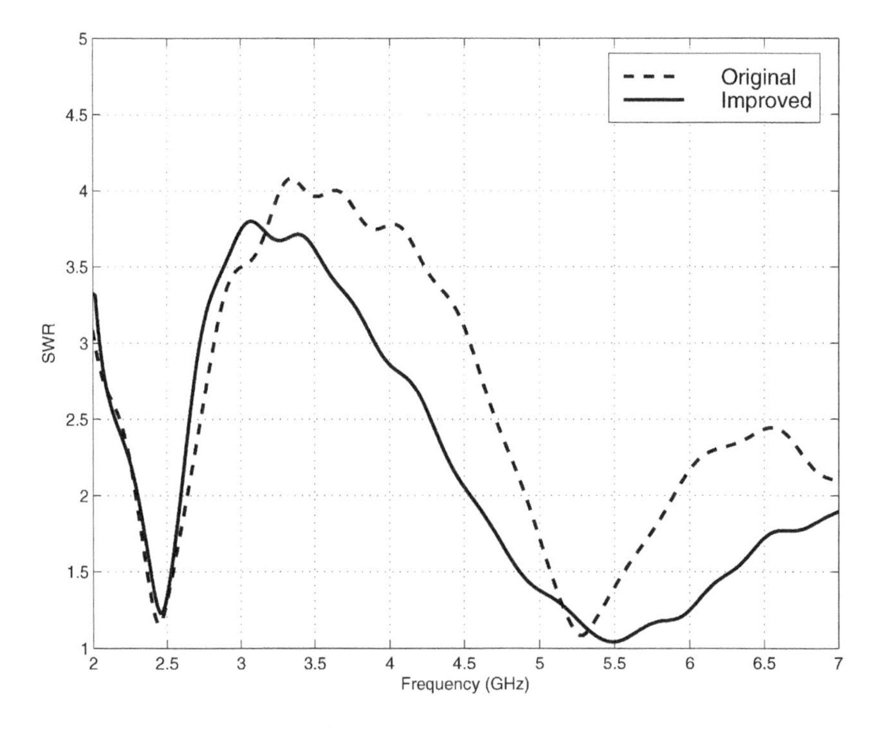

Figure 4.30 Comparison of measured SWR of the original and improved Hitachi Cable. (From [50]. Reproduced by permission of Hitachi Cable, Ltd.)

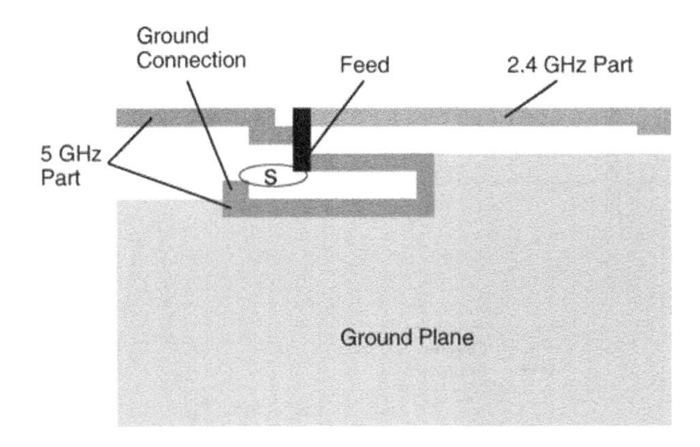

Figure 4.31 Improved Hitachi Cable antenna. (H. Tate, private communication. Reproduced by permission of Hitachi Cable, Ltd.)

Figure 4.30, it is clear that adding the second radiator for the 5 GHz band widens the antenna bandwidth, but it has little effect on the 2.4 GHz band.

Figure 4.32 shows photos of the original and improved Hitachi Cable antennas. The overall dimensions of both types of antennas are the same, that is, 31.0 (W) × 30.5 (H) × 0.2 (T) mm.

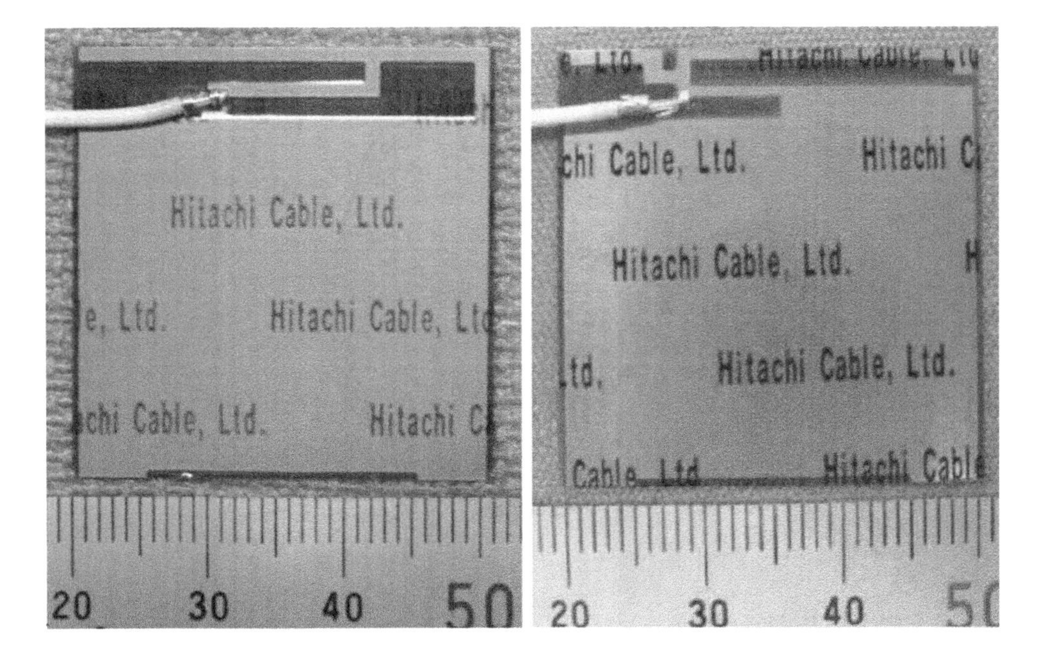

Figure 4.32 Photos of the Hitachi Cable original and improved antennas. (From [50]. Reproduced by permission of Hitachi Cable, Ltd.)

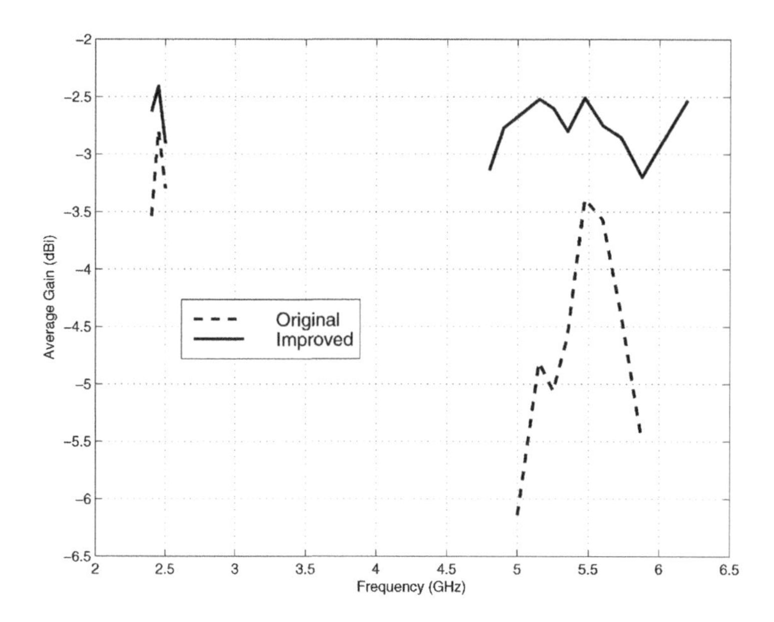

Figure 4.33 Comparison of measured average gain of the original and improved Hitachi Cable antennas. (From [50]. Reproduced by permission of Hitachi Cable, Ltd.)

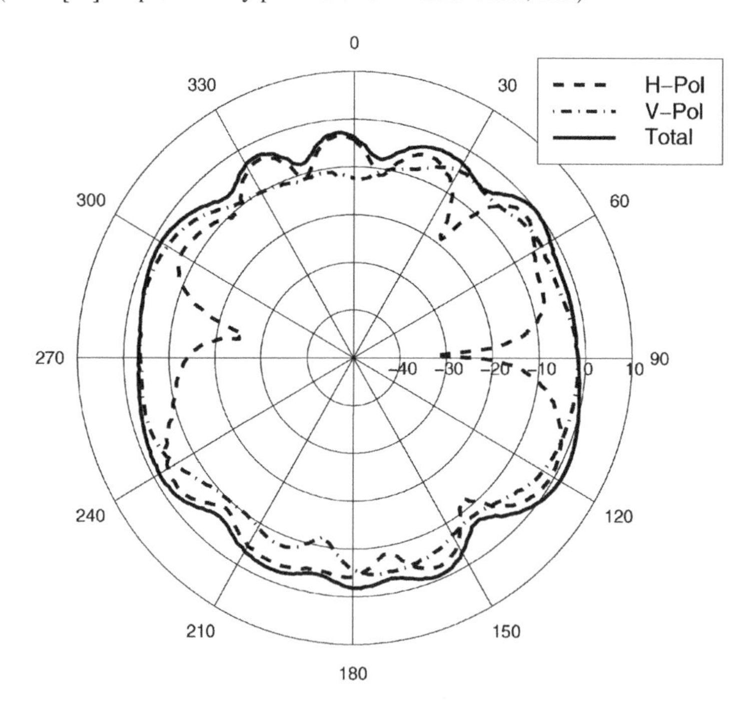

Figure 4.34 Measured radiation patterns of the improved Hitachi Cable at 2.45 GHz. (From [50]. Reproduced by permission of Hitachi Cable, Ltd.)

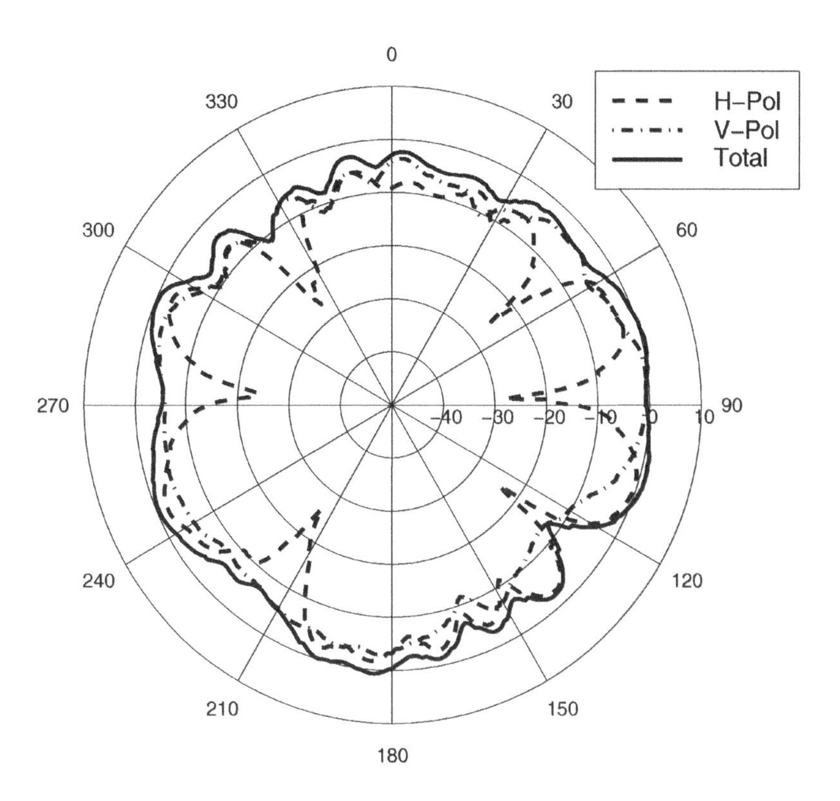

Figure 4.35 Measured radiation patterns of the improved Hitachi Cable at 5.25 GHz. (From [50]. Reproduced by permission of Hitachi Cable, Ltd.)

Figure 4.33 shows the measured average gain of the original and improved antennas. Although the second radiator is primarily used for improving the antenna performance at the 5 GHz band, we can also see about 0.5 dB gain improvement for the 2.4 GHz band. This is mainly due to the 5 GHz loop location and the feed point changes. The improvement in the 5 GHz band is tremendous in both gain value and gain flatness versus frequency.

Figure 4.34 and 4.35 show the measured radiation patterns of the improved antenna at 2.45 and 5.25 GHz, respectively. In both cases, the antenna has strong horizontal and vertical polarizations. And the overall radiation patterns are almost omnidirectional.

One variation of the original Hitachi Cable antenna design is also used in laptops [58]. Figure 4.36 shows a sketch of the antenna. Instead of a rectangular loop, this antenna uses a twisted loop in the shape of an '8'. The antenna is used in either a stamped sheet metal

Figure 4.36 Nissei antenna.

or PCB form. Comparing to the original Hitachi Cable antenna, this twisted loop actually helps the antenna performance in both 2.4 and 5 GHz bands.

4.9 Remarks on WLAN Antenna Design and Evaluations

Two performance parameters were used to define integrated antennas for laptop applications. One is SWR, and the other is the average antenna gain (see Section 4.3.3). Based upon link budget models and system requirements, the integrated antennas should have better than a 2:1 SWR bandwidth, wide enough to cover both the 2.4 and 5 GHz bands to ensure a wireless system with reliable, high data rate performance over a useful range or coverage area. The antenna should have average gain values similar to that of an isotropic radiator. The average gain value can be used in a communication link budget model to predict system level performance such as throughput, reliability, and range. The antenna polarization is not a critical parameter for laptop applications since laptops are primarily used in indoor environment where there is high scattering of signals. As one would expect, the best location for integrated antennas in laptops is on the laptop display as high as possible. But using this location forces a design tradeoff between the antenna's 'visibility' and the necessity of a lossy feed cable, since wireless cards are usually placed in the base of a laptop. The ultimate system cost, time to market, and performance are consciously traded off for each application. There is no one solution that will meet the needs of every laptop, much less every portable device. Current state of the art modeling tools are not accurate predictors of real system performance, but can be used to provide adequate estimates of the design for use with traditional cut-and-try methods.

As expected, the integrated wireless system provides much better performance than the PC card version system. Performance, convenience, and mechanical strength ensure the integrated wireless dominates the WLAN market in laptops.

Since the radio card is usually in the base of a laptop and the antenna is on the top of display, the feeding cable length tends to be long, more than 50 cm in most implementations. The coaxial cable used for the integrated wireless has a very small diameter, around 1.1 mm to allow routing through hinges, so the cable has more than 5 dB/m loss at 5 GHz. As a result, the cable loss will be more than 3 dB for the integrated wireless in the 5 GHz band. A loss of 3 dB is very costly from a wireless performance perspective. Therefore, more studies are needed for the 5 GHz wireless implementation.

4.10 Antennas for Wireless Wide Area Network Applications

The WLAN has become very popular, and WLAN devices have been integrated into almost all new laptop computers. However, WLAN connectivity is primarily constrained in hot spots such as airports, school campuses, hotels or homes. With the success of the WLAN and the convenience it brings to business travelers in hot spots, it is becoming increasingly important that this connectivity and mobility be maintained even when on the road and away from such WLAN hot spots. A wireless wide area network (WWAN) card for laptops providing connectivity via cellular networks has been shown to be effective and very useful to business travelers. With higher speed 3G phones finally appearing on the market, adoption of WWAN technology will continue to grow. Laptop manufacturers have already started to integrate the WWAN devices into their laptop computers and this trend will continue into the future.

Integrating WLAN antennas into laptop computers has been a daunting task for laptop makers especially due to shrinking laptop computer sizes. Integrating WWAN antennas which are larger than three times WLAN antennas into laptops will be extremely difficult. Many dualband, even quadband, antenna structures have been proposed for cellphone applications [59–64]. However, all these antenna structures can not be used for laptop applications due to the laptops' unique form factor. To begin this investigation, we will utilize what we learned from integrating WLAN subsystems into laptops.

4.10.1 INF Antenna Height Effects on Bandwidth

The INF antenna is widely used in mobile applications due to its simple design, flexibility, low cost and reliable performance [6, 59–62]. Many antenna types for portable applications are variations of the INF antenna; and many dualband or even triband antennas are extensions of the INF antenna [6, 63, 64]. An antenna made from a PCB or stamped from sheet metal is especially popular in laptop applications due to its easy integration and laptops, unique form factor [6]. It is well known that the INF antenna performance parameters, such as efficiency and impedance bandwidth, depend on the antenna height, while the resonant frequency of the antenna primarily depends on the antenna length. So it is important to study the relationships between antenna impedance bandwidth and the antenna height related to laptop applications, more specifically to WWAN applications.

Figure 4.37 shows the antenna layout on a 26×29.5 single-sided 0.059 in thick FR4 PCB attempting to represent the laptop LCD display [65]. The antenna locations and relevant dimensions are shown in Figure 4.38. The 65/95 dimension means that the length is 95 mm for the 824–894 MHz band and 65 mm for the 1850–1990 MHz PCS band. For a given height H, the length L and feed point FD have to be adjusted to ensure the antenna is either resonating at 859 MHz or 1920 MHz, the center of each band. Table 4.5 lists the parameters for this study.

Figure 4.39 shows the measured SWR for antennas at the 800 MHz band for different heights H. Note that the actual antenna height is $H + 2$ mm. Its clear from this figure that the antenna bandwidth increases as the parameter H increases. The curve in Figure 4.40 indicates the relationship between the antenna 2:1 SWR bandwidth and the height parameter H. The horizontal dashed line indicates the bandwidth (70 MHz) needed for the application. For the 800 MHz application, the minimum value for H has to be greater than 11 mm to cover the band. However, 11 mm is too large for most laptop applications. If 3:1 SWR bandwidth is required, such as used for cellphone applications, 8 mm for height H will be adequate.

Figure 4.41 shows the measured SWR for antennas at the PCS band for different heights H. Again, the actual antenna height is also $H + 2$ mm. Its clear from this figure that the antenna bandwidth increases as the parameter H increases. The curve in Figure 4.42 indicates the relationship between the antenna 2:1 SWR bandwidth and the height parameter H. The horizontal dashed line indicates the bandwidth (140 MHz) needed for the application. For the PCS application, the H value above 3 mm will provide the needed antenna bandwidth with good margin. An overall 5 mm antenna height is reasonable for most laptop constructions. So integrating a PCS band antenna into laptops should be relatively easy.

Figure 4.43 shows the measured radiation patterns at 1910 MHz, about the center (1920 MHz) of the PCS band for $H = 4$ mm on the horizontal plane. Note that 1920 MHz

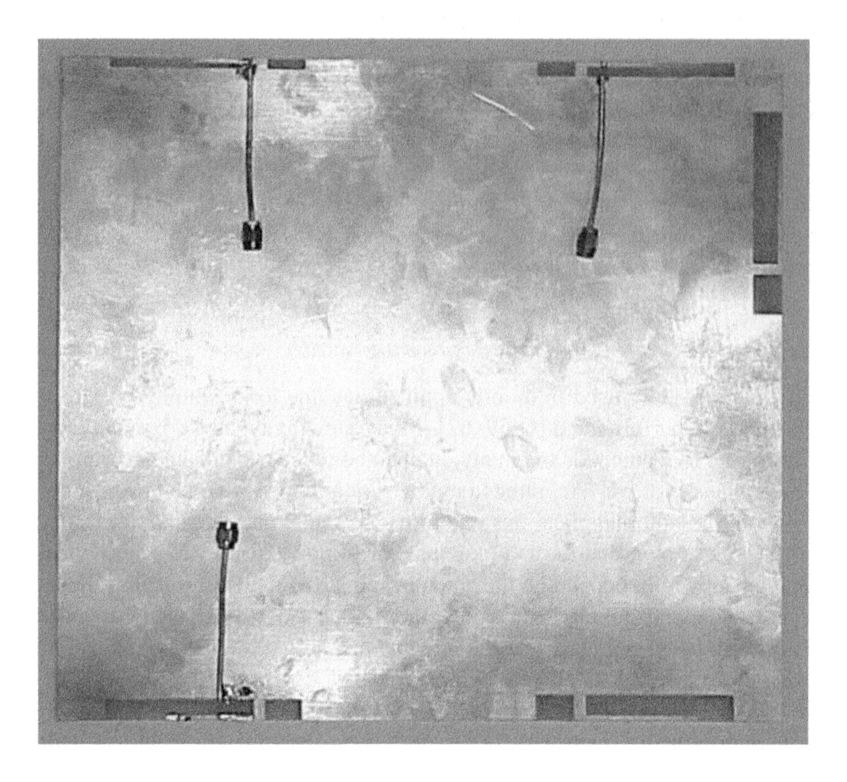

Figure 4.37 WWAN antennas implemented on FR4 PCB. (From [65]. Reproduced by permission of © IEEE.)

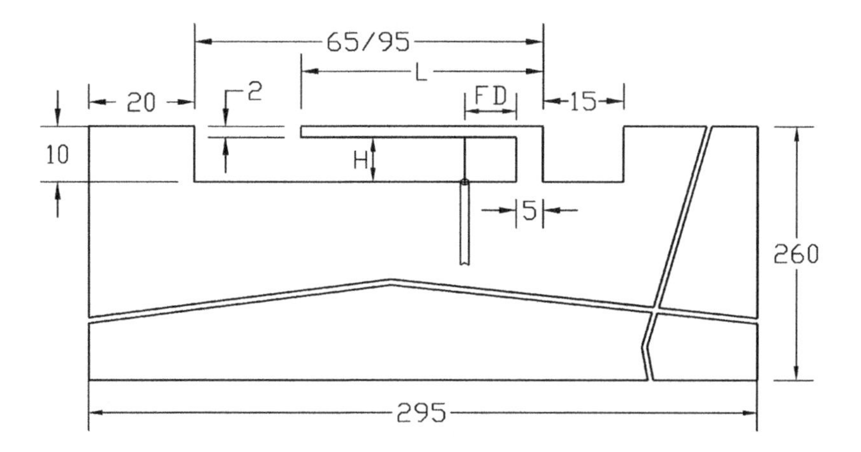

Figure 4.38 WWAN antenna locations and dimensions on a laptop display-size PCB (unit in mm). (From [65]. Reproduced by permission of © IEEE.)

Table 4.5 Antenna length and feed point for a given height.

H (mm)	824–894 MHz		1850–1990 MHz	
	L (mm)	FD (mm)	L (mm)	FD (mm)
2	65.0	3.5	30.5	2.5
4	66.0	4.0	34.5	6.5
6	66.0	4.0	41.5	13.5
8	68.5	6.0		
10	69.5	8.0		

(From [65]. Reproduced by permission of © IEEE.)

was not measured due to the frequency increment selected in the measurement setup. As can be seen, both polarizations are strong, and the overall radiation pattern is nearly omnidirectional. Figure 4.44 shows the measured overall radiation patterns through the PCS band. The overall radiation patterns do not change much through the PCS band. The radiation patterns are less sensitive to the H parameter. Radiation patterns in the 800 MHz band were not measured, but should be similar to those of the PCS band.

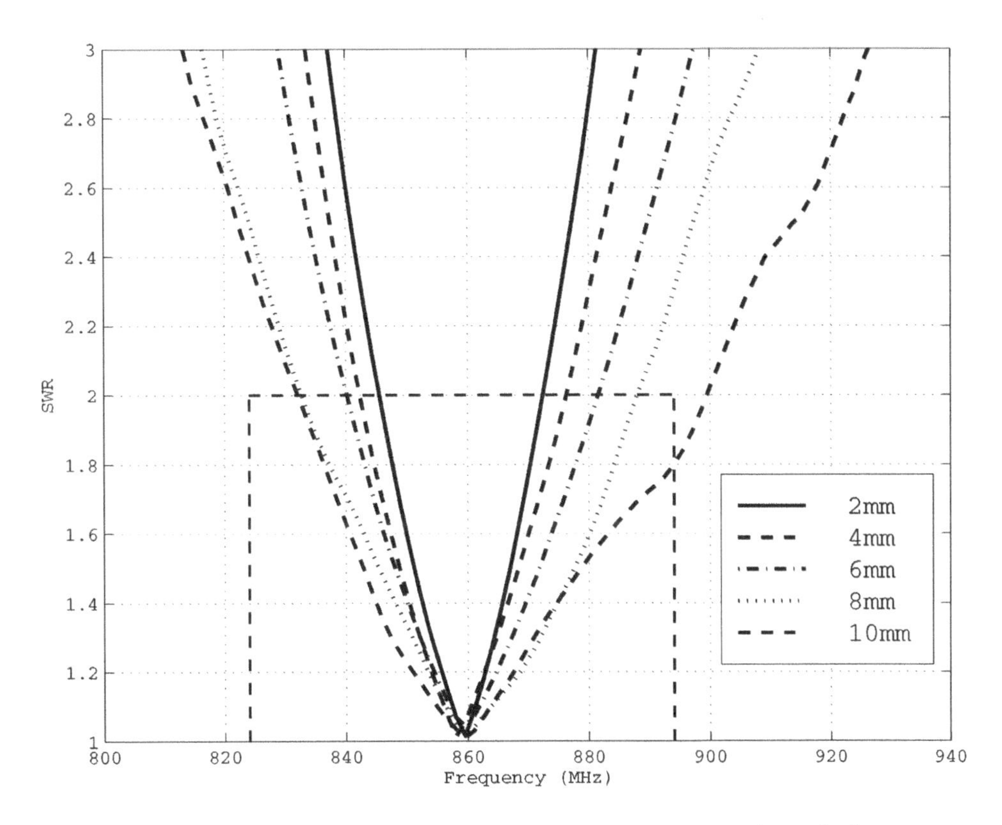

Figure 4.39 Measured SWR at the 800 MHz band for different heights H. (From [65]. Reproduced by permission of © IEEE.)

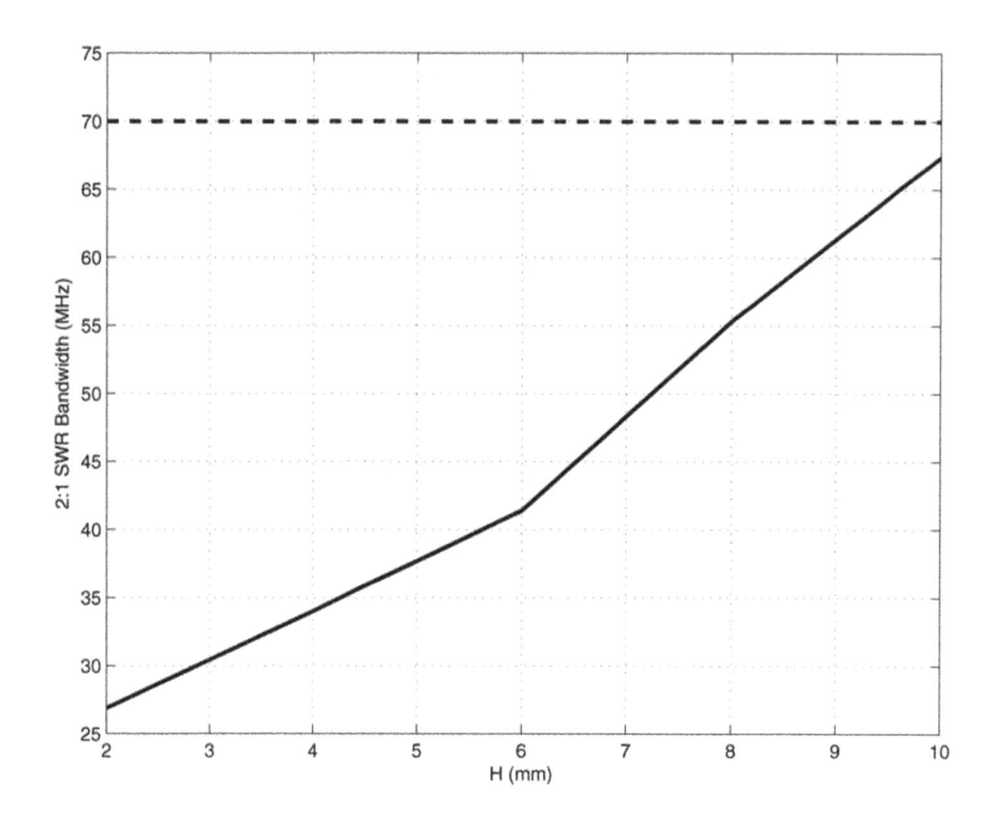

Figure 4.40 Measured 2:1 SWR bandwidth at the 800 MHz band for different heights H. (From [65]. Reproduced by permission of © IEEE.)

The measurement results indicate the antenna SWR bandwidth increases nearly linearly with the antenna height. Increasing the antenna height increases the antenna area, resulting in reduced antenna Q value, which is inversely proportional to the antenna bandwidth. So for practical applications, the best way to increase antenna bandwidth is to maximize the antenna area (or volume for 3D antenna structures) within a given physical constraint. For the 2D INF antenna implementation, as studied here, the height parameter has to be at least 11 mm to cover the 800 MHz band, while 3 mm covers the PCS band with a good margin.

4.10.2 A WWAN Dualband Example

Hitachi Cable designed a WWAN antenna, as shown in Figure 4.45, based on its thin film technology and WLAN antenna structure. Since the WWAN size is much larger than those of WLAN antennas, both antenna designers and laptop mechanical and industrial designers have to compromise. For all WLAN applications, the antennas are completely hidden in the the laptop display, so consumers do not see any bumps on the display surface. However, hiding the WWAN antennas completely in display is impossible. Therefore laptop mechanical and industrial designers allow the WWAN antenna to 'bump out' the display surface as much as

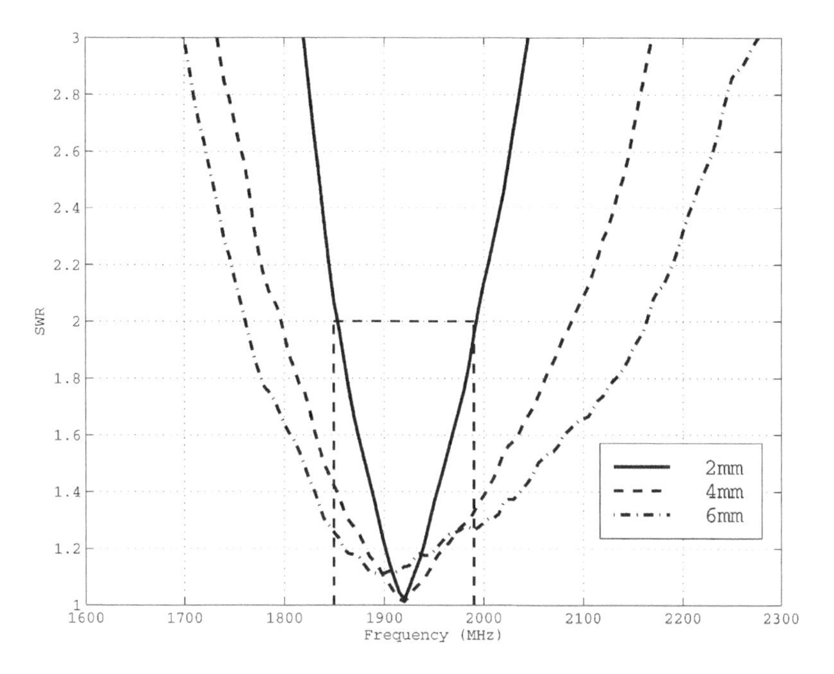

Figure 4.41 Measured SWR in the PCS band for different heights H. (From [65]. Reproduced by permission of © IEEE.)

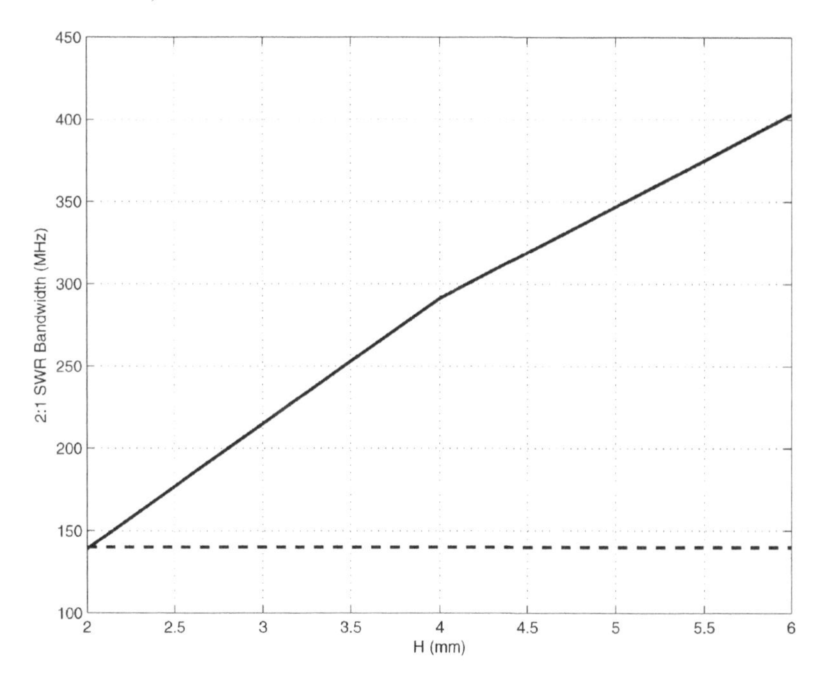

Figure 4.42 Measured 2:1 SWR bandwidth at the PCS band for different heights H. (From [65]. Reproduced by permission of © IEEE.)

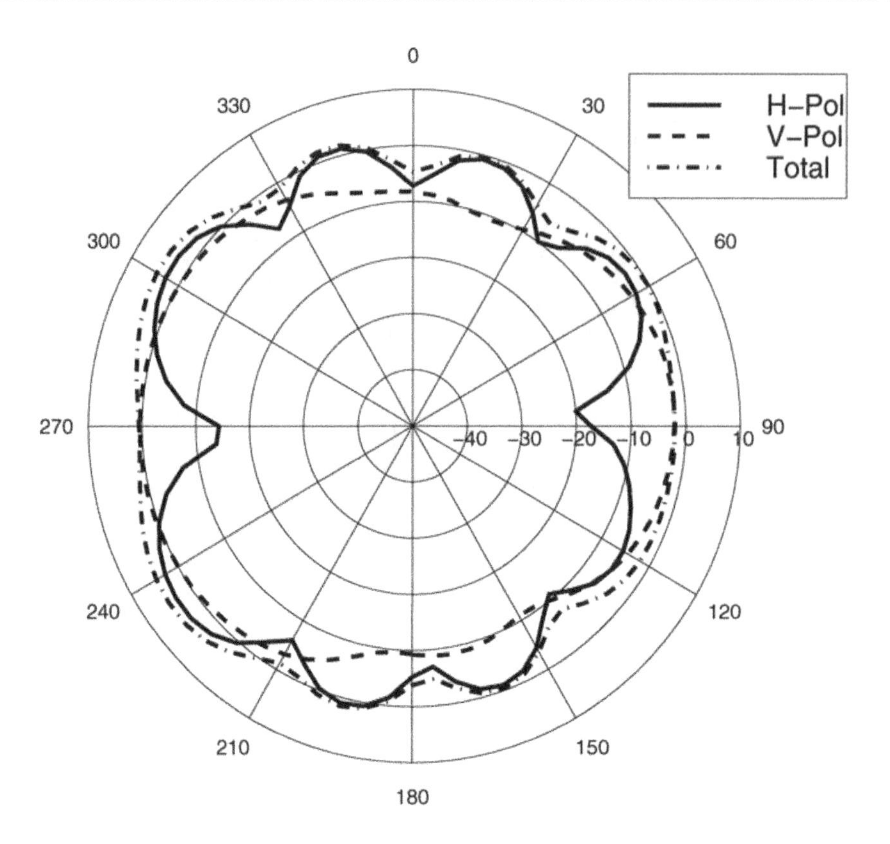

Figure 4.43 Measured radiation patterns at 1910 MHz on the horizontal plane. (From [65]. Reproduced by permission of © IEEE.)

5 mm. Figure 4.45 shows the principles of the Hitachi Cable design. One resonator is used for the 800 and 900 MHz band, referred as the low band. Similar to its WLAN antenna, two resonators are used for the DCS, PCS and UMTS bands from 1710 to 2170 MHz for world-wide applications. Two quarter-wavelength resonators are folded to reduce the resonator height. Due to its small size, the WLAN antenna resonators are above the LCD panel surface to ensure a high efficiency while completely hidden in the display (see Figure 4.1). For the WWAN antennas however, the antenna efficiency is maintained by extruding the antenna outside the display cover surface. A photo of the actual design is shown in Figure 4.46.

Figure 4.47 shows the measured SWR of the antenna in a laptop. It is clear the antenna covers all cellular bands world-wide. Figures 4.48 and 4.49 show the measured radiation patterns of the WWAN antenna in the laptop through the low and high bands, respectively. In the low band, the antenna radiation patterns begin to deteriorate when approaching 824 MHz, the low end of the 800 MHz band. However the radiation patterns do not change much through the 1710–2170 MHz range. In both cases, the radiation patterns are almost omnidirectional.

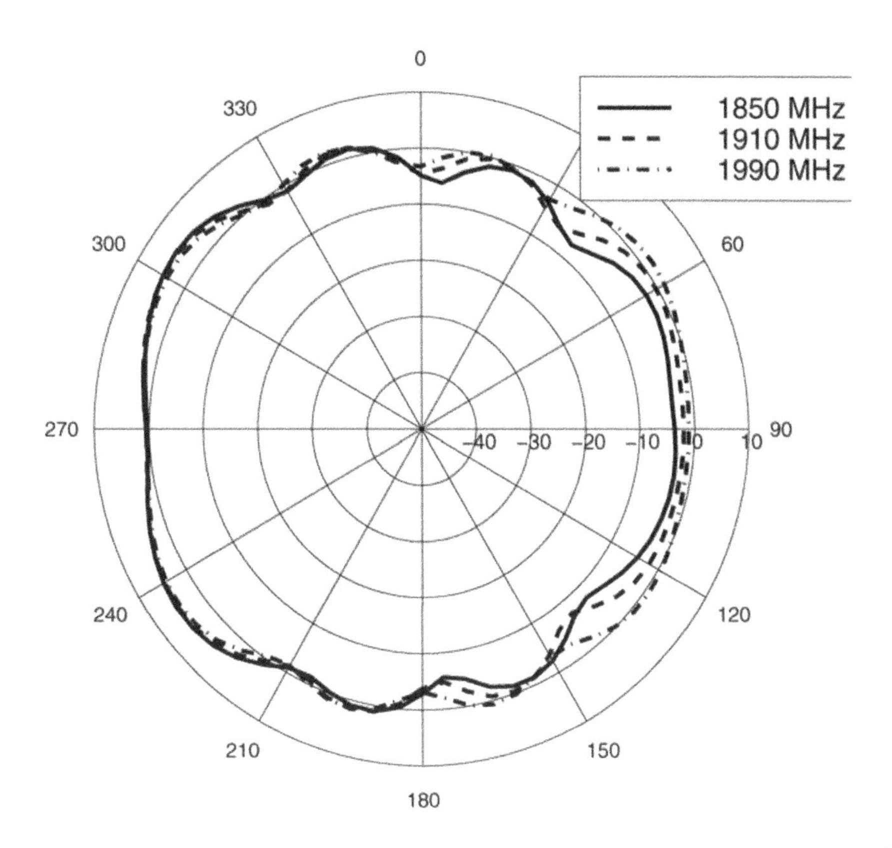

Figure 4.44 Measured radiation patterns in the PCS band on the horizontal plane. (From [65]. Reproduced by permission of © IEEE.)

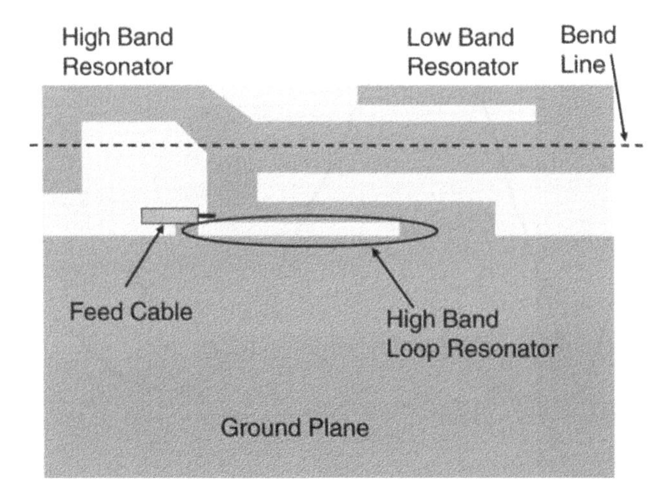

Figure 4.45 Principles of Hitachi Cable WWAN antenna design.

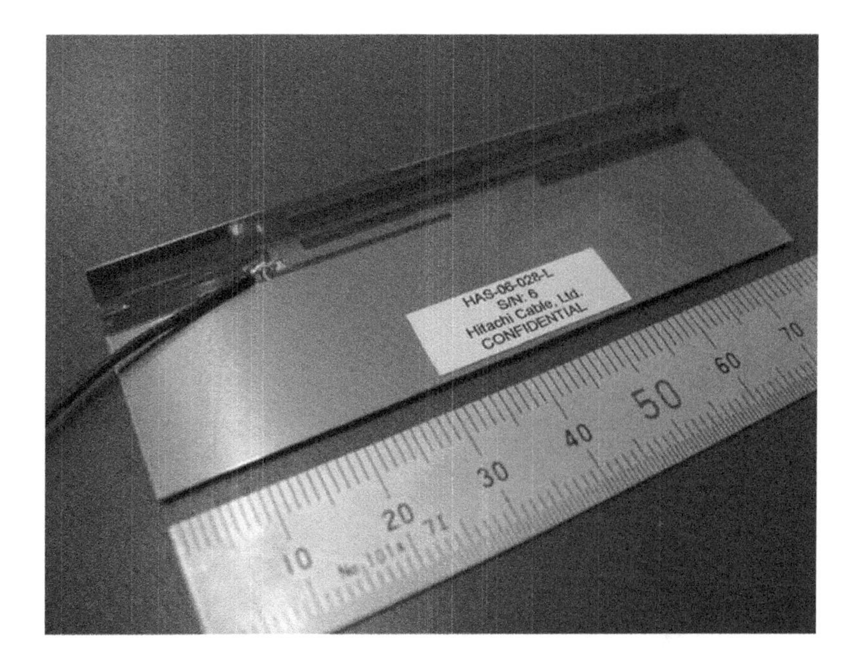

Figure 4.46 Hitachi Cable WWAN antenna. (H. Tate, private communication. Reproduced by permission of Hitachi Cable, Ltd.)

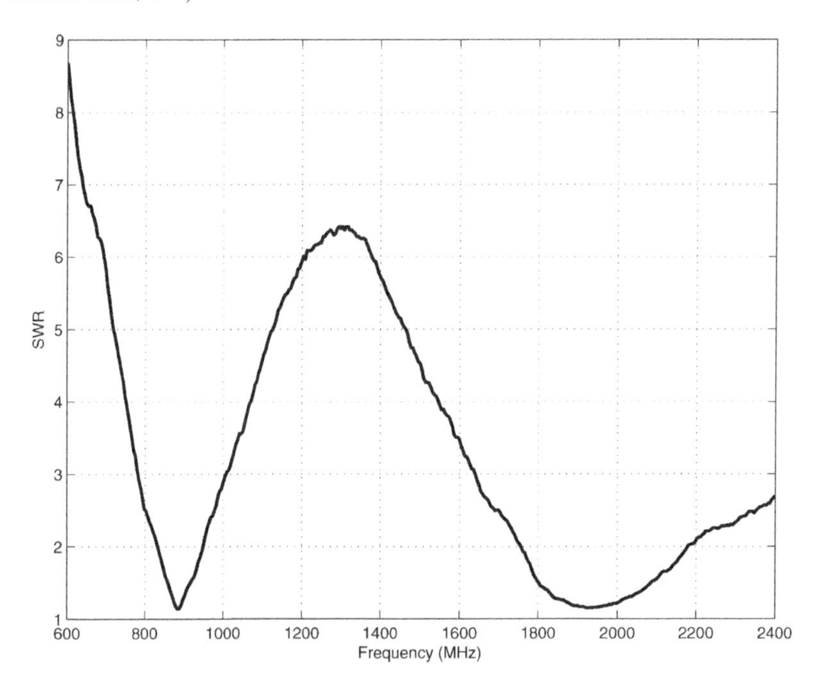

Figure 4.47 Measured SWR of the Hitachi Cable WWAN antenna. (H. Tate, private communication. Reproduced by permission of Hitachi Cable, Ltd.)

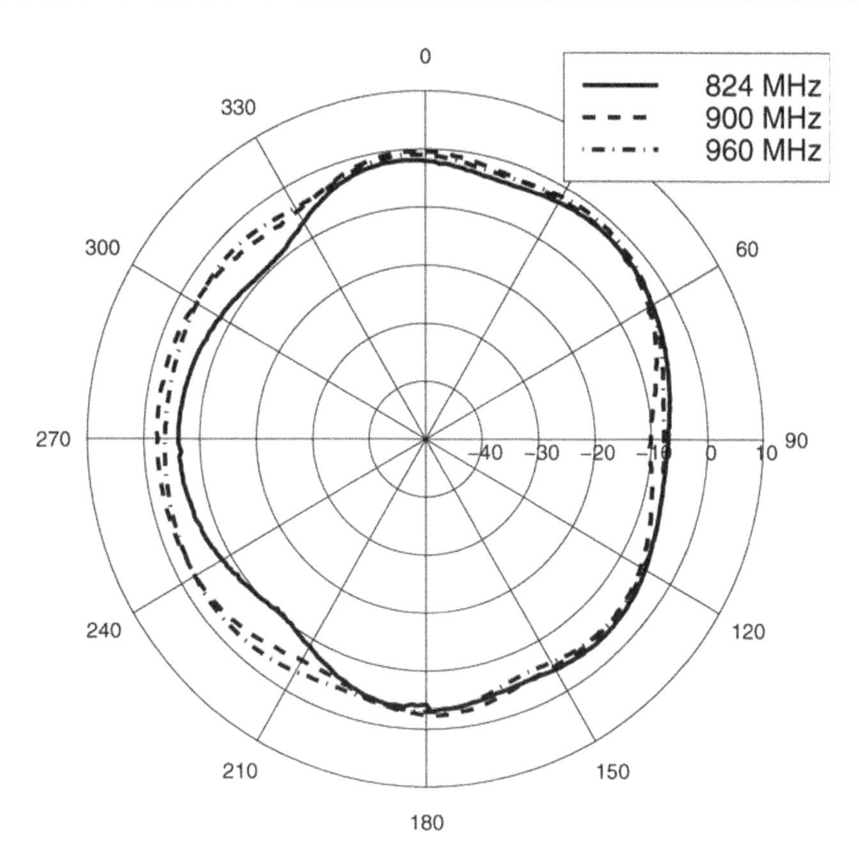

Figure 4.48 Measured radiation patterns of the Hitachi Cable WWAN antenna in the 800/900 MHz band. (H. Tate, private communication. Reproduced by permission of Hitachi Cable, Ltd.)

4.11 Ultra-Wide Band Antennas

The technology based on the ultra-wide band (UWB) concept has recently attracted much interest, especially since the release by the FCC of an extremely large frequency range for unlicensed commercial applications. Wireless systems based on the UWB technology have been explored and developed extensively [66]. UWB has been considered as a promising technology for short range wireless communications, especially for wireless USB applications at 480 Mbps. One of the challenges in UWB systems is to design antennas that can cover the ultra-wide bandwidth of 3.1–10.6 GHz yet are small enough to reside inside portable devices.

Among broadband antennas, planar designs have been considered as the most suitable candidates for portable UWB devices due to their simple mechanical structures and broadband characteristics [67–71]. However, the size of the usual planar antennas is still too large for many portable devices, such as laptop computers, where very thin and low-profile antennas are required. For example, the design reported by Suh *et al.* [71] has a very good impedance match within a 7:1 frequency range, but the antenna, with a height greater than 25 mm, is too large for laptop computer UWB applications. Another design challenge is the significant

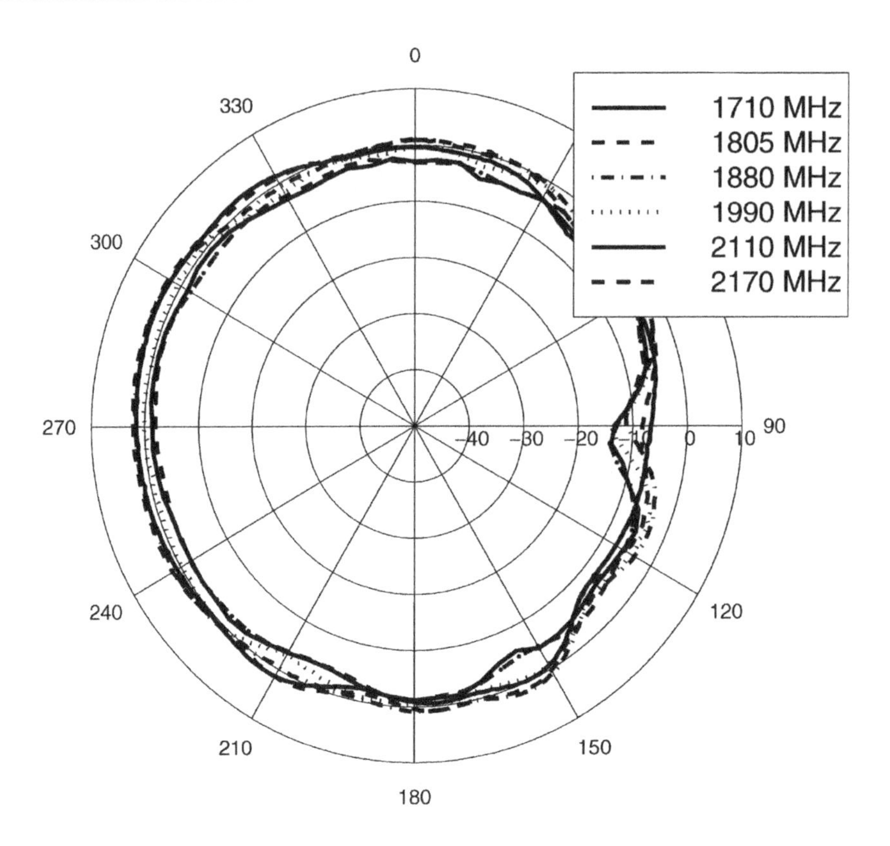

Figure 4.49 Measured radiation patterns of the Hitachi Cable WWAN antenna through the high band. (H. Tate, private communication. Reproduced by permission of Hitachi Cable, Ltd.)

proximity effects of the lossy display on the radiation efficiency of UWB antennas, since the antennas are usually embedded into the display of a laptop computer. The lossy metal objects in the display not only affect the impedance match, but also absorb the energy leaked from the RF feed cable as well as radiated from the antenna. Therefore the lossy objects result in significant reduction of the antenna efficiency. Thus the design must be carefully examined for practical laptop computer applications.

A typical laptop computer has three embedded antennas, one for Bluetooth™ and two for WLAN. (The latest laptops have five antennas, including two for WWAN applications.) Two antennas, usually the WLAN antennas, are typically embedded in the laptop display cover. Since laptop computers are getting smaller and thinner, it is very difficult to implement or embed more antennas, such as UWB antennas, into laptop computers. Therefore, it will be very useful to replace a WLAN antenna with an antenna that covers both WLAN and UWB bands while maintaining the existing WLAN antenna size and performance.

This section investigates a planar embedded dualband antenna that covers both 2.4 GHz and UWB bands suitable for laptop computer applications [72]. The proposed antenna is formed by a partial elliptic with a tail on one side of a PCB and a strip on the back side of the PCB. The back side strip is connected to the front structure with a metal strip on the side or a via the PCB. The antenna has a low profile and can be easily embedded into the display of a laptop computer. This simple structure made from common materials is very cost effective. The impedance and radiation performance of the antenna integrated into the lossy display of a laptop are taken into account. The SWR, maximum and average gain as well as radiation patterns of the embedded antenna are examined experimentally.

The proposed design is also suitable for other portable devices. However, the effects of environments, where the antennas are installed, on the performance of the embedded antennas should be carefully re-examined.

4.11.1 Description of the UWB Antenna

This antenna can be considered as a planar variation of the monopole antenna with one branch [40, 73]. The fat or widened monopole with smooth circular or elliptic feed is used to cover the UWB band. To further reduce the antenna height, it is necessary to top load the monopole. Symmetrical top loading is not as effective as the asymmetrical or tailed loading to reduce the monopole height. This monopole height reduction process is shown in Figures 4.50(a)–(c). To cover the 2.4 GHz WLAN band, another monopole or resonator is required. This secondary monopole can be a branch of the main monopole so only one feed is required. Figures 4.50(d)–(e) show the formation of the secondary monopole. Even though a PCB side connection is used here, a plated through hole (or via) close to the top edge of the PCB can also be used.

A dualband antenna with a metal strip was etched on a thin PCB (Roger4003, $\epsilon_r = 3.38$ and 0.0025 loss tangent at 10 GHz) of dimensions of 45 mm × 13 mm × 20 mil as shown in Figure 4.51. All dimensions were obtained by the simulation of the proposed antenna in

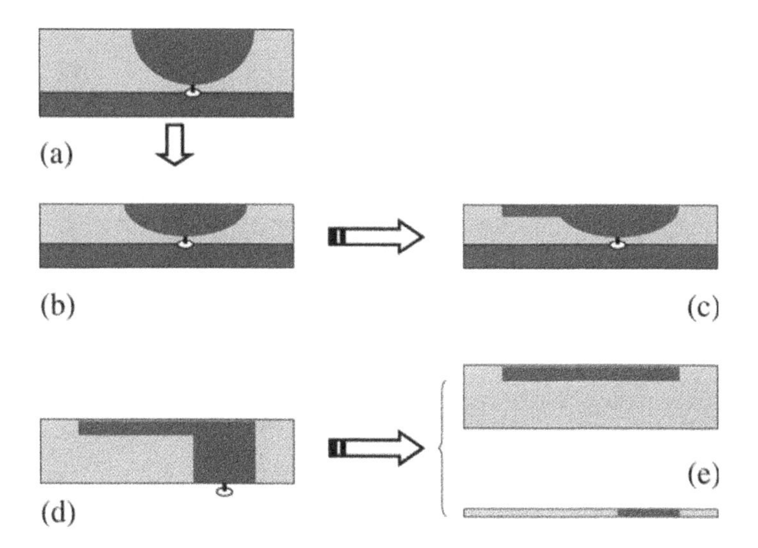

Figure 4.50 UWB dualband evolution. (From [72]. Reproduced by permission of © IEEE.)

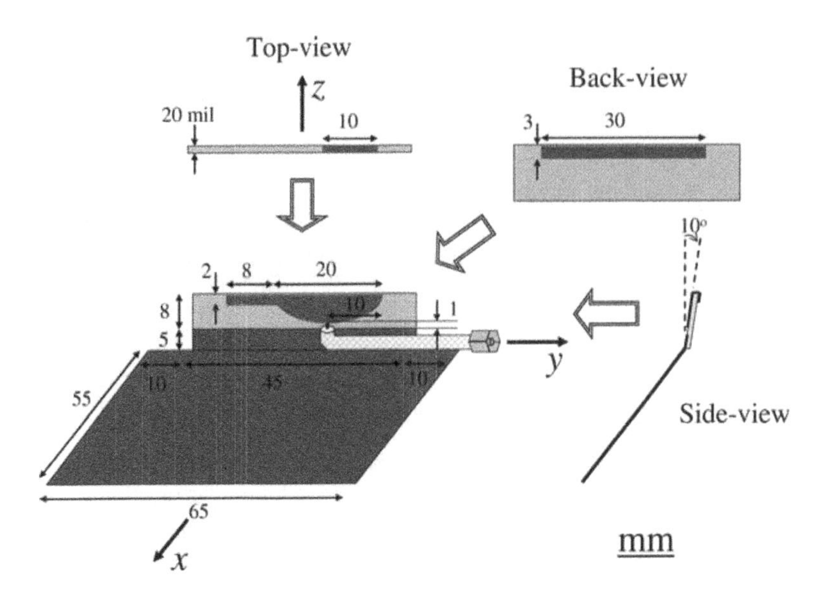

Figure 4.51 Geometry of proposed planar antenna. (From [72]. Reproduced by permission of © IEEE.)

Figure 4.52 A photo of the dualband UWB antenna prototype in a laptop display. (From [72]. Reproduced by permission of © IEEE.)

free space, and optimized experimentally for the antenna embedded into the lossy display of a laptop computer. The electromagnetic simulator used was IE3D (Zeland Software, Inc.), which is based on the method of moments. The height of the radiator portion is only 8 mm. The 5 mm × 45 mm metal strip is used for the ground connection and to improve antenna efficiency when embedded in a laptop display. Rectangular sheet metal or metal foil is used for the antenna ground plane. Due to the laptop display structure, the radiator portion is usually perpendicular to the ground plane. A ground plane of size 55 mm × 65 mm was electrically connected to the bottom edge of the 5 mm × 45 mm metal strip. The use of the ground plane is to ease the change in the performance of the antenna installed in a varying environment. The PCB tilts at an angle of 10° from the z-axis to fit the shape of the display cover edge as shown in Figures 4.51 and 4.52.

A 50 Ω coaxial cable excites the bottom of the partial elliptic with a feed gap of 1 mm through the ground plane to reduce the random effect of the feeding cable. By changing the feed gap, the impedance match can be adjusted. Usually, the feed gap is around 1 mm for the 3.1–10.6 GHz UWB band. Thus, the overall height of the antenna is about 13 mm.

The antenna can be embedded into the display of a laptop computer in several ways. To reduce the proximity effects of human body and support, such as the lap or a desk, the

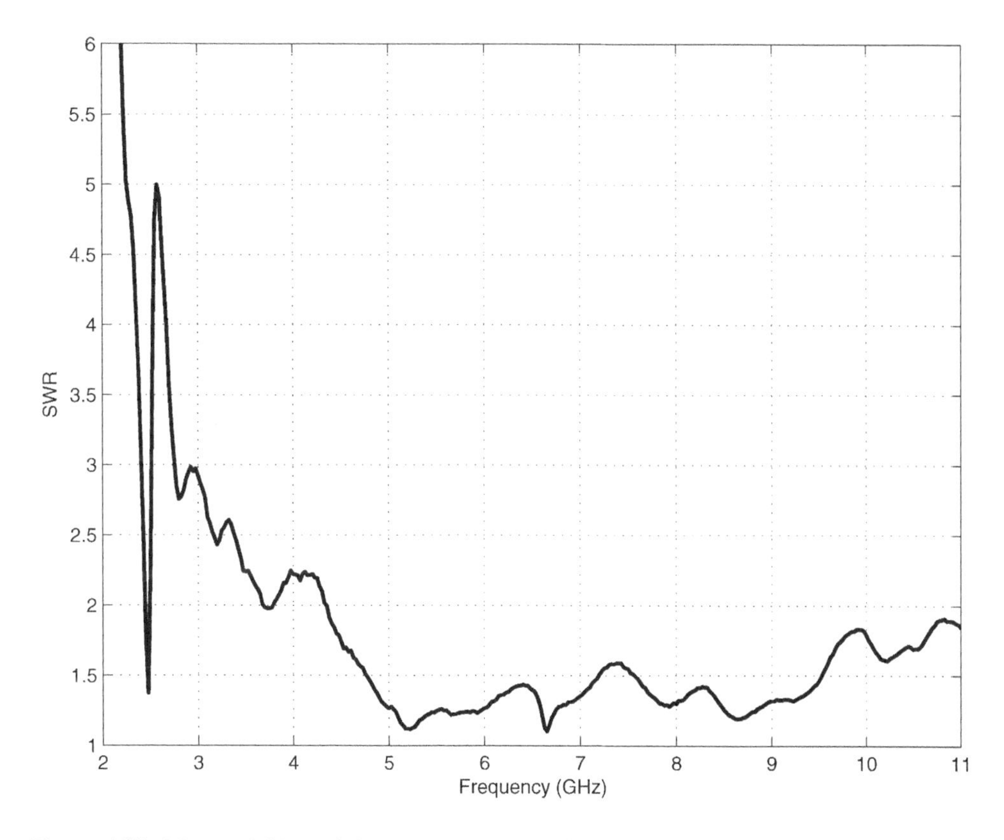

Figure 4.53 Measured SWR of the antenna embedded in a laptop display. (From [72]. Reproduced by permission of © IEEE.)

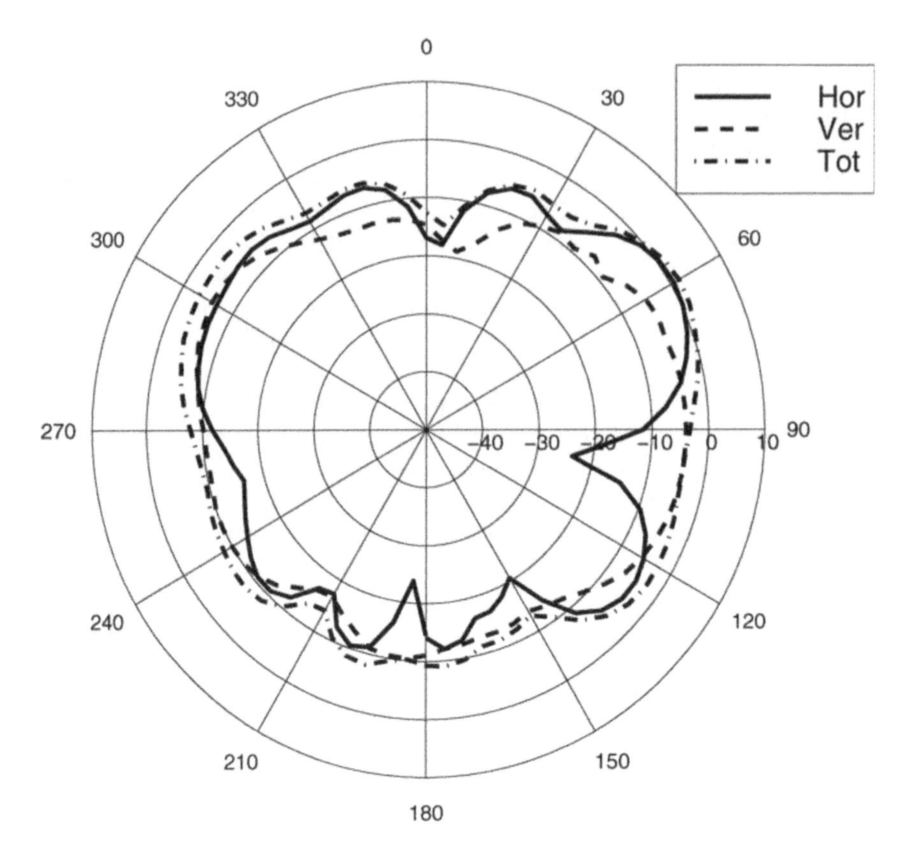

Figure 4.54 Measured radiation patters of the antenna embedded in a laptop display at 2.5 GHz. (From [72]. Reproduced by permission of © IEEE.)

antennas are mainly placed in the display frame or cover of a laptop computer, either on the left/right vertical edge or the top horizontal edge, very close to the lossy LCD panel. To be embedded into the display, the desirable design should be low profile and thin.

In this design, the antenna was placed on the top edge of the display cover. The metal LCD panel is lossy and usually leads to severe undesired ohmic losses. To keep the radiation effective, the top edge of the display cover has a 50 mm × 8 mm cutout. Then, the antenna was installed centrally in the cutout as displayed in Figure 4.52.

A lossy conducting LCD panel about 5 mm thick is installed very closely to the antenna with about 3 mm separation from the feed point at the bottom of the antenna. The proximity of the conducting LCD panel to some degree affects the impedance matching, particularly at the low frequencies. The 55 mm × 65 mm ground plane of the antenna is between the LCD panel and the display cover. The ground plane is electrically connected to the cover. Thus, the lossy cover contributes to the loss as well. In the test, the RF feeding cable went through the metal cover via a small hole. The investigation has shown that the LCD panel significantly affected both impedance matching and efficiency of the antenna embedded into the display as mentioned above.

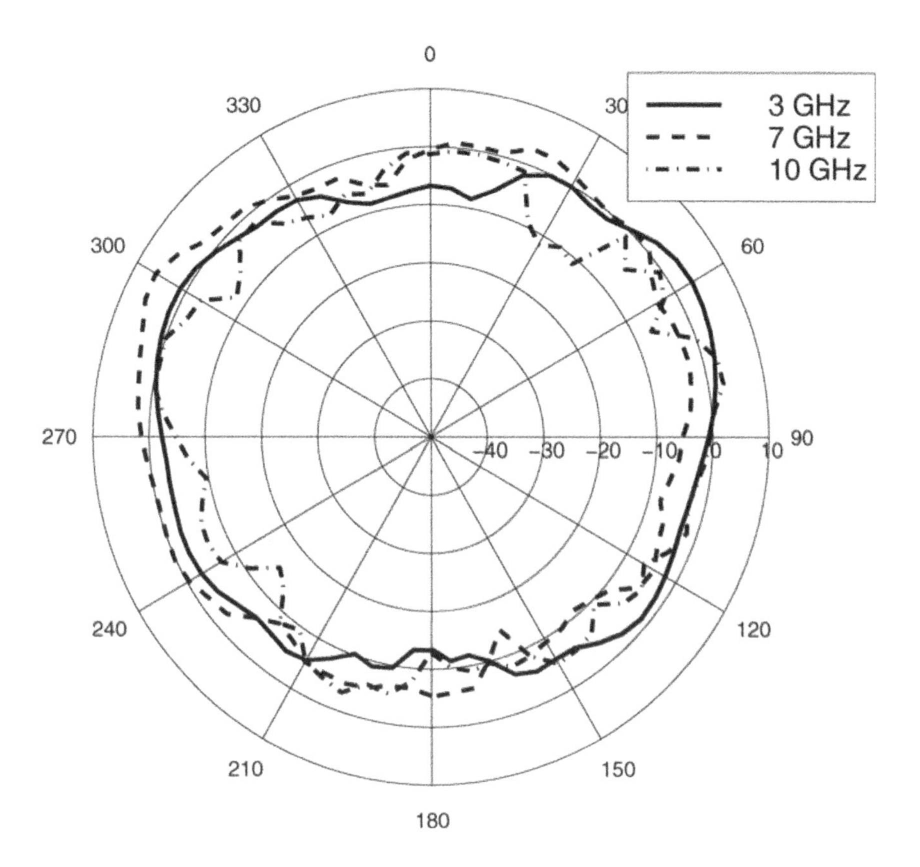

Figure 4.55 Measured radiation patters of the antenna embedded in a laptop display at 3, 7 and 10 GHz. (From [72]. Reproduced by permission of © IEEE.)

4.11.2 UWB Antenna Measurement Results

Both the impedance and radiation performance of an antenna embedded into the display of a laptop computer were simulated and examined experimentally.

The input impedance of the antenna was first measured. Compared with the antenna in free space, the antenna embedded into the display frame with a lossy conducting LCD panel has a broader bandwidth. It is inferred that the lossy cover and LCD panel somewhat reduce the Q of the antenna system so that a broad well-matched bandwidth can be attained. However, such losses will reduce the radiation efficiency and to some degree be harmful to antenna performance.

Figure 4.53 shows the measured SWR of the antenna across the 2–11 GHz frequency range. The antenna has enough 3:1 SWR bandwidth to cover the 2.4 GHz ISM band (2.4–2.5 GHz) and the 3.1–10.6 GHz UWB band.

Next, the radiation performance of the proposed antenna including gain and radiation patterns was measured in an anechoic chamber.

Figure 4.54 shows the measured radiation patterns at 2.5 GHz on the horizontal plane when the laptop was open 90°. The 2.5 GHz radiator is basically an inverted-L element, so the antenna has both horizontal and vertical polarizations. The overall radiation pattern is basically omnidirectional.

Figure 4.55 shows the measured radiation patterns at 3, 7 and 10 GHz across the UWB on the horizontal plane when the laptop was open 90°, respectively. The radiation patterns do not change much across the bands. The average gain is about 0 dBi, quite adequate for all standards. The laptop effects on the radiation patterns are obvious.

The measured results demonstrate that the proposed planar antenna for the UWB is basically a variation of the monopole, and features the monopole-like radiation patterns, which are quite consistent across the UWB band. The dips of the radiation patterns have been observed in the direction of computer users (z-direction, referring to Figure 4.51). The higher the frequency is, the higher the gain is.

References

[1] J. Geier, *Wireless LANs: Implementing Interoperable Networks.* Indianapolis, IN: Macmillan Technical Publishing, 1999.

[2] J. Bray and C. Sturma, *Bluetooth: Connect without Cables.* Upper Saddle River, NJ: Prentice Hall, 2001.

[3] J. Ross, *The Book of Wi-Fi.* San Francisco: No Starch Press, 2003.

[4] D. Liu, E. Flint and B. Gaucher, Integrated antennas for ThinkPads – design and performance. *The IBM Fourth Annual Personal Systems Institute Symposium,* Raleigh, NC, October 17–18, 2000.

[5] D. Liu, E. Flint and B. Gaucher, Integrated laptop antennas – design and evaluations. *Proceedings of the IEEE Antennas and Propagation Society International Symposium,* Vol. 4, pp 56–59, San Antonio, TX, June 16–21, 2002.

[6] D. Liu, B. Gaucher, E. Flint, T. Studwell and H. Usui, Developing integrated antenna subsystems for laptop computers. *IBM Journal of Research and Development,* vol. 47 (2003), pp 355–367.

[7] Wireless LAN security – an industry outlook, June 2005. http://www.researchandmarkets.com/reports/302681/.

[8] A. Huotari, A Comparison of 802.1 1a and 802.1 1b wireless LAN standards. The Linksvs Group, Inc., May 1, 2002.

[9] W. Sun, Optimizing WLAN performance with MIMO calls for careful analysis. *Wireless Net DesignLine,* January 2, 20006. http://www.wirelessnetdesignline.com/howto/175800500.

[10] V.K. Jones, G. Raleigh and R. van Nee, MIMO answers high-rate WLAN call. *EE Times,* December 31, 2003. http://www.eetimes.com/in focus/communications/OEG2003 123 1S0008.

[11] S.H. Wildstrom, Deluxe laptops that work anywhere. *Business Week,* p. 24, September 17, 2001.

[12] T.S. Rappaport, *Wireless Communications – Principles and Practice.* Upper Saddle River, NJ: Prentice Hall, 1996.

[13] C. Soras, M. Karaboikis, G. Tsachtsiris and V. Makios, Analysis and design of an inverted-F antenna printed on a PCMCIA card for the 2.4 GHz ISM band. *IEEE Antennas and Propagation Magazine*, 44 (2002), pp 37–44.

[14] J. Guterman, A.A. Moreira and C. Peixeiro, Omnidirectional wrapped microstrip antenna for WLAN applications in laptop computers. *Proceedings of the IEEE Antennas and Propagation Society International Symposium,* Vol. 3B, pp 301–304, Washington DC, July 2005.

[15] J. Guterman, A.A. Moreira and C. Peixeiro, Integration of omnidirectional wrapped microstrip antennas into laptops. *Antennas and Wireless Propagation Letters,* 5 (2006), pp 141–144.

[16] F.M. Caimi and G. O'Neill, Antenna designs for notebook computers: pattern measurements and performance considerations. *Proceedings of the IEEE Antennas and Propagation Society International Symposium,* Vol. 4A, pp 247–250, Washington, DC, July, 2005.

[17] R.R. Ramirez and F. DeFlaviis, Triangular microstrip patch antennas for dual mode 802.1 1a,b WLAN applications. *Proceedings of the IEEE Antennas and Propagation Society International Symposium,* Vol. 4, pp 44–47, San Antonio, TX, June 2002.

[18] C.A. Balanis, *Antenna Theory – Analysis and Design.* New York: Harper & Row, 1982.

[19] D. Liu and B. Gaucher, Performance analysis of inverted-F and slot antennas for WLAN applications. *Proceedings of the IEEE Antennas and Propagation Society International Symposium,* Vol. 2, pp 14–17, Columbus, OH, June 2003.

[20] J.T. Bernhard, Analysis of integrated antenna positions on a laptop computer for mobile data communications. *Proceedings of the IEEE Antennas and Propagation Society International Symposium,* Vol. 4, pp. 2210–2213, Montreal, July 1997.
[21] D.A. Strohschein and J.T. Bernhard, Evaluation of a novel integrated antenna assembly for mobile data networks using laptop computers. *Proceedings of the IEEE Antennas and Propagation Society International Symposium,* Vol. 4, pp. 1962–1965, Atlanta, GA, June 1998.
[22] K. Ito and T. Hosoe, Study of the characteristics of planar inverted F antenna mounted in laptop computers for wireless LAN. *Proceedings of the IEEE Antennas and Propagation Society International Symposium,* Vol. 2, pp. 22–25, Columbus, OH, June 2003.
[23] K.L. Wong, *Planar Antennas for Wireless Communications.* Hoboken, NJ: John Wiley & Sons, Inc., 2003.
[24] K.L. Wong, L.C. Chou and C.M. Su, Dual-band flat-plate antenna with a shorted parasitic element for laptop applications. *IEEE Transactions on Antennas and Propagation,* 53 (2005), 539–544.
[25] C.M. Su and K.L. Wong, Narrow flat-plate antenna for 2.4 GHz WLAN operation. *IEE Electronics Letters,* 39 (2003), 344–345.
[26] K.L. Wong, L.C. Chou and C. Wang, Integrated wideband metal-plate antenna for WLAN/WMAN operation for laptops. *Proceedings of the IEEE Antennas and Propagation Society International Symposium,* Vol. 4A, pp. 235–238, Washington, DC, July 2005.
[27] R. Bancroft, Development and integration of a commercially viable 802.1 1a/b/g HiperLan/WLAN antenna into laptop computers. *Proceedings of the IEEE Antennas and Propagation Society International Symposium,* Vol. 4A, pp. 231–234, Washington, DC, July 2005.
[28] Y. Ge, K.P. Esselle and T.S. Bird, Small quad-band WLAN antenna. *Proceedings of the IEEE Antennas and Propagation Society International Symposium,* Vol. 4B, pp. 56–59, Washington, DC, July 2005.
[29] J. Yeo, Y.J. Lee and R. Mittra, A novel dual-band WLAN antenna for notebook platforms. *Proceedings of the IEEE Antennas and Propagation Society International Symposium,* Vol. 2, pp. 1439–1442, Monterey, CA, June 2004.
[30] S. Rogers, J. Scott, J. Marsh and D. Lin, An embedded quad-band WLAN antenna for laptop computers and equivalent circuit model. *Proceedings of the IEEE Antennas and Propagation Society International Symposium,* Vol. 3, pp. 2588–2591, Monterey, CA, June 2004.
[31] S.H. Yeh and K.L. Wong, Dual-band F-shaped monopole antenna for 2.4/5.2 GHz WLAN application. *Proceedings of the IEEE Antennas and Propagation Society International Symposium,* Vol. 4, pp. 72–75, San Antonio, TX, June 2002.
[32] Y.J. Cho, Y.S. Shin and S.O. Park, An internal PIFA for 2.4/5 GHz WLAN applications. *Proceedings of the Asia-Pacific Microwave Conference (APMC2005),* Vol. 4, December 2005.
[33] N. Behdad and K, Sarabandi, A compact dual-/multi-band wireless LAN antenna. *Proceedings of the IEEE Antennas and Propagation Society International Symposium,* Vol. 2B, pp. 527–530, Washington DC, July 2005.
[34] K.L. Wong, L.C. Chou and C.M. Su, Dual-band flat-plate antenna with a shorted parasitic element for laptop applications. *IEEE Transactions on Antennas and Propagation,* 53 (2005), 539–544.
[35] C.M. Su, W.S. Chen and K.L. Wong, Metal-plate shorted T-shaped monopole for internal laptop antenna for 2.4/5 GHz WLAN operation. *Proceedings of the IEEE Antennas and Propagation Society International Symposium,* Vol. 2, pp. 1943–1946, Monterey, CA, June 2004.
[36] H. Okado, A 2.4 and 5 GHz dual band antenna. *Proceedings of the IEEE Antennas and Propagation Society International Symposium,* vol. 3, pp. 2596–2598, Monterey, CA, June 2004.
[37] H. Chen, J. Chen, P. Cheng and Y. Lin, Microstrip-fed printed dipole antenna for 2.4/5.2 GHz WLAN operation. *Proceedings of the IEEE Antennas and Propagation Society International Symposium,* Vol. 3, pp. 2584–2587, Monterey, CA, June 2004.
[38] J. Jan, L. Tseng, W. Chen, Y. Cheng, Printed monopole antennas stacked with a shorted parasitic wire for Bluetooth and WLAN applications. *Proceedings of the IEEE Antennas and Propagation Society International Symposium,* Vol. 3, pp. 2607–2610, Monterey, CA, June 2004.
[39] D. Liu, B. Gaucher and T. Hildner, A dualband antenna for WLAN applications. *Proceedings of the First IEEE International Workshop on Antenna Technology: Small Antennas and Novel Metamaterials,* pp. 201–204, Singapore, March 2005.
[40] D. Liu and B. Gaucher, A branched inverted-F antenna for dual band WLAN applications. *Proceedings of the IEEE Antennas and Propagation Society International Symposium,* Vol. 3, pp. 2623–2626, Monterey, CA, June 2004.

[41] D. Liu and B. Gaucher, A new multiband antenna for WLAN/cellular applications. *Proceedings of the IEEE 60th Vehicular Technology Conference,* Vol. 1, pp. 243–246, Los Angeles, September 2004.

[42] D. Liu and B. Gaucher, A dual band antenna for WLAN applications. *Proceedings of JINA2004,* pp. 436–437, Nice, November 2004.

[43] D. Liu, B. Gaucher and E. Flint, A new dual-band antenna for ISM applications. *Proceedings of the IEEE 56th Vehicular Technology Conference,* Vol. 2, pp. 937–940, Vancouver, September 2002.

[44] D. Liu, A dual-band antenna for 2.4 GHz ISM and 5 GHz UNII applications. *Proceedings of the URSI International Symposium on Electromagnetic Theory,* pp. 344–346, Victoria, Canada, May 2001.

[45] Y. Wang and S. Chung, A new dual-band antenna for WLAN applications. *Proceedings of the IEEE Antennas and Propagation Society International Symposium,* Vol. 3, pp. 2611–2614, Monterey, CA, June 2004.

[46] B.S. Collins, V. Nahar, S.P. Kingsley and S.Q. Zhang, A dual-band hybrid dielectric antenna for laptop computers. *Proceedings of the IEEE Antennas and Propagation Society International Symposium,* Vol. 3, pp. 2619–2622, Monterey, CA, June 2004.

[47] D. Liu and B. Gaucher, A triband antenna for WLAN applications. *Proceedings of the IEEE Antennas and Propagation Society International Symposium,* Vol. 2, pp. 18–21, Columbus, OH, June 2003.

[48] S. Fujio and T. Asano, Dual band coupled floating element PCB antenna. *Proceedings of the IEEE Antennas and Propagation Society International Symposium,* vol. 3, pp. 2599–2602, Monterey, CA, June 2004.

[49] M. Karikomi, Parasitic element excitation of a dual-frequency printed dipole antenna. *Proceedings of the Electronic Information Communication Society National Conference,* B-73, 1989.

[50] K. Fukuchi, Y. Yamamoto, K. Sato, R. Sato and H. Tate, Wide-band wireless LAN antenna for IEEE 802.11 a/b/g. *Hitachi Cable Review,* no. 23, August 2004.

[51] D. Liu, Analysis of a closely-coupled dual band antenna. *Proceedings of the Wireless Communications Conference,* pp. 86–89, Boulder, CO, August 1996.

[52] B.J. Rubin and S. Daijavad, Radiation and scattering from structures involving finite-size dielectric regions. *IEEE Transactions on Antennas and Propagation,* 38 (1990), 1863–1873.

[53] M. Ikegaya, T. Sugiyama and H. Tate, Dual band film type antenna for mobile devices. *Proceedings of the North American Radio Science Meeting,* AP/URSI B Session 133.8, Columbus, OH, June 2003.

[54] M. Ikegaya, T. Sugiyama, S. Takaba, S. Suzuki, R. Komagine and H. Tate, Film type antenna for mobile devices for 2.4 GHz range. *Hitachidensen,* no. 21, 2002.

[55] M. Ikegaya, T. Sugiyama, S. Takaba, S. Suzuki, R. Komagine and H. Tate, Development of film type antenna for mobile devices. *Hitachi Cable Review,* no. 21, August 2002.

[56] K. Fukuchi, T. Ogawa, M. Ikegaya, H. Tate and K. Takei, Small and thin structure plate type wideband antenna (3 GHz–6 GHz) for wireless communications. *Proceedings of the IEEE Antennas and Propagation Society International Symposium,* Vol. 3, pp. 2615–2618, Monterey, CA, June 2004.

[57] K. Fukuchi, K. Sato, H. Tate and K. Takei, Film type wide-band antenna for next-generation communication systems at frequency range from 2.3 to 6 GHz. *Hitachi Cable Review,* no. 24, August 2005.

[58] T. Ito, H. Moriyasu and M. Matsui, A small antenna for laptop applications. *Proceedings of the Second IEEE International Workshop on Antenna Technology: Small Antennas and Novel Metamaterials,* pp. 233–236, White Plains, NY, March 2006.

[59] H. Haruki and A. Kobayashi, The inverted-F antenna for portable radio units. *Digest of IECE Japan,* p. 613, 1982.

[60] K. Hirasawa and M. Haneishi, *Analysis, Design, and Measurement of Small and Low-Profile Antennas.* Boston: Artech House.

[61] T. Taga and K. Tsunekawa, Performance analysis of a built-in planar inverted-F antenna for 800 MHz hand portable radio units. *IEEE Journal of Selected Areas in Communications*, 5 (1987), 921–929.

[62] P. Salonen, L. Sydänheimo, M. Keskilammi and M. Kivikoshi, Planar inverted-F antenna for wearable applications. *Proceedings of the 3rd International Symposium on Wearable Computers,* pp. 95–98, San Francisco, October 1999.

[63] Z. Li, Y. Rahmat-Samii and T. Kaiponen, Bandwidth study of a dual band PIFA on a fixed substrate for wireless communication. *IEEE Transactions on Antennas and Propagation,* pp. 22–27, January 2003.

[64] P. Song, P.S. Hall, H. Ghafouri-Shiraz and D. Wake, Triple band planar inverted F antenna. *Proceedings of the IEEE Antennas and Propagation Society International Symposium,* Vol. 2, pp. 908–911, Orlando, FL, July 1999.

[65] D. Liu and B. Gaucher, The inverted-F antenna height effects on bandwidth. *Proceedings of the IEEE Antennas and Propagation Society International Symposium,* Vol. 2A, pp. 367–370, Washington, DC, July 2005.

[66] Federal Communications Commission, *First Order and Report, Revision of Part 15 of the Commission's Rules Regarding UWB Transmission Systems,* FCC 02-48, April 22, 2002.
[67] Z.N. Chen, X.H. Wu, N. Yang and M.Y.W. Chia, Planar square monopoles for UWB radio systems. In *Proceedings of the IEEE Asia-Pacific Microwave Conference,* pp. 1632–1635, Korea, 2003.
[68] X.H. Wu, Z.N. Chen and N. Yang, Planar diamond antenna in UWB radio systems. *Proceedings of the IEEE Asia-Pacific Microwave Conference,* pp. 1620–1623, Korea, 2003.
[69] N. Yang, Z.N. Chen and X.H. Wu, Study of circular planar monopoles for UWB radio systems. *Proceedings of the IEEE Asia-Pacific Microwave Conference,* pp. 1628–1631, Korea, 2003.
[70] Y. Zhang, Z.N. Chen and M.Y.W. Chia, Effects of finite ground plane and dielectric substrate on planar dipoles for UWB applications. *Proceedings of the IEEE Antennas and Propagation Society International Symposium,* Vol. 3, pp. 2512–2515, Monterey, CA, June 2004.
[71] S.Y. Suh, W.L. Stutzman and W.A. Davis, A new ultrawideband printed monopole antenna: the planar inverted cone antenna (PICA). *IEEE Transactions on Antennas and Propagation,* 52 (2004), 1361–1364.
[72] Z.N. Chen, D. Liu and B. Gaucher, A planar dualband antenna for 2.4 GHz and UWB laptop applications. *Proceedings of the IEEE 63rd Vehicular Technology Conference,* Melbourne, May 2006.
[73] D. Liu, A multi-branch monopole antenna for dual-band antenna cellular applications. *Proceedings of the IEEE Antennas and Propagation Society International Symposium,* Vol. 3, pp. 1578–1581, Orlando, FL, 1999.

5

Antenna Issues in Microwave Thermal Therapies

Koichi Ito
Faculty of Engineering, Chiba University

Kazuyuki Saito
Research Center for Frontier Medical Engineering, Chiba University

5.1 Microwave Thermal Therapies

5.1.1 Introduction

Microwave applications have assumed considerable importance in medicine [1] because they are effective in the reduction of the mental and physical burden borne by patients [2]. Such applications are of three types. First, thermal treatments which use microwave energy as a source of heat. Second, diagnosis and information gathering inside the human body (e.g. by computerized tomography and magnetic resonance imaging) and non-invasive temperature measurement inside the human body [3, 4]. It has not switched x-rays, ultrasound and magnetic fields are also studied strenuously. Third, the gathering of medical information on the human body from outside the body, and information transmission [5]. Techniques in this category are considered to be an extension of communication technologies. Therefore, this chapter describes the characteristics of antennas for microwave thermal therapies.

5.1.2 Classification by Therapeutic Temperature

In recent few decades, various types of microwave antennas for thermal therapy have been investigated. There are two major types of treatments, which can be classified depending on the therapeutic temperature (Figure 5.1). Hyperthermia is one of the modalities for cancer treatment, utilizing the difference in thermal sensitivity between tumor and normal tissue [6]. Here, the tumor must be heated up to the therapeutic temperature of 42–45 °C without

Antennas for Portable Devices Zhi Ning Chen

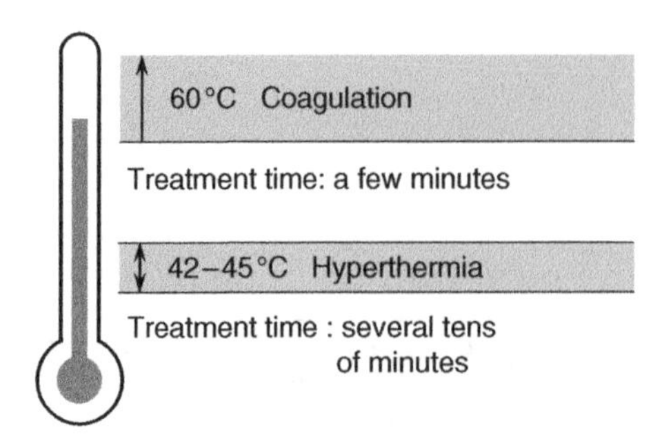

Figure 5.1 Therapeutic temperatures.

overheating the surrounding normal tissues. Moreover, the therapeutic effect of other cancer treatments such as radiation therapy and chemotherapy can be enhanced by using them together with hyperthermia.

Microwave coagulation therapy (MCT) has been used mainly for the treatment of small tumors such as hepatocellular carcinoma [7]. In the treatment, a thin microwave antenna is inserted into the tumor, and the microwave energy provided by the antenna heats up the tumor to produce the coagulated region including the cancer cells. The therapeutic temperature of this treatment is above 60 °C.

In particular, this chapter focuses on antennas for interstitial microwave hyperthermia. However, the interstitial heating technique (described later) can be also be applied to MCT.

5.1.3 Heating Schemes

The size and shape of the target tumors are various. Therefore, it is impossible to treat different kinds of tumor with a unique type of antenna. Figure 5.2 shows the various types of antennas used. Antennas may be external (radiator and contact antennas) or internal (intracavitary and interstitial antennas).

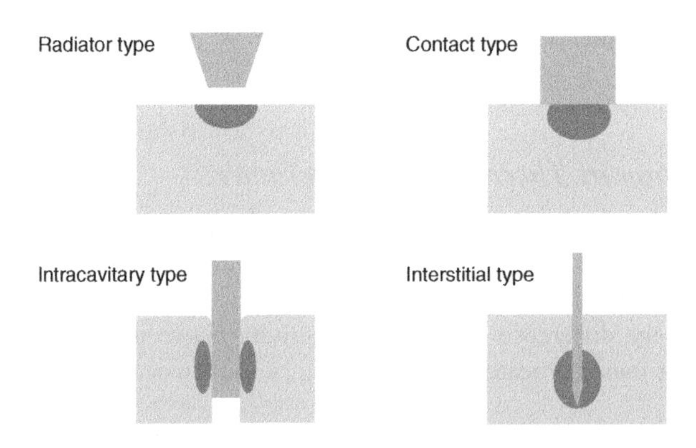

Figure 5.2 Various types of antennas for the treatment of tumors.

In particular, this chapter focuses on the antennas for interstitial microwave hyperthermia and MCT, both for thermal treatment of cancer. However, the results of this chapter can also be applied more widely to the improvement of antennas for the treatment of cardiac catheter ablation [8], the treatment of benign prostatic hypertrophy [9], and so on, because the antennas for such treatments and the antennas for cancer treatment have some common characteristics (e.g. very thin diameter). Moreover, the results of this chapter will give suggestions not only for the characteristics of the antenna inside the human body but also for the behavior of electromagnetic waves inside lossy media.

5.2 Interstitial Microwave Hyperthermia

5.2.1 Introduction and Requirements

First of all, the structure of the antenna for interstitial heating must be thin. Typically, its diameter should be less than 2–3 mm. It is dependent on the target of the treatment. Interstitial microwave hyperthermia is used for the treatment of large-volume, deep-seated tumors. In the treatment, a thin microwave antenna is inserted into the tumor and heated up by microwave energy. The antenna usually employed is as an array applicator[1] allowing the insertion of several elements into the tissue. Such antennas can be utilized as an adjuvant therapy to interstitial radiation therapy, by using the same catheter (Figure 5.3). In the system of Figure 5.3, first, thin microwave antennas such as the coaxial-slot antenna are inserted into the catheter. After heating, only the antennas are removed from the catheters. Then, a radiation source such as iridium 192 is automatically inserted into the catheter with a high

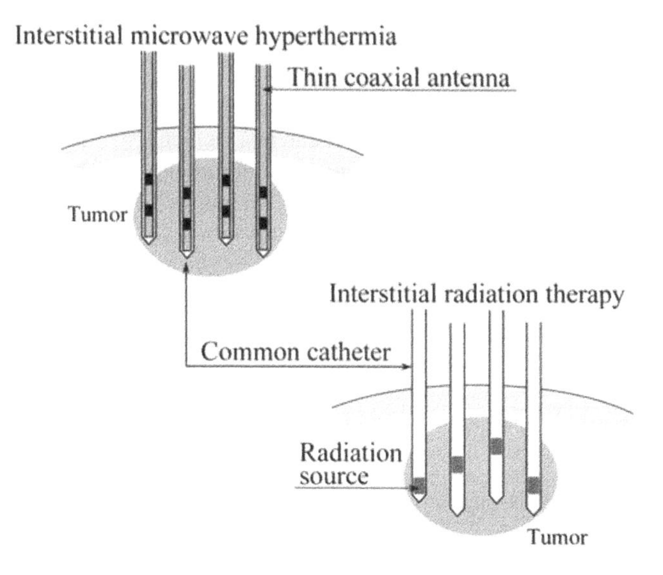

Figure 5.3 Combination of interstitial microwave hyperthermia and interstitial radiation therapy.

[1] The term 'applicator' indicates the part of the heating system that is directly applied to the human body.

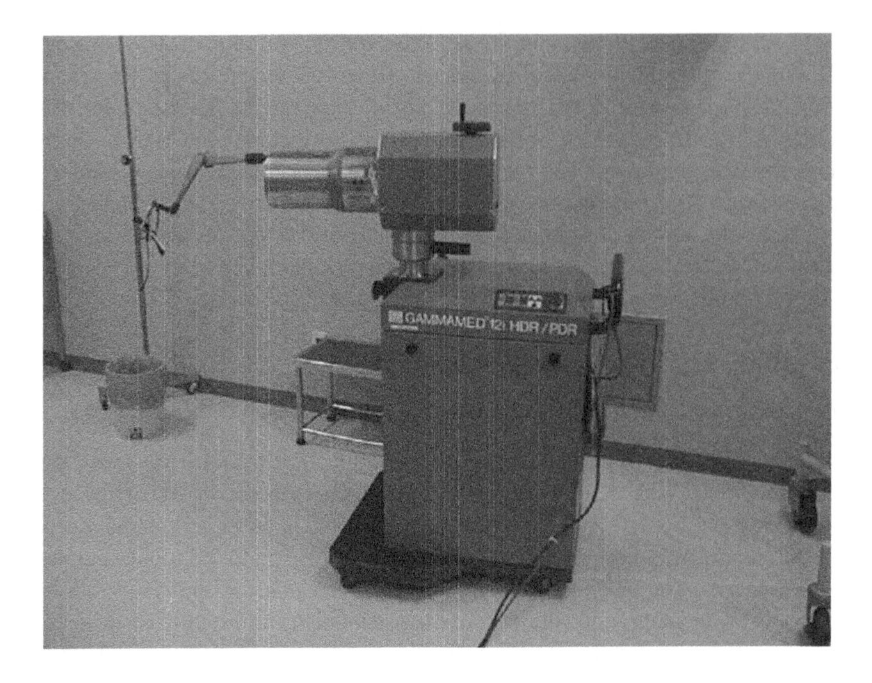

Figure 5.4 Treatment system for interstitial radiation therapy (high dose rate afterloading system).

dose rate afterloading system. Figure 5.4 shows the high dose rate afterloading system for interstitial radiation therapy.

In the antennas used in treatment, it is possible to change the heating pattern in the perpendicular direction of the antenna axis by varying the number of elements and the antenna insertion points. Control of the heating pattern in the longitudinal direction of the antenna axis is realized by changing the structure of the antenna elements while keeping the structure thin. This is especially important for treatment of brain tumors described in Section 5.4.

An example of an interstitial heating system is shown in Figure 5.5. The temperature in the tumor is monitored by means of inserted temperature sensors, and the output power

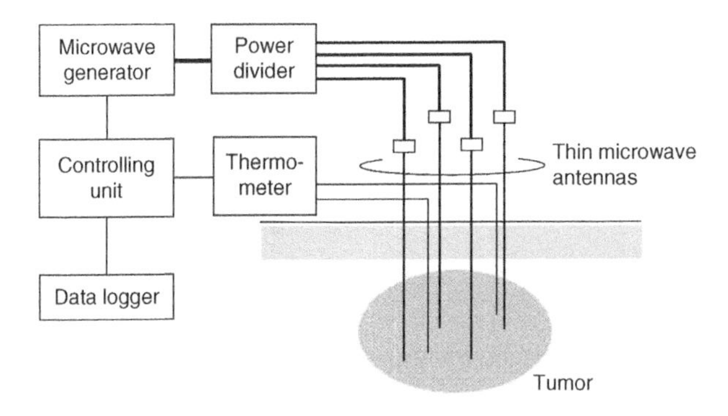

Figure 5.5 Interstitial heating system.

of the microwave generator is modified by feedback control. The microwave power output from the generator feeds the antennas through a power divider.

5.2.2 Coaxial-Slot Antenna

Several types of interstitial antennas have been developed and reported [10, 11]. Figure 5.6 shows typical antennas for interstitial heating. The authors have been studying a coaxial-slot antenna (Figure 5.6(f) or 5.6(g)) for such application. Figure 5.7 and Table 5.1 show the basic structure and an example of the structural parameters of the antenna. This antenna is composed of a thin semi-rigid coaxial cable. Some ring slots are cut on the outer conductor of the thin coaxial cable and the tip of the cable is short-circuited. The antenna is inserted into a catheter made of polytetrafluoroethylene for hygiene reasons. The operating frequency is 2.45 GHz, which is one of the Industrial, Scientific, and Medical (ISM) frequencies. From our previous investigations, it is clear that the coaxial-slot antenna with two slots, which are set so that L_{ts} and L_{ls} are equal to 20 mm and 10 mm, respectively, generates a localized heating region only around the tip of the antenna [13]. Therefore, we employ these structural parameters for the antenna in this chapter.

5.2.3 Numerical Calculation

5.2.3.1 Procedure of Calculation

In the antennas for telecommunications and broadcasting, input impedance, radiation pattern, and radiation efficiency are among the important factors for performance evaluation. In contrast, in antennas for thermal treatment, it is the specific absorption rate (SAR) and the temperature distribution in the body that are the important criteria. Here, the numerical calculation of the temperature distribution is described.

Figure 5.8 shows the flowchart for computer simulation for calculating the temperature distribution around the coaxial-slot antenna inside the tissue. First, we calculate the electric

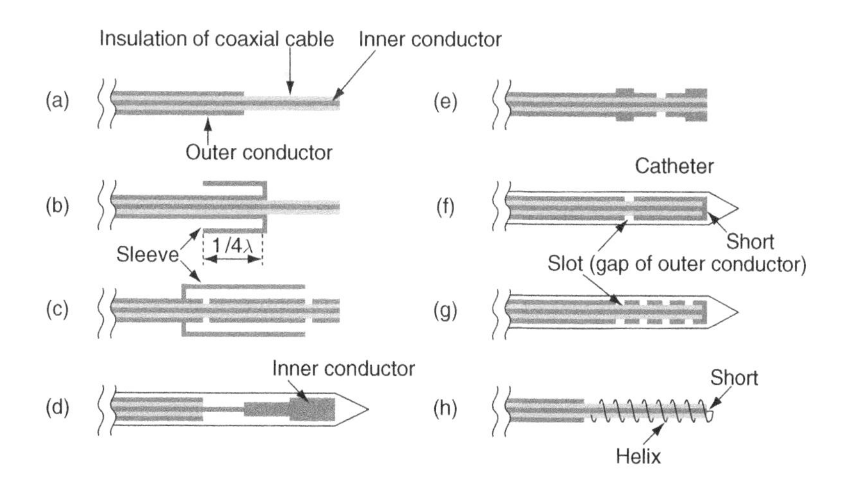

Figure 5.6 Configuration of various antennas (around the tip) for interstitial heating [12].

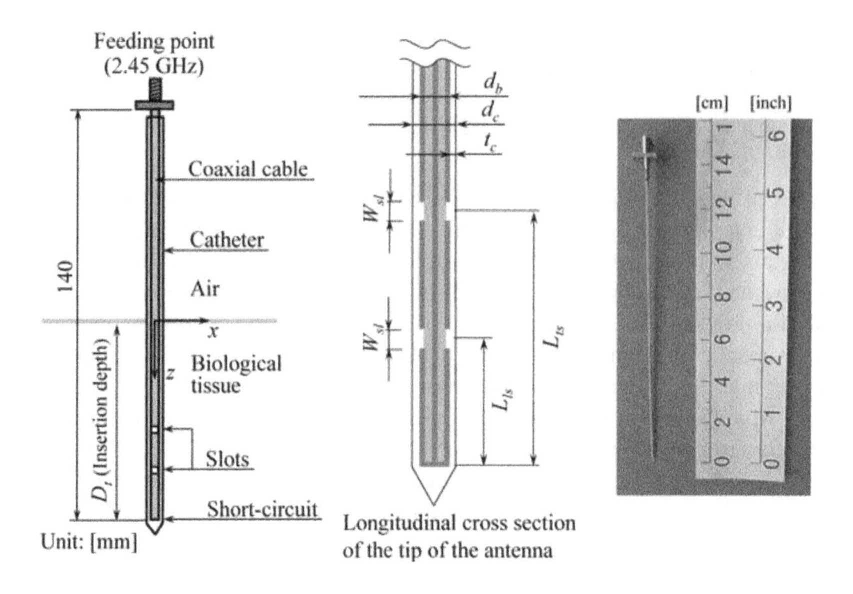

Figure 5.7 Basic structure of the coaxial-slot antenna.

Table 5.1 Structural parameters of the coaxial-slot antenna with two slots.

d_b (diameter of the antenna) [mm]	1.19
d_c (external diameter of the catheter) [mm]	1.79
t_c (thickness of the catheter) [mm]	0.3
L_{ts} (length from the tip to the center of the slot close to the feeding point) [mm]	20.0
L_{ls} (length from the tip to the center of the slot close to the tip) [mm]	10.0
W_{sl} (width of the slot) [mm]	1.0
ε_{rc} (relative permittivity of the catheter)	2.6

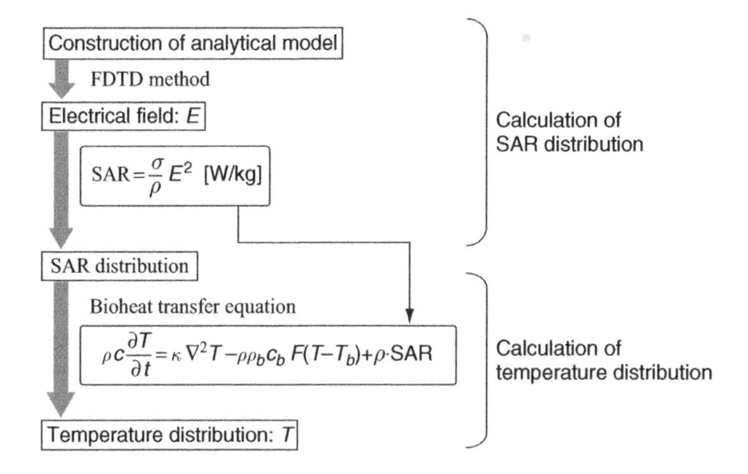

Figure 5.8 Procedure for temperature calculation.

field around the antenna using the finite difference time domain (FDTD) method. Next, we calculate the SAR distribution around the antenna:

$$\mathrm{SAR} = \frac{\sigma}{\rho} E^2 \quad [\mathrm{W/kg}], \tag{5.1}$$

where σ is the conductivity of the tissue [S/m], ρ the density of the tissue [kg/m^3] and E the electric field (rms) [V/m]. The SAR takes a value proportional to the square of the electric field around the antennas and is equivalent to the heating source generated by the electric field in the tissue.

Finally, in order to obtain the temperature distribution in the tissue, we numerically analyze the bioheat transfer equation including the obtained SAR distribution by using the finite difference method (FDM).

5.2.3.2 FDTD Calculation for Electromagnetic Field

Figure 5.9 shows the FDTD calculation modeling for the coaxial-slot antenna. To model the coaxial-slot antenna precisely, a very fine mesh model, which takes up considerable computer memory and is time-consuming, is needed because the antenna is very thin. Therefore, basically, a rectangular antenna cross-section is employed instead of a transverse cross-section for the calculations for array applicators shown in Figure 5.16. Moreover, a staircasing approximation model is employed for example analyses of input impedance of the

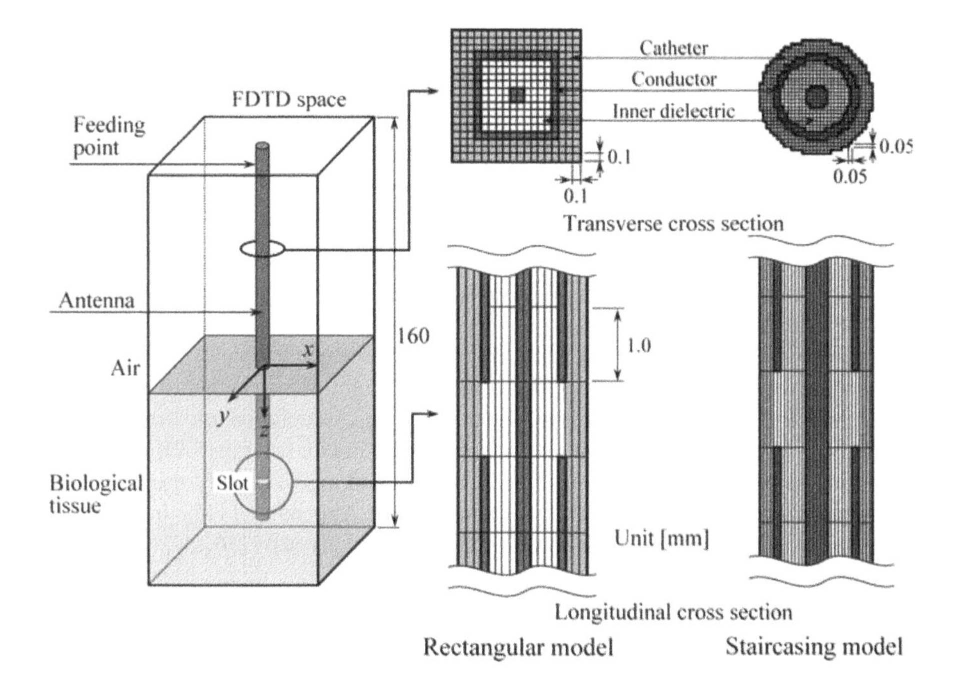

Figure 5.9 FDTD calculation models of the coaxial-slot antenna.

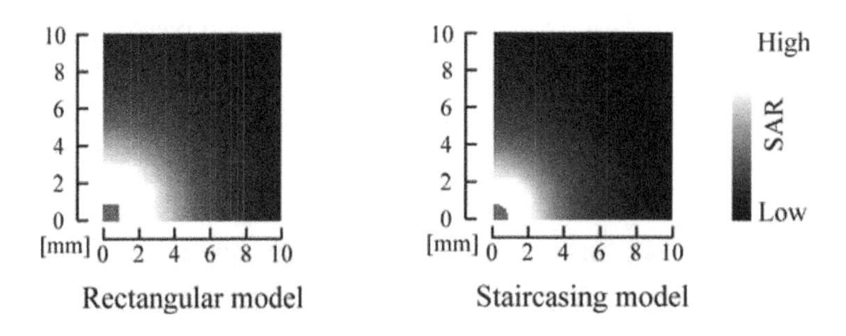

Figure 5.10 SAR distributions at the corner of the calculation model.

single antenna. In addition, Figure 5.10 shows the calculated results of the SAR distributions near the antenna by the FDTD calculation method for both the 'rectangular model' and 'staircasing model' (only a quarter region is shown due to structural symmetry in the *x*–*y* plane). From this figure, almost the same SAR distributions are observed in both models.

In order to calculate the SAR or temperature distributions, a steady-state analysis is performed, feeding a sinusoidal electric field between the inner and the outer conductors of the coaxial cable. The outer boundary condition of the FDTD space is the Mur first-order condition. In the analysis, non-uniform grids are employed and small-sized grids are used only for the antenna.

5.2.3.3 Temperature Analysis

We can calculate the temperature distribution inside the biological tissue by solving the bioheat transfer equation given by [14]:

$$\rho\, c\, \frac{\partial T}{\partial t} = \kappa\, \nabla^2 T - \rho\, \rho_b\, c_b\, F\ (T - T_b) + \rho \cdot \mathrm{SAR}, \tag{5.2}$$

where T is the temperature [°C], t time [s], ρ density [kg/m^3], c specific heat [J/kg·K], κ thermal conductivity [W/m·K], ρ_b the density of the blood [kg/m^3], c_b the specific heat of the blood [J/kg·K], T_b the temperature of the blood [°C] and F the blood flow rate [m^3/kg·s]. The first, second and third terms on the right-hand side of (5.2) denote the thermal conduction, the heat dissipation by the blood flow and the heat generation by the electric field, respectively. We assumed that the temperature of the blood T_b is equal to the initial temperature of the tissue. We substituted the SAR obtained by the previous electromagnetic calculation into the third term on the right of (5.2). The FDM was used in a numerical calculation for solving (5.2). We employed the same grids as those of the electromagnetic analysis and analyzed only inside the tissue. The finite difference approximation of (5.2) is shown in the Appendix of [15]. In this case, the stability criterion of the temperature analysis is assumed to be given by [16]:

$$\Delta t \leq \frac{2\rho\, c}{\rho\, \rho_b\, c_b\, F + 4\kappa \left(\frac{1}{\Delta x^2} + \frac{1}{\Delta y^2} + \frac{1}{\Delta z^2}\right)} \tag{5.3}$$

where Δt [s] is the time step for temperature analysis and Δx, Δy, Δz [m] are minimum cell sizes. From (5.3) it is clear that we must choose a small time step when κ increases. In general, κ in the conductor of the antenna is very large compared with that in the biological tissue. However, we cannot choose a very small time step for practical calculation. Therefore, we assumed that the heat transfer in the conductor is small because the conductor has very small volume. Under this assumption, we replaced the antenna with an object having the same material as that of the catheter.

5.2.4 Performance of the Coaxial-Slot Antenna

5.2.4.1 SAR Distributions

Figure 5.11 shows the calculated SAR distributions around the antenna. Here, the coaxial-slot antenna is inserted into liver tissue ($\varepsilon_r = 43.03$, $\sigma = 1.69\,\text{S/m}$ at 2.45 GHz [17]), for instance. The observation plane is the longitudinal (x–z) plane that includes the antenna. Here, the calculated results for a coaxial-slot antenna with a single slot ($L_{ts} = 10\,\text{mm}$) and with two slots ($L_{ts} = 20\,\text{mm}$, $L_{ls} = 10\,\text{mm}$) are shown for comparison. In this case, the antenna insertion depth was set as 70 mm. From Figure 5.11(a), the high SAR region exists only around the tip of the antenna. On the other hand, in Figure 5.11(b), we can observe that the high SAR region goes a long way in the antenna insertion direction for the coaxial-slot antenna with a single slot. By comparing both results, we can observe that a more localized SAR profile around the tip of the antenna is achieved by use of the coaxial-slot antenna with two slots.

Figure 5.12 shows the measured SAR distributions around these two antennas. The SAR distributions are measured by the thermographic method [18]. Figure 5.13 illustrates the measurement system. In this system, first, the antenna is placed between the pre-cut phantoms and a high power microwave energy (e.g., several tens of watts) is supplied to the antenna.

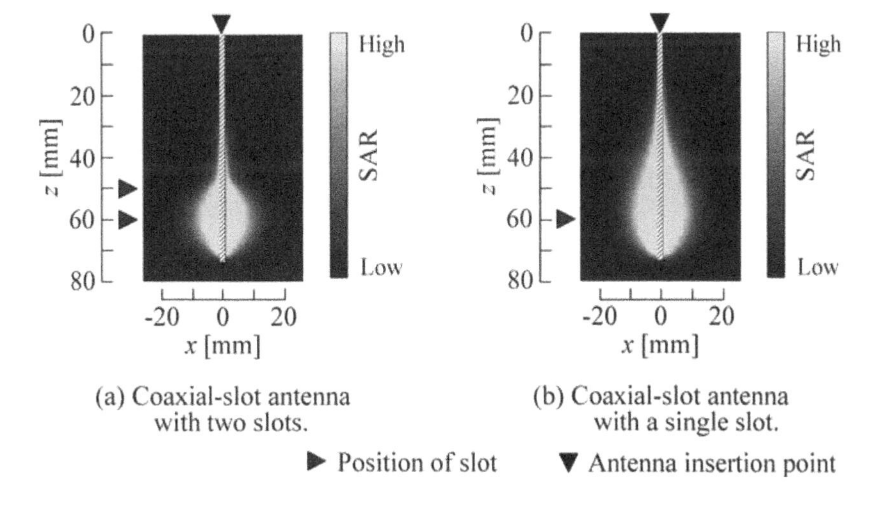

Figure 5.11 Calculated SAR distributions of coaxial-slot antennas.

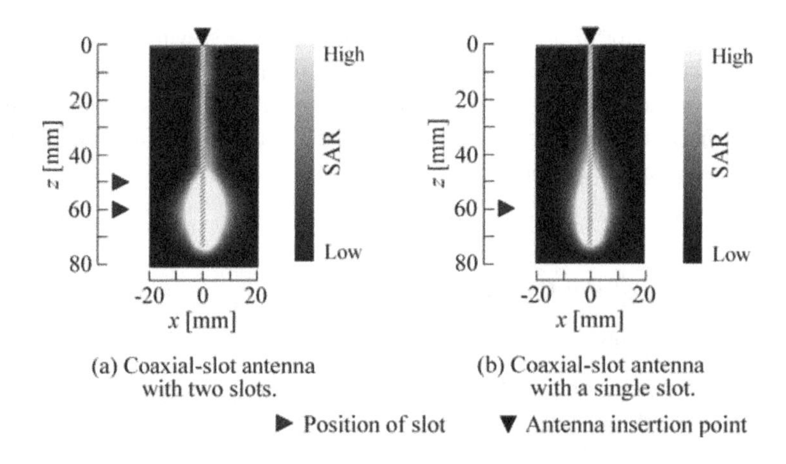

Figure 5.12 Measured SAR distributions of coaxial-slot antennas.

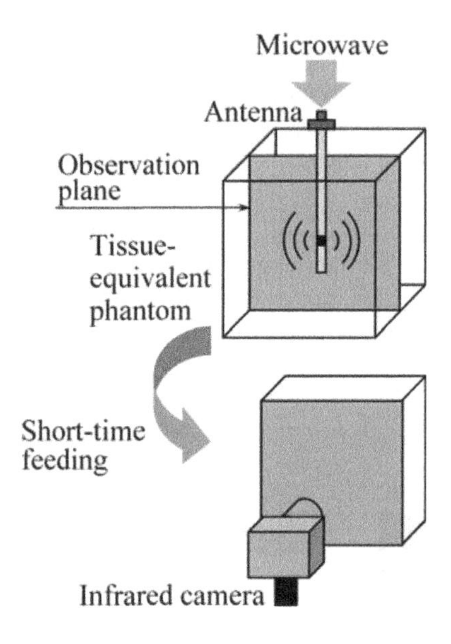

Figure 5.13 SAR measurement by the thermographic method.

After a short-time feeding, the inner surface of the phantom is observed by the infrared camera. Here, the SAR distribution is estimated by

$$\mathrm{SAR} = c\frac{\Delta T}{\Delta t}\ [\mathrm{W/kg}] \tag{5.4}$$

where c is the specific heat of the phantom [J/kg·K], Δt the feeding time [s], and ΔT the temperature rise [K]. From Figure 5.12, a localized heating pattern is observed by employing the coaxial-slot antenna with two slots in the experiment.

5.2.4.2 Dependency of the SAR on the Antenna Insertion Depth

In the actual treatments, not only generating a localized heating region but also having the antenna insertion depth independent of the heating pattern is very important. Therefore, we investigated the dependence on the insertion depth (D_t) of the coaxial-slot antenna with two slots (L_{ts} =20 mm, L_{ls} =10 mm) and a single slot (L_{ts} =10 mm) for comparison. Figure 5.14 shows the calculated SAR profiles on the SAR observation line, which is located at a distance of 3 mm away from the center of the antenna in the x–z plane, under various values of D_t. Here, D_t was changed from 30 mm to 70 mm every 20 mm. From Figure 5.14(a), the SAR profiles of the antenna with a single slot depend on the insertion depth. Here, in the case of D_t =30 mm, the value of the SAR around $z = 0$ (equivalent to surface of the patient) is approximately 25 % of the peak value. Therefore, it is considered that if this kind of antenna is employed for treatment, we cannot perform the treatment completely, because the normal tissue around the surface would overheat. However, from Figure 5.14(b), we can observe that the SAR profiles are independent of the antenna insertion depth. In particular, the value of the SAR around $z = 0$ for D_t = 30 mm is less than 5 % of the peak value. This result is of a great use for interstitial heating as we do not have to take the insertion depth into consideration.

5.2.4.3 Electric Current Distributions on the Antennas

In order to understand the mechanism of generation of a localized heating region by the coaxial-slot antenna with two slots, we calculated the electric current distribution on the antenna. Figure 5.15 shows the calculated current distributions on coaxial-slot antennas with

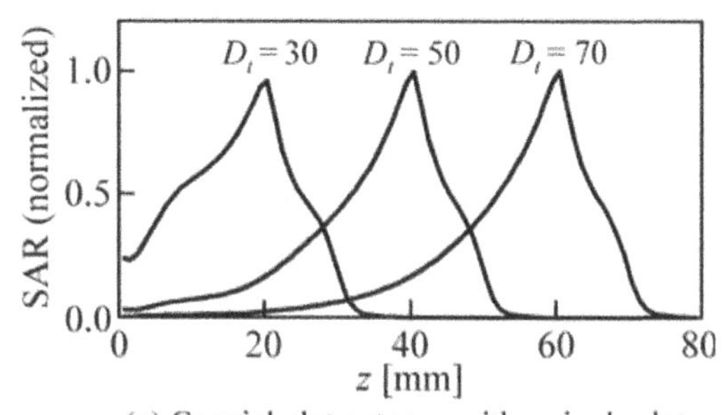

(a) Coaxial-slot antenna with a single slot.

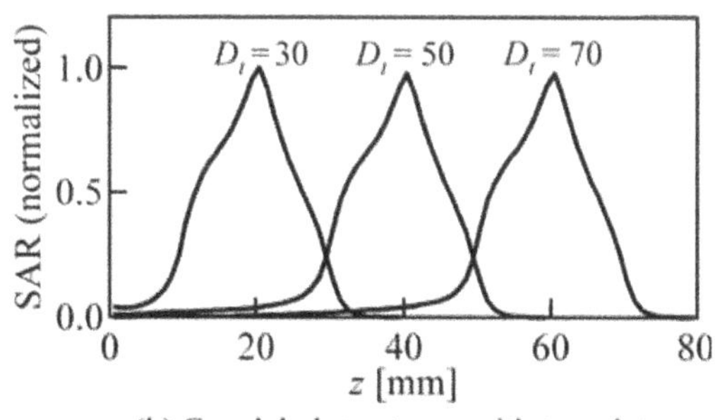

(b) Coaxial-slot antenna with two slots.

Figure 5.14 SAR distributions under various insertion depths.

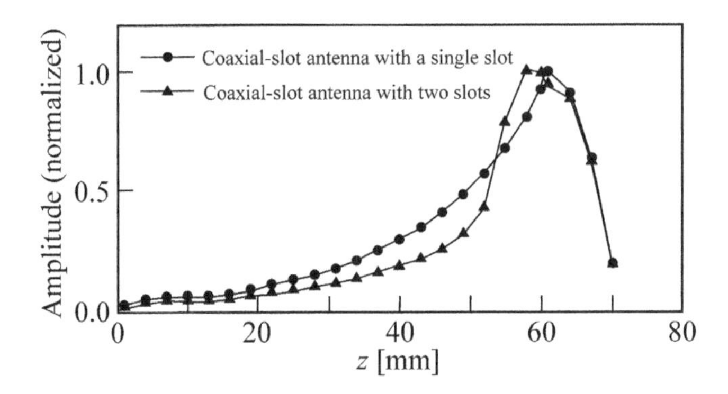

Figure 5.15 Electric current distributions on the antennas.

two slots ($L_{ls} = 10$ mm, $L_{ts} = 20$ mm) and a single slot ($L_{ts} = 10$ mm) at $D_t = 70$ mm. The coaxial-slot antenna with a single slot has a monotonous change in the current distribution from $z = 0$ to $z = 60$ mm (position of the slot). On the other hand, the current distribution of the coaxial-slot antenna with two slots is localized around a region where 50 mm $< z <$ 70 mm. This tendency is similar to that of the SAR distribution shown in Figure 5.14(b) (for D_t =70 mm). It is assumed that the localized current distribution generates a localized SAR profile around the antenna.

5.2.5 Temperature Distributions Around the Antennas

5.2.5.1 Calculation Models

In the actual treatments, the single antenna or the array applicators are chosen depending on the tumor size. Therefore, the temperature distributions around the single antenna and the array applicators are calculated. Figure 5.16 shows the calculation models for the single coaxial-slot antenna and the array applicators with their temperature observation planes discussed in the next section. The antenna spacing and the antenna insertion depth are 20 mm and 30 mm, respectively, considering the actual treatments described later. In addition, parameters of biological tissues for the calculations are listed in Table 5.2.

5.2.5.2 Results of Calculations

Figures 5.17–5.19 show the calculated temperature distributions in the observation planes, which are defined in Figure 5.16, for the single antenna and the array applicators composed of two or four antennas, respectively. In order to simulate the situation of the actual treatments, in the calculations the net input power of the antenna is 5.0 W for each element and the heating time is 600 s. In addition, the initial temperature of the biological tissue is 37 °C.

From Figure 5.17, in the case of the single antenna, a spherical heating region is generated around the tip of the antenna. The size of the effective heating region (the region higher than 42 °C) in the z direction is almost the same as L_{ts} (length from the tip of the antenna to the upper slot).

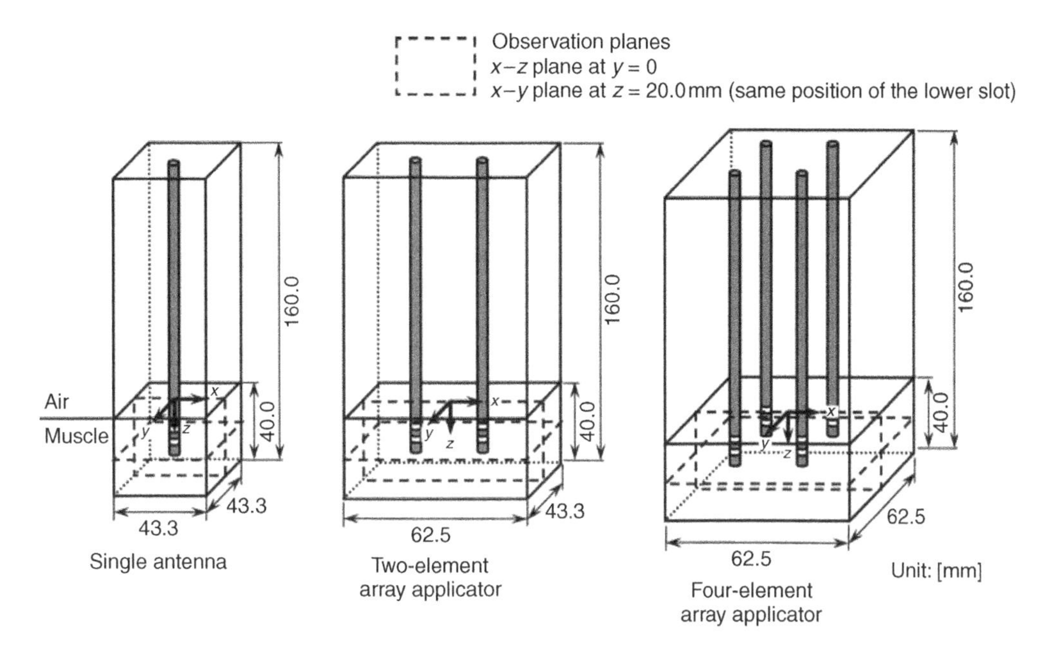

Figure 5.16 Calculation models for temperature distributions.

Table 5.2 Parameters of biological tissues for the calculations.

Electrical properties (muscle, at 2.45 GHz)		
Relative permittivity ε_r	47.0	
Conductivity σ[S/m]	2.21	
Thermal properties	Muscle	Blood
Specific heat c [J/kg·K]	3500	3960
Thermal conductivity κ [W/m·K]	0.60	—
Density ρ [kg/m^3]	1020	1060
Blood flow (muscle)		
Blood flow rate $F \times 10^{-6}$ [m^3/kg·s]	8.30	

In Figure 5.18, we can observe the heating region not only around each antenna element but also at the center of two antenna elements ($x = y = 0$ in the x–y plane). It is assumed that the heating region around the center of two antenna elements in the x–y plane is generated by the enhancement of electric fields caused by each antenna element. The size of the effective heating region in the z direction is also almost the same as L_{ts}.

As shown in Figure 5.19, when using the array applicator composed of four antenna elements, the heating region is generated inside the region of the array applicator. Therefore, when the tumor is located inside this heating region, it can be completely heated. The size of the effective heating region in the z direction is also almost the same as L_{ts}.

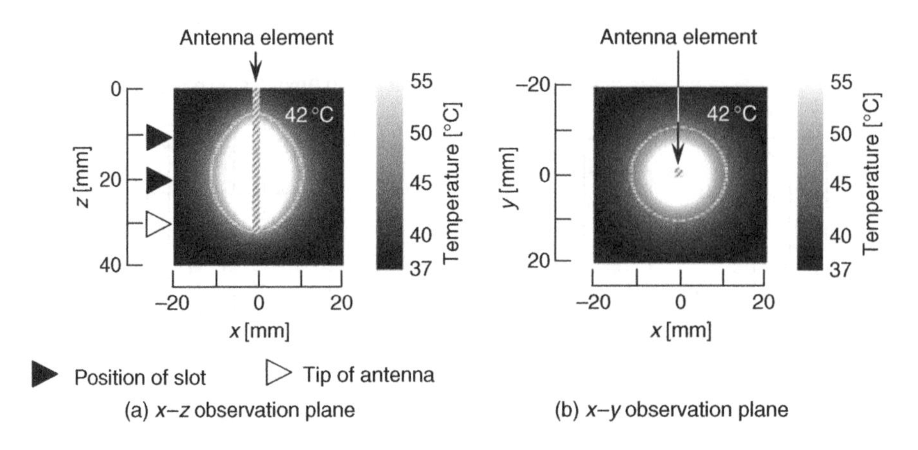

Figure 5.17 Calculated temperature distributions of a single antenna.

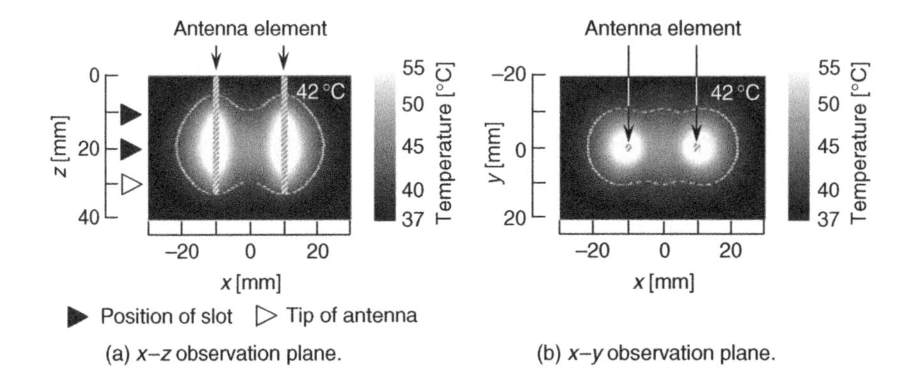

Figure 5.18 Calculated temperature distributions of a two-element array applicator.

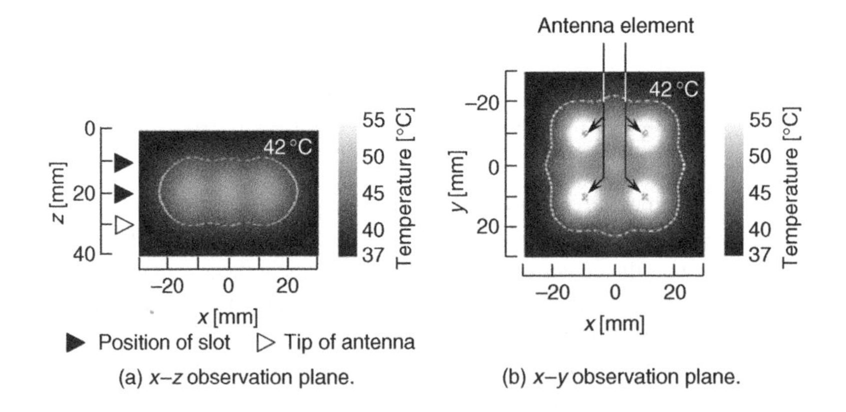

Figure 5.19 Calculated temperature distributions of a four-element array applicator.

In all cases, we cannot observe a hot spot around the surface of the tissue, which is placed at $z = 0$ in the x–z observation plane. Hence, it is considered that injuries to the skin are well suppressed by use of the proposed antennas.

5.3 Clinical Trials

5.3.1 Equipment

Figure 5.20 shows the treatment system. This consists of a microwave generator (Alfresa Pharma, AZM-520), a microwave power reflection meter (Rohde & Schwarz, NRT and NRT-Z44), and a thermometer (Laxtron, M-790). We do not have detailed specifications for the power dividers (order made, one 2-divider and two 4-dividers, up to 8 ports in total). Here, the microwave generator is sold for MCT and is appropriate for hyperthermic treatments. In all cases, we measure the temperature in and around the tumor at up to four points. Figure 5.21 explains the positions of the antennas and the temperature probes in the horizontal (x–y) plane where the size of the tumors is a maximum. During the treatment, we especially focus on the temperature of the reference point (probe no. 2 in Figure 5.21), which is placed at the border of the tumor, and control the output power of the microwave generator by checking the temperature at this point. In order to heat the tumor sufficiently, the minimum temperature of the reference point is maintained at 42 °C.

5.3.2 Treatment by Use of a Single Antenna

Here, the clinical trial by use of a single antenna is introduced. The patient was a female aged 60 with a tumor (gingival carcinoma postoperative, neck node recurrence) in her right

Figure 5.20 Equipment for treatments.

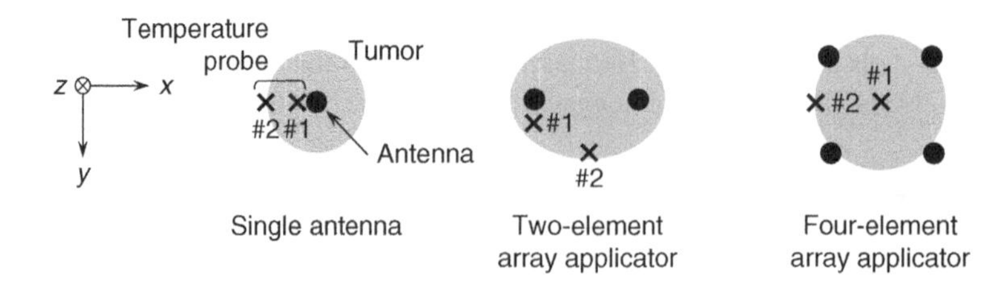

Figure 5.21 Positions of the antennas and the temperature probes. (Only two important temperature measurement points are shown.)

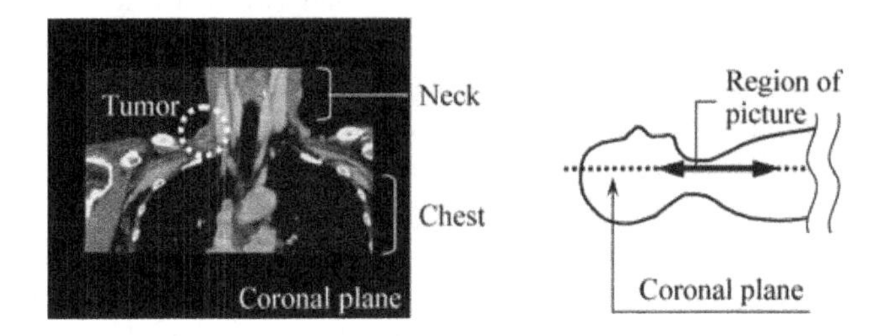

Figure 5.22 Patient undergoing treatment by X-ray Computer Tomography.

shoulder. Figure 5.22 shows the patient, including the targeted tumor, undergoing treatment by X-ray computer tomography (CT). The diameter of the targeted tumor is approximately 25 mm and the antenna insertion depth D_t is 27 mm. The shape of the heating pattern of the coaxial-slot antenna with two slots is independent of the insertion depth, as described in the previous section. Therefore, the antenna is useful for heating relatively shallow tumors as in this case.

Figure 5.23(a) shows the positions of the antenna and the thermosensors. In this case, we employed three thermosensors in and around the targeted tumor for reliable heating. Here, the sensors #1, #2, and #3 are located near the antenna, outside the targeted tumor, and directly under the skin, respectively. Figure 5.23(b) is a photograph of the patient during the treatment.

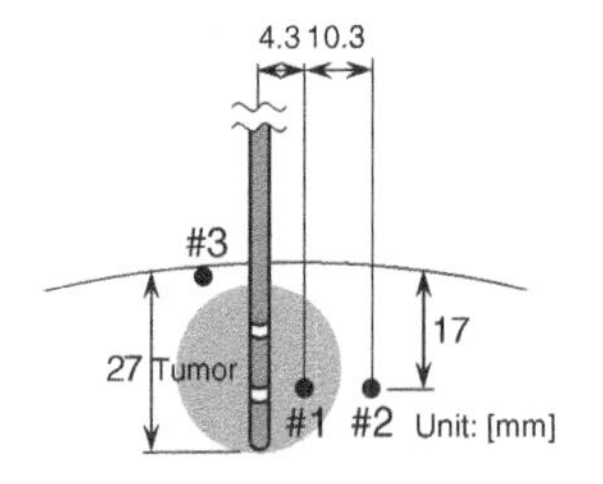

(a) Temperature measurement points 1–3.

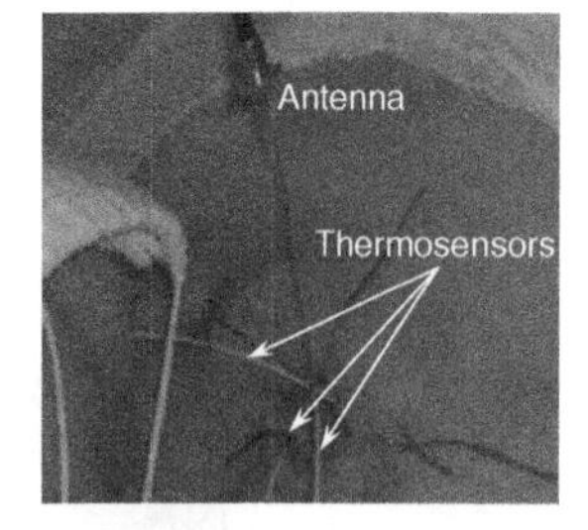

(b) Photograph during the treatment.

Figure 5.23 Positions of the antenna and thermosensors.

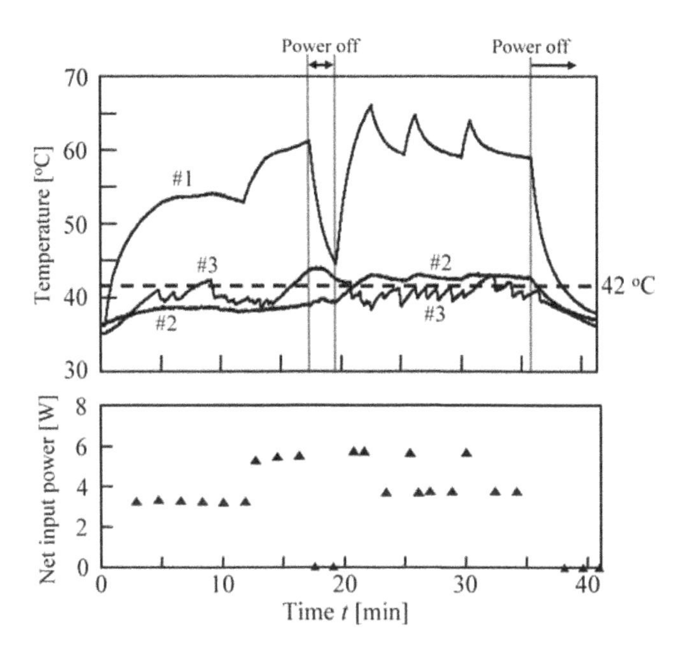

Figure 5.24 Transition of the temperature and net input power of the antenna. At $t > 20$ min, the skin around the antenna insertion point was cooled down by a physiological salt solution to avoid the overheating of the normal tissue. Therefore, the temperature of sensor #3 at $t > 20$ min is lower than that of sensor #2.

Figure 5.24 shows the transitions of the temperature and the net input power of the antenna. As mentioned earlier, we chose sensor #2 as a temperature reference point and controlled the output power of the microwave generator so that the minimum temperature of sensor #2 is 42 °C. Although sensor #2 was placed outside of the targeted tumor, the minimum temperature of sensor was 42 °C at the steady state (20 min $< t <$ 35 min). Therefore, in this case, we may say that the targeted tumor was completely covered by the therapeutic temperature. Moreover, we could observe a high temperature in sensor #1. Therefore, coagulation necrosis could also be expected inside the tumor.

5.3.3 Treatment by Use of an Array Applicator

In the case of treatment of large-volume tumor, the array applicator is employed. Here, the treatment by the four-element array applicator is introduced. The patient was a male aged 61 with a tumor (esophagus cancer postoperative, supraclavicular lymph node recurrence) in his right shoulder. Figures 5.25 shows the positions of the antennas and the thermosensors and a photograph of the patient during the treatment. The antennas were placed so that the targeted tumor was surrounded. Figure 5.26 shows the feeding system for the treatment. The output power is divided into four by means of a power divider. Therefore, the divided microwave powers all have the same amplitude and phase.

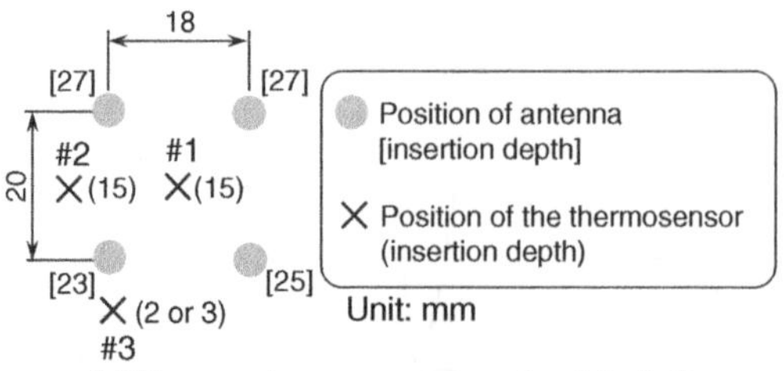

(a) Temperature measurement points 1–3.

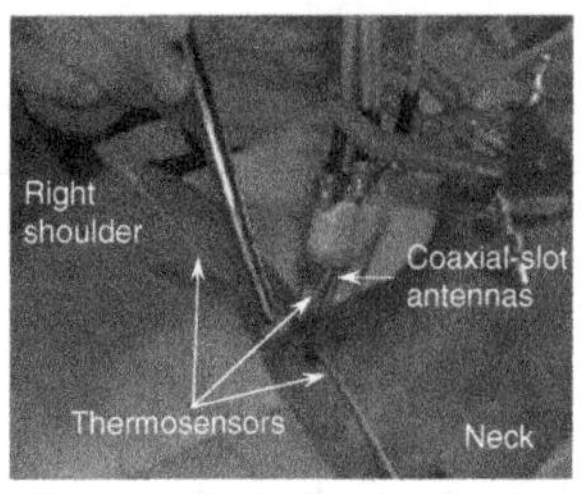

(b) Photograph during the treatment.

Figure 5.25 Position of the antennas and thermosensors.

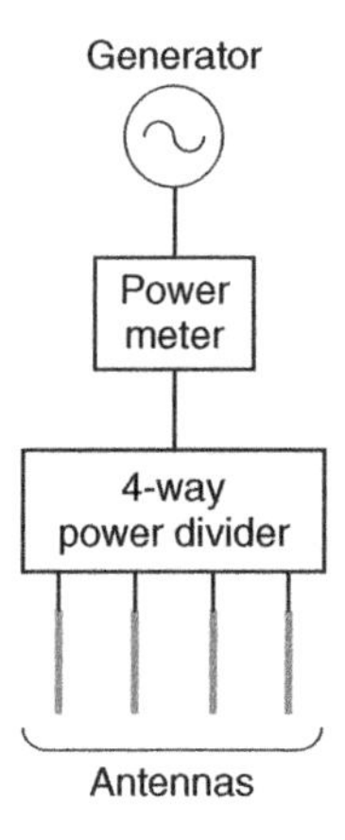

Figure 5.26 Feeding system for treatment.

Here, we employed three thermosensors in and around the targeted tumor for reliable heating, sensors #1, #2, and #3 being located at the center of the array applicator, outside the targeted tumor, and directly under the skin, respectively.

Figure 5.27 shows the temperature transitions. The net input power of the array applicator was approximately 18.0 W (we did not change the output power of the microwave generator). Although sensor #2 was placed outside of the targeted tumor, the minimum temperature of the sensor was 42 °C at steady state (5 min $< t <$ 29 min). Therefore, in this case, we may say that the targeted tumor was completely covered by the therapeutic temperature. In addition,

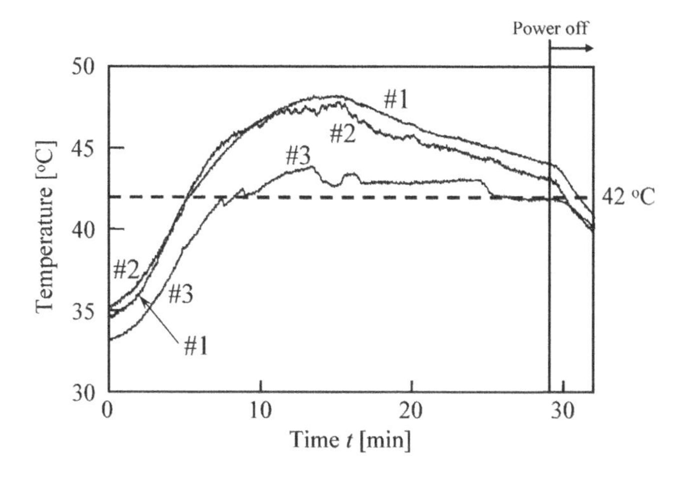

Figure 5.27 Temperature transitions.

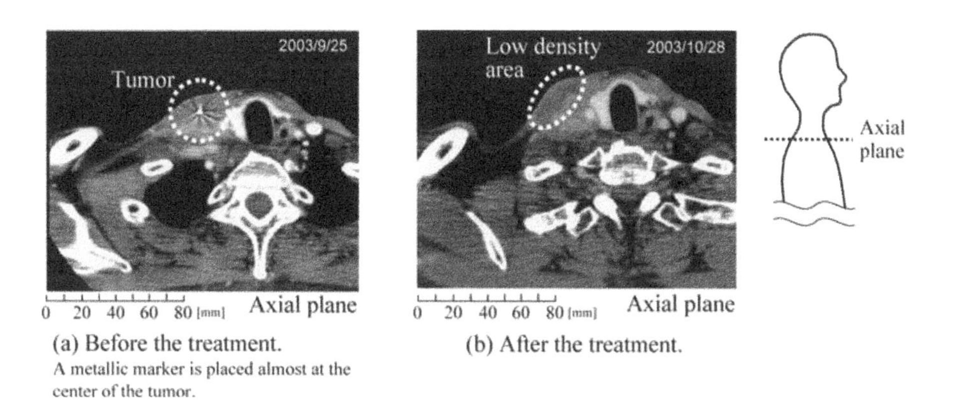

Figure 5.28 Pictures of the patient by X-ray CT.

Figure 5.28 shows tomograms of the patient before and after the treatment. In Figure 5.28(b) we can observe a low density area, which is the area of necrosis, at the tumor. At present we do not observe any re-expansion of the tumor.

5.3.4 Results of the Treatment

Table 5.3 summarizes the results of the treatments realized by the coaxial-slot antennas. The treatments were efficient for all heated sites. Moreover, no critical side effects, including injury of skin, occurred. Therefore, it is considered that we could effectively heat the tumor without any problems in all cases.

Table 5.3 Results of the treatment.

	Case 1	Case 2	Case 3	Case 4
Age	59	89	62	77
Sex	Female	Female	Male	Female
Primary	Gingiva	Soft palate	Esophagus	Lip
Pathology	Squamous cell carcinoma	Adenocarcinoma	Squamous cell carcinoma	Squamous cell carcinoma
Heated site	Supraclavicular node recurrence	Primary lesion	Supraclavicular node recurrence	Submandibular node recurrence
Number of antennas	1	1	2 or 4	1
Follow-up time	7.3 months	1 year 7 months	1 year 7 months	4 months
Result	Died from other metastasis No evidence of tumor growth in heated site	Alive without disease	Alive without disease	Alive No evidence of tumor growth

5.4 Other Applications

5.4.1 Treatment of Brain Tumors

It is thought that interstitial hyperthermia is an effective tool for the treatment of brain tumors, because it is difficult to treat such tumors by surgical operation, radiation therapy, and so on. At present, the radio frequency interstitial heating system is used for the treatment (Figure 5.29 (a)). In this system, one or a few needle type electrodes are inserted into the targeted tumor and an external electrode is needed. The current between the needle type electrode and the external electrode heats up the tumor. The heating region in the perpendicular direction of the electrode axis may be insufficient in some cases. By employing the microwave heating technique, the heating region may be expanded. In addition, the external electrode may not be necessary (Figure 5.29(b)).

The control of the heating region is more important than in the other cases. In order to generate a controllable heating region in the direction of the antenna axis, we employ a coaxial-dipole antenna [19] and improve its structure. Figure 5.30 and Table 5.4 show the basic structure of the coaxial-dipole antenna and its structural parameters, respectively. In this structure, the size of the heating region in the direction of antenna axis is controlled by changing the length of the sleeves ($2L_d$). In addition, impedance matching is realized by adjusting the distance between the tip of the feeding line and the slot position (L_{ts}). Figure 5.31 illustrates the SAR distributions around the coaxial-dipole antenna by changing the $2L_d$ length (the SAR observation line is the same as in Figure 5.14). In Figure 5.31 controllable SAR distributions depending on the length of the sleeves are observed.

5.4.2 Intracavitary Microwave Hyperthermia for Bile Duct Carcinoma

This section describes the coaxial-slot antenna for intracavitary heating for treatment of bile duct carcinoma. The bile duct is seated in a deep region of the body and some thick blood vessels lie near the bile duct. Therefore, bile duct carcinoma is generally difficult to treat by surgical operation. In such cases, a multidisciplinary approach, which combines radiation therapy, chemotherapy, thermal therapy, and so on, is expected to improve the quality of

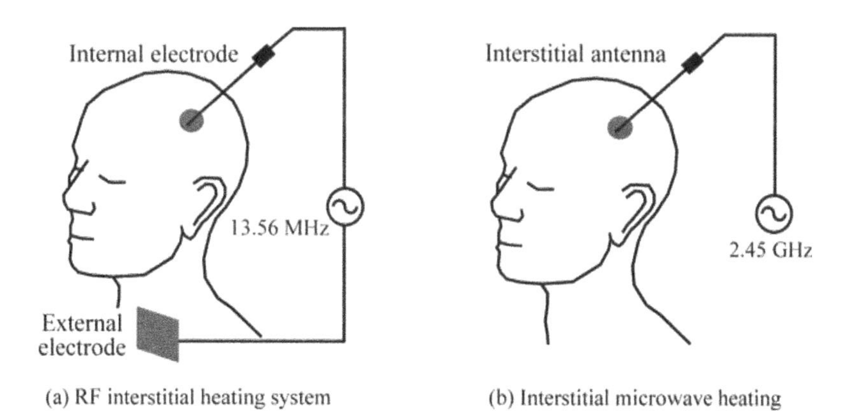

Figure 5.29 Treatment of brain tumors.

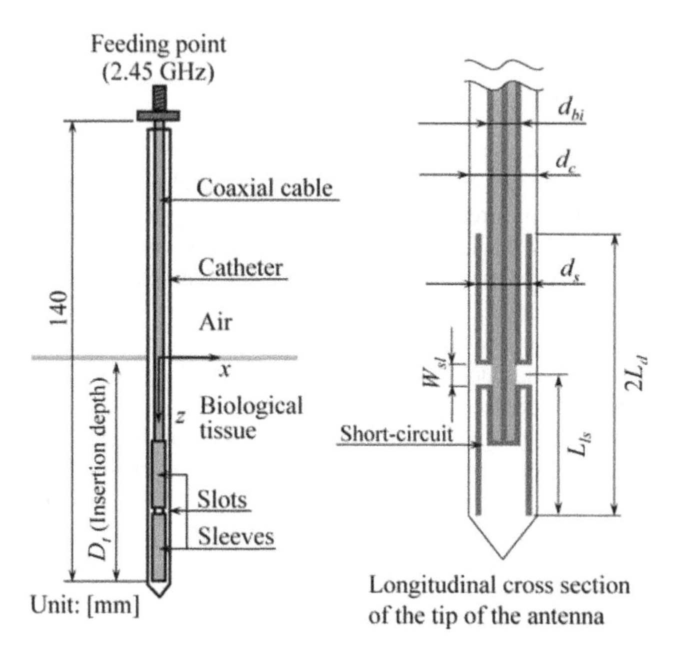

Figure 5.30 Basic structure of the coaxial-dipole antenna.

Table 5.4 Structural parameters of the coaxial-dipole antenna.

d_{bi} (diameter of the coaxial cable) [mm]	1.19
d_{bs} (diameter of the sleeves) [mm]	1.79
d_c (external diameter of the catheter) [mm]	0.3
L_{ts} (length from the tip to the center of the slot close to the feeding point) [mm]	20.0
W_{sl} (width of the slot) [mm]	1.0
ε_{rc} (relative permittivity of the catheter)	2.6

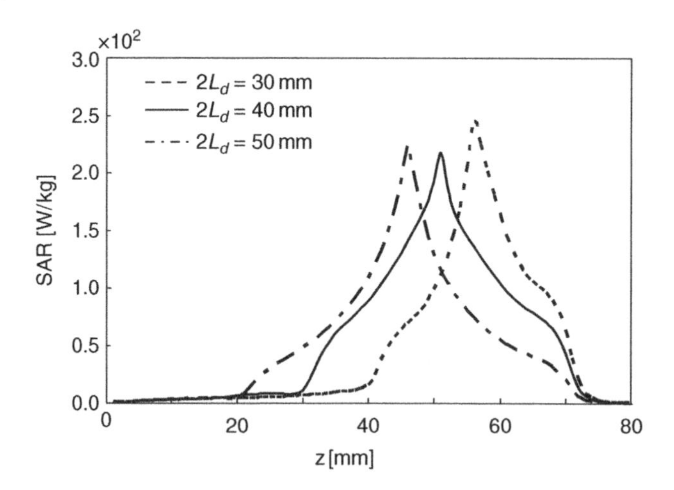

Figure 5.31 SAR distribution around the coaxial-dipole antenna by changing the length of the sleeves (normalized by 1.0 W incident power).

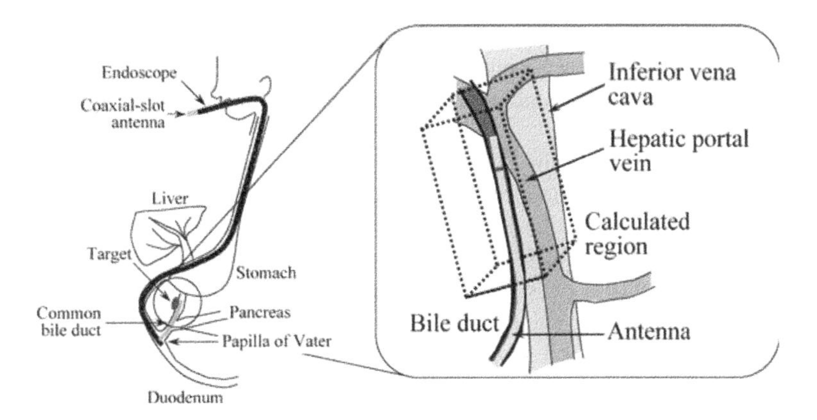

Figure 5.32 Scheme of the treatment for bile duct carcinoma by intracavitary microwave heating. The calculated region is explained in the text.

life of the patient. For these reasons, investigations of the antenna for such a treatment are important.

Figure 5.32 shows the scheme of the treatment. In the treatment, the endoscope is first inserted into the duodenum and a long and flexible coaxial-slot antenna is inserted into the forceps channel of the endoscope, which is used to insert the tool for surgical treatment. Finally, the antenna is guided to the bile duct through the papilla of Vater, which is located in the duodenum, and is inserted into the bile duct.

Figure 5.33(a) shows the structure of the proposed coaxial-slot antenna for this treatment. The basic structure of the antenna is the same as in Figure 5.7, except that this antenna

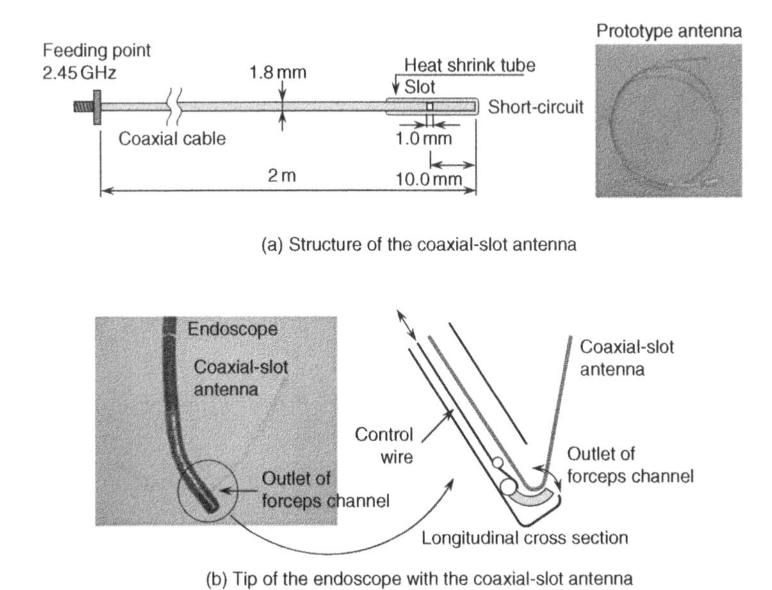

Figure 5.33 Structure of the coaxial-slot antenna for treatment of bile duct carcinoma.

is fabricated using a commercially available flexible coaxial cable. We confirmed that the fabricated prototype antenna can be inserted into the forceps channel without any problems (Figure 5.33(b)). The antenna is bent at the outlet of the channel by a barb whose bend angle is controllable by the operator.

In order to check the possibility of heating the bile duct carcinoma, we numerically calculated the temperature distribution around the tip of the antenna. Figure 5.34 shows the calculation model based on the realistic human model developed at Brooks Air Force Laboratories [20]. The location of the calculated region is indicated in Fig. 5.32. In order to construct this model, a region including the bile duct is extracted. In Figure 5.34, only the bile duct is indicated inside the calculated region, though there are a few other organs such as liver, stomach, duodenum, and small intestine. The physical properties of the biological tissues are listed in Table 5.5. In addition, in Figure 5.34, two temperature observation planes are shown. Here, observation plane #1 is the x–y plane at $z = 50$ mm (including the slot) and observation plane #2 is the x–z plane at $y = 0$.

In the hyperthermic treatment, it is important to heat the tumor to more than the therapeutic temperature. Therefore, temperature measurement in and around the tumor is needed for a reliable treatment. In general, some thin thermosensors such as fiber-optic thermosensors are employed for the measurement [13]. However, in this case, it is difficult to place any sensors non-invasively especially inside the tumor. In addition, it is clear that the region close to the antenna is heated to a very high temperature. This high temperature will cause lesions of the bile duct. In order to avoid the high temperature, if we choose a low input power of the antenna, the size of the therapeutic heating region will not be appropriate.

In order to solve these problems, an on–off feeding control is employed. In our preliminary investigations, we defined the parameters of the on–off feeding as follows:

- Interval: 2 s (on-feeding 2 s, off-feeding 2 s)
- Input power of the antenna: 15.0 W (This value includes the loss of the actual antenna cable. The actual radiated power from the antenna is less than this value.)
- Allowed maximum temperature: 60 °C.

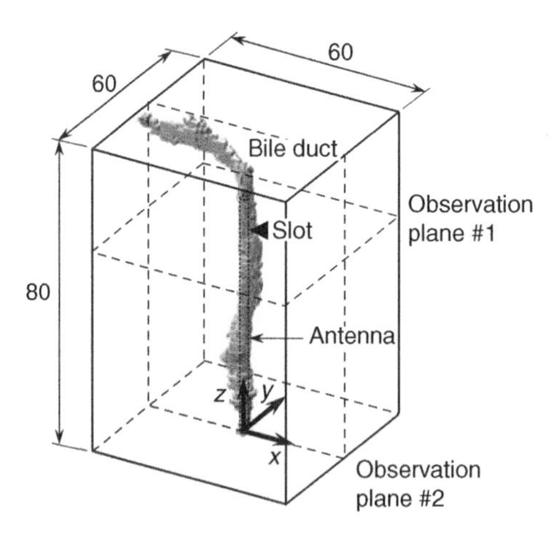

Figure 5.34 Calculation model based on realistic human model. Dimensions in millimetres.

Table 5.5 Physical properties of the biological tissues [20–22].

	Relative permittivity	Conductivity [S/m]	Density [kg/m^3]	Specific heat [J/kg·K]	Thermal conductivity [W/m·K]	Blood flow rate [m^3/kg·s]
Air (internal)	1.0	0.0	—	—	—	—
Bile	68.4	2.80	1010	3960	0.500	0.0
Fat	5.3	0.10	916	2300	0.220	5.00×10^{-7}
Lymph	57.2	1.97	1040	3960	0.515	7.50×10^{-6}
Muscle membrane	42.9	1.59	1040	3500	0.600	8.30×10^{-6}
Muscle	52.7	1.74	1047	3500	0.600	8.30×10^{-6}
Stomach	62.2	2.21	1050	3500	0.527	6.67×10^{-6}
Glands	57.2	1.97	1050	3500	0.600	0.0
Blood vessel	42.5	1.44	1040	3500	0.600	0.0
Liver	43.0	1.69	1030	3600	0.497	1.67×10^{-5}
Gall bladder	57.6	2.06	1030	3500	0.600	8.30×10^{-6}
Bone (cortical)	44.8	2.10	1038	1300	0.436	4.20×10^{-7}
Cartilage	38.8	1.76	1097	1300	0.436	4.20×10^{-7}
Ligaments	43.1	1.68	1220	3500	0.600	8.30×10^{-6}
Intestine (large)	53.9	2.04	1043	3500	0.600	1.33×10^{-5}
Intestine (small)	54.4	3.17	1043	3500	0.600	1.67×10^{-5}
Pancreas	57.2	1.97	1045	3500	0.441	1.00×10^{-5}
Blood	58.3	2.54	1058	3960	—	—
Kidneys	52.7	2.43	1050	3890	0.539	6.25×10^{-5}
Bone marrow	5.3	0.10	1040	1300	0.436	4.20×10^{-7}
Bladder	18.0	0.69	1030	3900	0.561	0.0
Bone (cancellous)	18.5	0.81	1920	1300	0.436	4.20×10^{-7}

Figure 5.35(a) shows the calculated temperature transitions. The temperature observation points are indicated in Figure 5.35(b). The calculated temperature transition without on–off feeding control is also shown for comparison. In Figure 5.35(a) we can observe that the maximum temperature at observation point #1 is less than 60 °C with on–off feeding control,

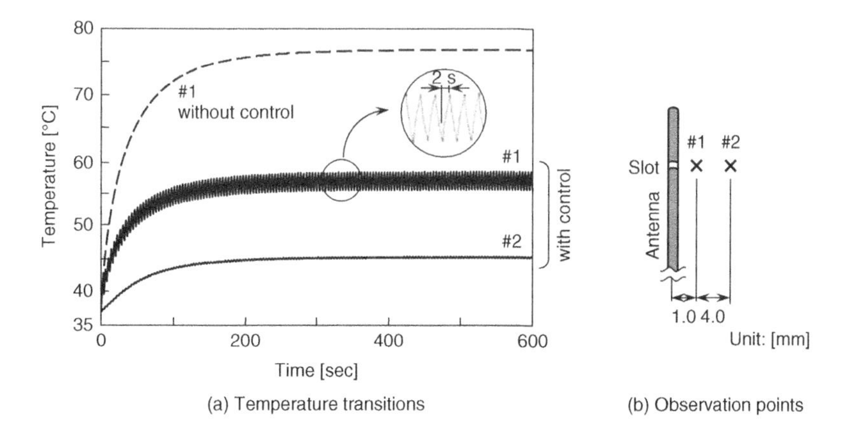

Figure 5.35 Calculated temperature transition.

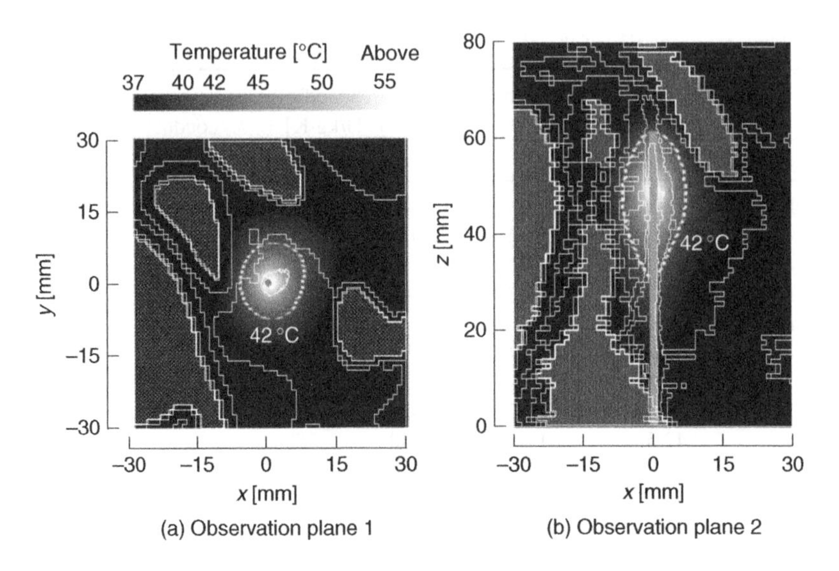

Figure 5.36 Calculated temperature distributions around the bile duct.

although the temperature at this point exceeds 70 °C without the control. Moreover, the temperature at observation point #2, which is placed 5.0 mm from the antenna axis, exceeds the lowest therapeutic temperature (42 °C).

Figure 5.36 shows the calculated temperature distributions in the observation planes defined in Figure 5.34. In this figure, the white dotted lines indicate 42 °C, which is the lowest temperature for the treatment. These temperature distributions, for which the minimum size of heating region by the on–off feeding control is chosen, are the results at the steady state. From Figure 5.36, the diameter of effective heating region (the region higher than 42 °C) is approximately 15 mm in the x–y plane. The heating pattern in the axial direction of the antenna can be controlled by shifting the antenna.

5.5 Summary

In recent years, various types of medical applications of microwaves have been widely investigated and reported. In particular, minimally invasive microwave thermal therapies using thin antennas are of a great interest. Among them are interstitial microwave hyperthermia and microwave coagulation therapy for medical treatment of cancer, cardiac catheter ablation for ventricular arrhythmia treatment, thermal treatment of benign prostatic hypertrophy. In this chapter, after describing the principle of hyperthermic treatment for cancer, some heating schemes using microwave techniques were explained. Then a coaxial-slot antenna, which is one of the thin coaxial antennas, and array applicators composed of several coaxial-slot antennas were introduced. Moreover, some fundamental characteristics of the coaxial-slot antenna and the array applicators, such as the SAR and temperature distributions around the antennas inside the human body, and the current distribution on the antenna, were described employing FDTD calculations and temperature computations inside the biological tissue by solving the bioheat transfer equation. Finally, some results of actual clinical trials using the proposed coaxial-slot antennas were explained from a technical point of view. Other therapeutic applications of coaxial-slot antennas, such as hyperthermic treatment for brain tumors and intracavitary hyperthermia for bile duct carcinoma, were also introduced.

References

[1] F. Sterzer, Microwave medical devices, *IEEE Microwave Magazine*, 3 (2002), 65–70.

[2] K. Ito, Medical applications of microwave. *Proceedings of the 1996 Asia-Pacific Microwave Conference*, Vol. 1, pp. 257–260, New Delhi, December 1996.

[3] S. Mizushina, H. Ohba, K. Abe, S. Mizoshiri, and T. Sugiura, Recent trends in medical microwave radiometry. *IEICE Transactions on Communications*, E-78B (1995), 789–798.

[4] J. Montreuil and M. Nachman, Multiangle method for temperature measurement of biological tissues by microwave radiometry. *IEEE Transactions on Microwave Theory and Techniques*, 39 (1991), 1235–1238.

[5] K. Shimizu, S. Matsuda, I. Saito, K. Yamamoto, and T. Hatsuda, Application of biotelemetry technique for advanced emergency radio system *IEICE Transactions on Communications*, E-78B (1995) 818–825.

[6] M.H. Seegenschmiedt, P. Fessenden, and C.C. Vernon (eds), *Thermoradiotherapy and Thermochemotherapy*. Berlin: Springer-Verlag, 1995.

[7] T. Seki, M. Wakabayashi, T. Nakagawa, T. Itoh, T. Shiro, K. Kunieda, M. Sato, S. Uchiyama, and K. Inoue, Ultrasonically guided percutaneous microwave coagulation therapy for small carcinoma. *Cancer*, 74 (1994), 817–825.

[8] P. Bernardi, M. Cavagnaro, J.C. Lin, S. Pisa, and E. Piuzzi, Distribution of SAR and temperature elevation induced in a phantom by a microwave cardiac ablation catheter. *IEEE Transactions on Microwave Theory and Techniques*, 52 (2004), 1978–1986.

[9] D. Despretz, J.-C. Camart, C. Michel, J.-J. Fabre, B. Prevost, J.-P. Sozanski, and M. Chivé, Microwave prostatic hyperthermia: interest of urethral and rectal applicators combination – Theoretical study and animal experimental results. *IEEE Transactions on Microwave Theory and Techniques*, 44 (1996), 1762–1768.

[10] J.C. Lin and Y.-J. Wang, Interstitial microwave antennas for thermal therapy. *International Journal of Hyperthermia*, 3 (1987), 37–47.

[11] L. Hamada, K. Saito, H. Yoshimura, and K. Ito, Dielectric-loaded coaxial-slot antenna for interstitial microwave hyperthermia: longitudinal control of heating patterns. *International Journal of Hyperthermia*, 16 (2000), 219–229.

[12] K. Ito and K. Furuya, Basics of microwave interstitial hyperthermia. *Japanese Journal of Hyperthermic Oncology*, 12 (1996), 8–21 (in Japanese).

[13] K. Saito, H. Yoshimura, K. Ito, Y. Aoyagi, and H. Horita, Clinical trials of interstitial microwave hyperthermia by use of coaxial-slot antenna with two slots. *IEEE Transactions on Microwave Theory and Techniques*, 52 (2004), 1987–1991.

[14] H.H. Pennes, Analysis of tissue and arterial blood temperatures in the resting human forearm. *Journal of Applied Physiology*, 1 (1948), 93–122.

[15] K. Saito, Y. Hayashi, H. Yoshimura, and K. Ito, Heating characteristics of array applicator composed of two coaxial-slot antennas for microwave coagulation therapy. *IEEE Transactions on Microwave Theory and Techniques*, 48, (2000), 1800–1806.

[16] J. Wang and O. Fujiwara, FDTD computation of temperature rise in the human head for portable telephones. *IEEE Transactions on Microwave Theory and Techniques*, 47, (1999), 1528–1534.

[17] C. Gabriel, Compilation of the dielectric properties of body tissues at RF and microwave frequencies. Brooks Air Force Technical Report AL/OE-TR-1996-0037. http://www.fcc.gov/fcc-bin/dielec.sh.

[18] Y. Okano, K. Ito, I. Ida, and M. Takahashi, The SAR evaluation method by a combination of thermographic experiments and biological tissue-equivalent phantoms. *IEEE Transactions on Microwave Theory and Techniques*, 48 (2000), 2094–2103.

[19] K. Iwata, K. Udagawa, M. S. Wu, K. Ito, and H. Kasai, A basic study of coaxial-dipole applicator for microwave interstitial hyperthermia. *Proceedings of the 12th Annual Meeting of the Japanese Society of Hyperthermic Oncology*, pp. 230–231, September 1995.

[20] http://www.brooks.af.mil/AFRL/HED/hedr/hedr.html.

[21] F.A. Duck, *Physical Properties of Tissue* New York: Academic, 1990.

[22] P.M. Van Den Berg, A.T. De Hoop, A. Segal, and N. Praagman, A computational model of the electromagnetic heating of biological tissue with application to hyperthermic cancer therapy. *IEEE Transactions on Biomedical Engineering*, 30 (1983), 797–805.

6

Antennas for Wearable Devices

Akram Alomainy and Yang Hao
Department of Electronic Engineering, Queen Mary, University of London, UK
Frank Pasveer
Healthcare Devices and Instrumentation, Philips Research, Netherlands

6.1 Introduction

Communication technologies are heading towards a future in which user-specified information is available on demand. In order to ensure the smooth transition of information from surrounding networks and from shared devices, computing and communication equipment needs to be body-centric. The antenna is an essential part of wireless body-centric networks. Its complexity depends on the radio transceiver requirements and also on the propagation characteristics of the surrounding environment. For the conventional long to short wave radio communication, conventional antennas have proven to be more than sufficient to provide desired performance, minimizing the restraints on cost and production time. On the other hand, for today's and tomorrow's communication devices, the antenna is required to perform more than one task – that is, it needs to operate at different frequencies to account for the increasing new technologies and services available to the user. Therefore, care is needed in designing antennas for body-worn devices, which are often hidden and small in size and weight.

This chapter briefly introduces wireless personal networks and the progression to body area networks (WBANs), highlighting the properties and applications of such networks. The main characteristics of wearable antennas, their design requirements and theoretical considerations are discussed. The effects of antenna parameters and types on radio channels in body-centric networks are demonstrated. To give a clear picture of practical considerations needed in antenna design for wearable devices deployed for commercial applications, a case study is

Antennas for Portable Devices Zhi Ning Chen

presented with detailed analyses and investigations of the antenna design and performance for healthcare sensors.

In order to understand the requirements for wearable antennas and the restrictions on antenna system deployment for body-centric networks, the main features of WBANs need to be introduced.

6.1.1 Wireless Body Area Networks

Body area networks (BANs) are a natural progression from the personal area network (PAN) concept, and they are wireless networks with nodes normally situated on the human body or in close proximity [1]. Advances in communication and electronic technologies have enabled the development of compact and intelligent devices that can be placed on the human body or implanted inside it, thus facilitating the introduction of BANs. High processing and complex BANs will be needed in the future to provide the powerful computational functionalities required for advanced applications. These requirements have led to increasing research and development activities in the area of WBAN applications for many purposes [2–7], with the main interest being in healthcare and wearable computers.

The idea of a body area network was initiated for medical purposes in order to keep continuous record of patients, health at all times. Sensors are placed around the human body to measure specified parameters and signals in the body, such as blood pressure, heart signals, sugar level, and temperature. As an extension to these sensors, base units can be deployed on or close to the human body to collect information or relay command signals to the various sensors in order to perform a desired operation. Figure 6.1 presents an illustration of the kind of BAN applied in healthcare services.

WBANs can be applied in many fields, such as:

- assistance to emergency services such as police, paramedics and fire fighters;
- military applications including soldier location tracking, image and video transmission and instant decentralized communications;
- augmented reality to support production and maintenance;
- access/identification systems by identification of individual peripheral devices;
- navigation support in the car or while walking;
- pulse rate monitoring in sports.

The ultimate WBAN should allow users to enjoy such applications with minimum interference, low transmission power and low complexity.

BANs have distinctive features and requirements that make them different from other wireless networks. This includes the additional restriction on electromagnetic pollution due to proximity to the human body which requires extremely low transmission power. The devices deployed within BANs have limited sources of energy due to their small size. Some devices are implanted in the body, which means that regular battery recharging is not a feasible option. Due to the large number of nodes, for specific applications, placed on the human body (which is a relatively small area), the interference is quite strong. In addition, the human body tissue is a lossy medium; hence the wave propagating within the WBAN faces large attenuation before reaching the specified receiver.

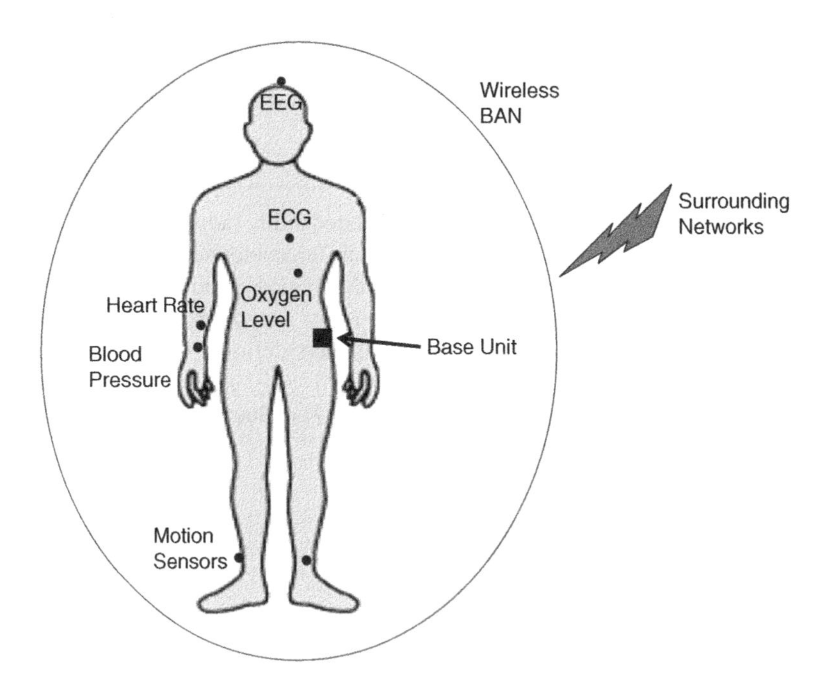

Figure 6.1 WBAN application in health care.

BANs have special network topologies and features determined by the human body. In comparison to indoor propagation channels, the permanent presence of the body leads to the derivation of deterministic radio channel models to be applied in designing efficient and reliable systems.

6.1.2 Antenna Design Requirements for Wireless BAN/PAN

Antennas play a vital role in defining the optimal design of the radio system, since they are used to transmit/receive the signal through free space as electromagnetic waves from/to the specified destination. However, the characteristics and behaviour of the antenna need to adhere to certain specifications set by the wireless standard or system technology requirements. This means that the transmitting and receiving frequency bands of the various units need to be justified accordingly. Another important parameter is the antenna gain that directly affects the power transmitted. Since there are restrictions on the level of power to which the human body can be exposed, the design of the antenna and the other RF components requires careful consideration.

In designing antennas for wearable and handheld applications, the electromagnetic interaction among the antennas, devices and the human body is an important factor to be considered. Various application dependent requirements necessitate thorough evaluation of different antenna configurations and also the effects of multi-path fading, shadowing, human body absorption, and so on. For the wireless body-centric network to be accepted by the public, wearable antennas need to be hidden and low profile. This requires a possible integration of these systems within everyday clothing.

6.1.2.1 Wearable Antenna Parameters

Conventional antenna parameters include impedance bandwidth, radiation pattern, directivity, efficiency and gain which are usually applied to fully characterize an antenna [8]. These parameters are usually presented within the classical situation of an antenna placed in free space. However, when the antenna is in or close to a lossy medium, such as human tissue, the performance changes significantly and the parameters defining the antenna need to be revisited and redefined.

In a medium with complex permittivity and non-zero conductivity, the effective permittivity ε_{eff} and conductivity σ_{eff} are usually expressed as

$$\varepsilon_{eff} = \varepsilon' - \frac{\sigma''}{\omega}, \tag{6.1}$$

$$\sigma_{eff} = \sigma' - \omega\varepsilon'', \tag{6.2}$$

where the permittivity and conductivity are composed of real and imaginary parts,

$$\varepsilon = \varepsilon' - \mathrm{j}\varepsilon'', \tag{6.3}$$

$$\sigma = \sigma' - \mathrm{j}\sigma''. \tag{6.4}$$

The permittivity of a medium is usually scaled to that of the vacuum for simplicity,

$$\varepsilon_r = \frac{\varepsilon_{eff}}{\varepsilon_0} \tag{6.5}$$

and ε_0 is given as 8.854×10^{-12} F/m.

The equations above indicate the differences between free space and lossy material, hence the imaginary part of the permittivity includes the conductivity of the material which defines the loss that is usually expressed as dissipation or loss tangent,

$$\tan\delta = \frac{\sigma_{eff}}{\omega\varepsilon_{eff}}. \tag{6.6}$$

The biological system of the human body is an irregularly shaped dielectric medium with frequency dependent permittivity and conductivity. The distribution of the internal electromagnetic field and the scattered energy depends largely on the body's physiological parameters, geometry as well as the frequency and the polarization of the incident wave.

Figure 6.2 shows measured permittivity and conductivity for a number of human tissues in the band 1–11 GHz. The results were obtained from a compilation study presented in [9, 10], which covers a wide range of different body tissues. Therefore, one major difference that can be identified directly when placing an antenna on a lossy medium, in this case the human body, is the deviation in wavelength value from the free space one. The effective wavelength

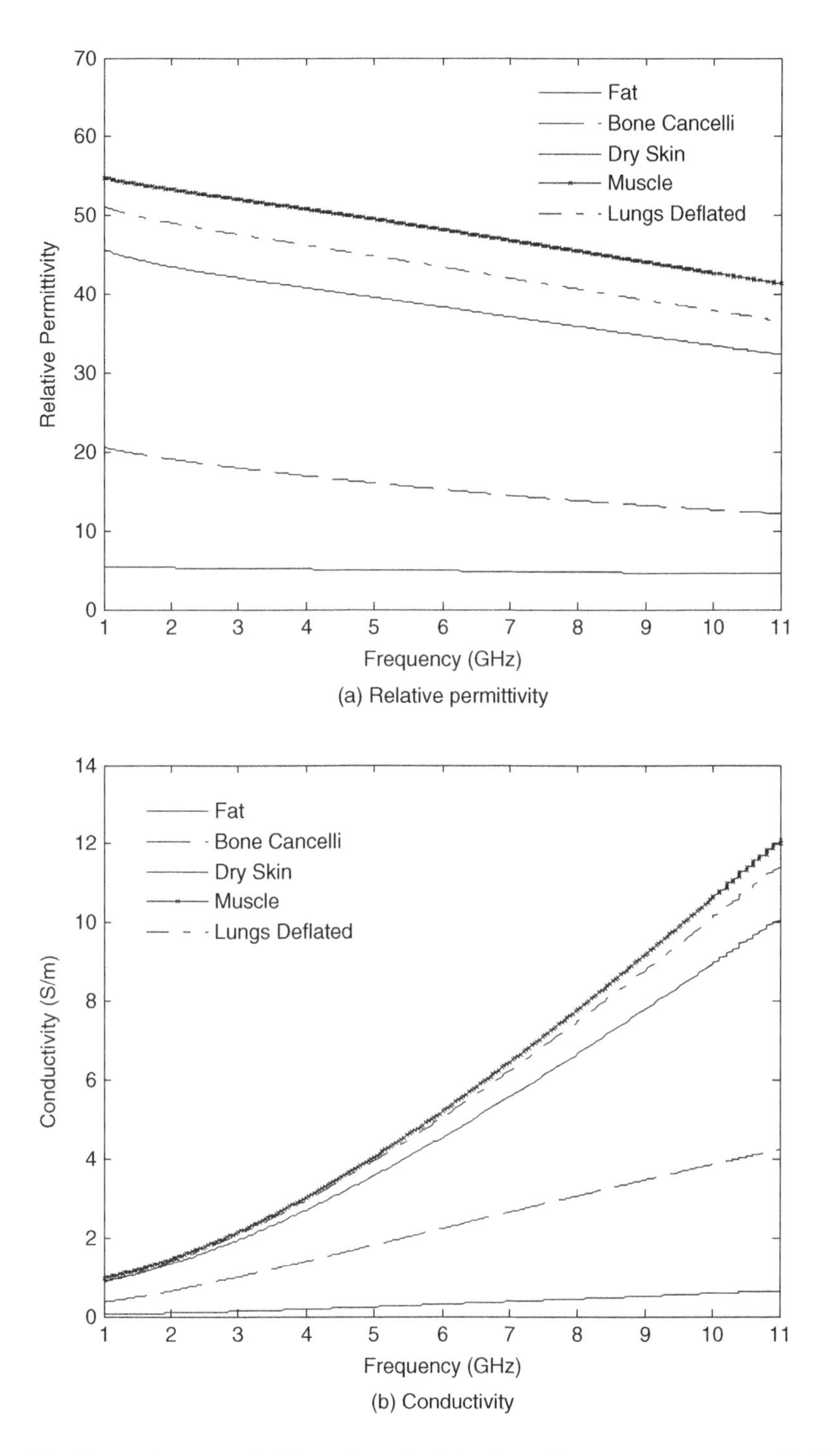

Figure 6.2 Human tissue permittivity and conductivity for various organs as measured in [9, 10].

λ_{eff} at the specified frequency will become shorter since the wave travels more slowly in a lossy medium

$$\lambda_{eff} = \frac{\lambda_0}{Re\left[\sqrt{\varepsilon_r - j\sigma_e/\omega\varepsilon_0}\right]} \tag{6.7}$$

Where λ_0 is the wavelength in free space. However, the effective permittivity as seen by the antenna depends on the distance between the antenna and the body and also on the location since the human electric properties are different for various tissue types and thicknesses. The general rule of thumb is that the further the antenna is from the body the closer its performance to that in free space. This also depends on the antenna type, its structure and the matching circuit.

Wire antennas operating in standalone modes and planar antennas directly printed on substrate will experience changes in wavelength and hence deviation in resonance frequency, depending on the distance from the body. On the other hand, antennas with ground planes or reflectors incorporated in their design will experience less effect when placed on the body from operating frequency and impedance matching factors independent of distance from the body.

An important factor in characterizing antennas is the radiation pattern and hence, gain and efficiency of the antenna. The antenna patterns and efficiency definitions are not obvious and cannot be directly derived from conventional pattern descriptors when the antenna is placed in or on a lossy medium. This is due to losses in the medium that cause waves in the far-field to attenuate more quickly and finally to zero.

Antenna efficiency is proportional to antenna gain [8],

$$G(\theta, \phi) = \eta D(\theta, \phi), \tag{6.8}$$

where η is the efficiency factor and $D(\theta, \phi)$ is the antenna directivity which is obtained from the antenna normalized power pattern P_n that is related to the far-field amplitude F,

$$D(\theta, \phi) = \frac{P_n(\theta, \phi)}{P_n(\theta, \phi)_{\text{average}}} = \frac{\left|\vec{F}(\theta, \phi)\right|^2}{\left|\vec{F}(\theta, \phi)\right|^2_{\text{average}}}. \tag{6.9}$$

The wearable antenna efficiency is different from that in free space, due to changes in antenna far-field patterns and also in the electric field distribution at varying distances from the body. However, the radiation efficiency of an antenna in either lossless or lossy medium can be generalized as

$$Efficiency_{\text{radiation}} = \frac{RadiatedPower}{DeliveredPower}. \tag{6.10}$$

An important quantity, which is in direct relation to antenna patterns and of great interest in wearable antenna designs, is the front–back ratio. This ratio defines the difference in power radiated in two opposite directions wherever the antenna is placed. The ratio varies depending on antenna location on the body and also on antenna structure. For example, the presence of the ground plane in a patch antenna reflects the electric field travelling backwards; hence the front–back ratio is not significantly different when placed in free space and on the body, which is not the case for conventional dipoles or monopoles with radiator parallel to the body.

6.1.2.2 Wearable Ultra-Wideband Antenna Requirements

The aforementioned parameters are conventionally used to describe antenna performance for narrowband systems. However, for wideband or ultra wideband (UWB) operations, additional parameters are needed to fully characterize the antenna, especially for wearable antennas. UWB short-range wireless communication, which covers the 3.1–10.6 GHz frequency band as defined by the Federal Communications Commission (FCC), is different from a traditional carrier wave system [11]. A UWB system sends very low power pulses, below the transmission noise threshold. In UWB communications, the antennas are significant pulse-shaping filters. Any distortion of the signal in the frequency domain causes distortion of the transmitted pulse shape, thereby increasing the complexity of the detection mechanism at the receiver.

For UWB antennas, an important additional criterion has to be taken into account, which is the dependence of antenna patterns on frequency. This criterion is considered essential in designing suitable UWB antennas due to the large relative bandwidth of UWB antennas; the variations of the antenna pattern over the frequency range considered are more distinct. In addition, the emission rules for UWB radiation specify that the power spectral density must be limited in each possible direction. The regulations enforce a limit on the emitted power in the frequency-angle domain [11].

The whole UWB radio system transfer function; in both frequency and time domain can be divided into three functions: the transmit antenna transfer function $H_{Tx}(f)$, the channel transfer function $H_{ch}(f)$ and the receiving antenna function $H_{Rx}(f)$. As presented in [12, 13], for a transmitting antenna, the transfer function expressed in frequency terms is the ratio of the vector amplitude of the radiated electric field at a point, P, to the complex amplitude of the signal input to the antenna as a function of frequency,

$$H_{Tx}(\omega, \theta, \Phi) = \frac{E_{rad}(\omega, \theta, \Phi)}{V_{in}(\omega)}. \tag{6.11}$$

For a receiving antenna, the transfer function expressed in frequency terms is the ratio of the complex amplitude response at an antenna output port to a source of emitted electric filed vector amplitude at a point, P,

$$H_{Rx}(\omega, \theta, \Phi) = \frac{V_{inc}(\omega)}{E_{inc}(\omega, \theta, \Phi)}. \tag{6.12}$$

The transmitting transfer function is the time derivative of the receiving transfer function; in other words the receiving transfer function is the integral of the time history of the radiation field. Hence, the ratio of the transmitting transfer function of an antenna to the receiving transfer function of the same antenna, is proportional to frequency [13],

$$H_{Tx}(\omega, \theta, \Phi) = \frac{j\omega}{C_o} H_{Rx}(\omega, \theta, \Phi). \tag{6.13}$$

where $\omega = 2\pi f$ is the operating frequency, C_o is the speed of light and (θ, ϕ) is the desired orientation, which in addition to traditional antenna characteristics will help in providing clearer picture of the antenna transfer function.

Due to the narrow pulses transmitted in UWB systems, impulse response and time domain characterization of the antenna is of great importance. The impulse response of a UWB antenna is also direction dependent, which urges the introduction of a spatial rms delay spread of the UWB antenna in addition to the conventional rms delay spread of the radio propagation channel. Another time domain UWB antenna parameter is the energy level enclosed by the time window of the radiated/received pulse.

One of the key parameters in correctly describing UWB antenna performance is pulse fidelity. In a time domain formulation, the fidelity between waveforms *x(t)* and *y(t)* is generally defined as a normalized correlation coefficient [14],

$$F = \left| \frac{\int_{-\infty}^{\infty} x(t)y(t-\tau)dt}{\sqrt{\int_{-\infty}^{\infty} |x(t)|^2 \int_{-\infty}^{\infty} |y(t)|^2}} \right|. \tag{6.14}$$

The fidelity factor, F, compares the shapes of the cross-correlated waveforms but not their amplitudes. In practice this factor is calculated for a given direction in space in order to fully characterize the spatial radiation properties of an antenna. The fidelity depends not only on the antenna characteristics, but also on the excitation pulse, thus it is also a system dependent parameter. The fidelity plays an important role in defining wearable antenna performance to determine several system parameters, especially in impulse radio systems.

6.2 Modelling and Characterization of Wearable Antennas

Fuelled by the idea of user-centric approach to future communication technology, many research projects have been initiated, under the rubric of smart clothing/textiles, to integrate antennas and RF systems into clothing, paying heed to size reduction and cost effectiveness [15, 16]. The effect of the human body on the operation of antennas located in close proximity has been investigated widely and thoroughly in the literature [17], including the absorption of energy within the body, the specific absorption rate (SAR) for proximate antennas [18] and the propagation on and off the body for use in mobile phones [19].

6.2.1 Wearable Antennas for BANs/PANs

6.2.1.1 Progress in Wearable Antennas

Although the main application of body-worn sensors is currently within the medical engineering community, the military has for sometime been investigating the potential applications of flexible and hidden wearable antennas for communication purposes, especially in the UHF and VHF bands. The key objective of these studies within the military engineering community is to design flexible, efficient, multi-function, multi-band and hidden antenna systems that can provide reliability and security for the soldier (the mobile communication base will be hidden to enemies and also protected from natural destructors). The US Army

Figure 6.3 Body-worn squad level antenna vest used for military applications [20].

Natick Soldier Center [20, 21] has transformed a rigid meander double-loop antenna into a wearable, flexible, textile based antenna that is compatible with the radio system used by soldiers in the battlefield and has been integrated into a vest (Figure 6.3). The vest provides a body conformal antenna that is visually hidden, protecting the soldier's location and life.

Another example of wearable antennas used in a military communication application is the Harris broadband body-worn dipole antenna operating in the 30–108 MHz band [22]. The antenna is omnidirectional, vertically polarized and can provide sufficient communication range; it is also ground independent which makes it configurable for any situation. In addition, the antenna is easily attached to the user's rucksack or vest. MegaWave has also developed a body-worn antenna integrated within the soldier's vest to provide radio and remote communication functionalities in different environment [23].

The application of wearable antennas in wireless communications for consumer technologies still provides the main market for the design and development of the body-worn antenna. Antennas are mainly designed to operate in the GSM/PCS bands and the unlicensed ISM band (2.4 GHz). Wearable antennas, as the name implies, are supposed to be integrated within the clothing or secured on the body. Another important factor to be considered when designing wearable antennas is the radiated power and absorption by the human body and the health risks as a consequence of increased exposure to radiated electromagnetic fields.

Salonen *et al.* [24, 25] explore the use of the planar inverted-F antenna (PIFA) as a wearable antenna that can be placed on the human body for maximum functionality and improved safety. They examine the possibility of achieving dual-band operation by simply adding a slot in the antenna design, and explore how the ground plane of the antenna provides a shield to the antenna, directing maximum power away from the body (Figure 6.4). They discuss the flexible PIFA antenna fabricated on a flexible substrate, enabling the placement of the antenna on the arm or conformal to the body, in [25]. They apply simple techniques

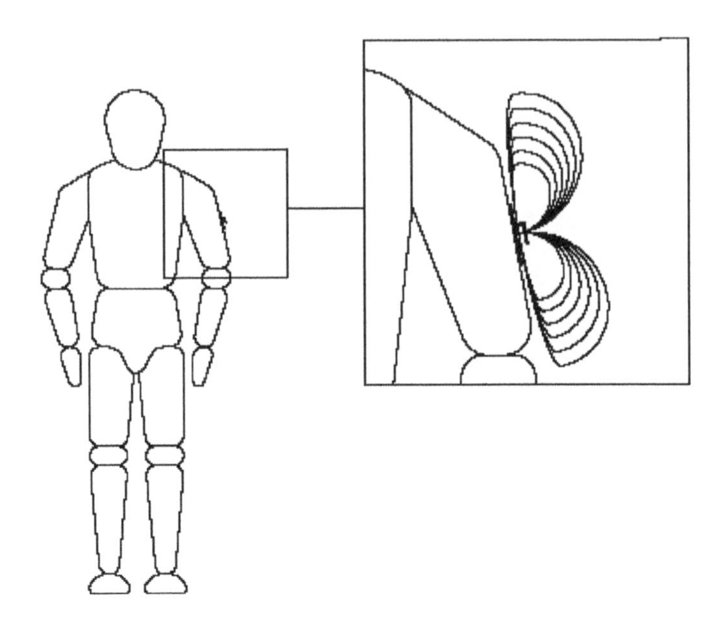

Figure 6.4 PIFA antenna placed on the arm as an optimum location to minimize health risks. The ground plane directs maximum power away from the body [24].

to provide wide-band operation, giving the system designers more freedom and potential applications for multi-functioning operation, in addition to possible array structure that will allow larger coverage area for the antenna with respect to restricted directionality.

Due to the increased demand for multi-frequency and multi-function antennas for use in smart clothing and future consumer communication technologies, fabric and textile antenna designs have received a vast amount of attention in recent years. Salonen *et al.* [26, 27] have addressed the design and development of a dual-band textile antenna for wearable applications. Fleece fabric was considered for antennas for GSM and WLAN bands. The study highlighted the need to understand accurately the sensitivity of the dielectric constant of textile materials. The fabricated dual-band textile antenna showed good agreement with predictions; however, the effect of conformal placement of the antenna on the body and the influence of non-planar textile surfaces on antenna performance have not been analysed in great detail.

The problem of addressing antenna performance when the design is conformal to the body, in other words the conventional antenna flat substrate design is deformed, has been tackled by Cibin *et al.* [28]. A flexible wearable E-shaped shorted PIFA antenna has been developed, fabricated and its performance validated. The antenna demonstrates a usable frequency range of 360–460 MHz in all expected field conditions, even when covered in wet soil. The antenna far-field gain radiation pattern is reasonably omnidirectional, with gain better by 15 dB than that of the body-worn half-wavelength dipole antenna. The antenna provides robustness and more comfort to the wearer, in addition to potential integration in clothing (Figure 6.5).

As a measure of antenna applicability to body-worn communication technologies, the amount of power absorbed by the human body and the risk to the human in the long run need to be examined and investigated thoroughly [29]. SAR levels of the power radiated by

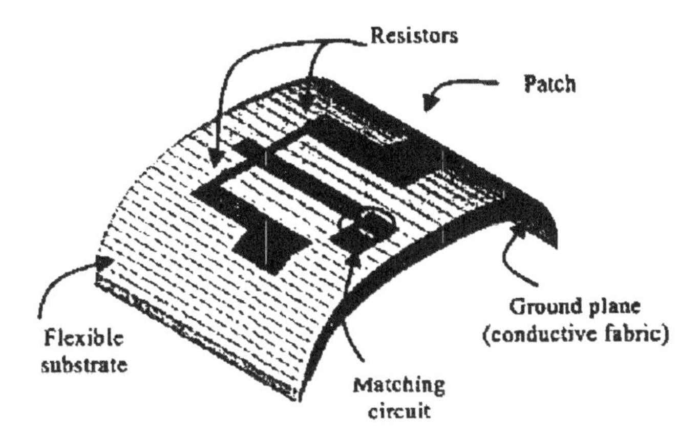

Figure 6.5 Configuration of conformal, flexible E-shaped antenna developed for wearable antenna applications [28].

the antenna when placed on or in close proximity to the body have been widely studied and explored in the literature [30–33].

Salonen *et al.* [31] demonstrated the effect of placing a textile patch antenna in the vicinity of the human head and the influence of the presence of human tissue on antenna impedance bandwidth and radiation patterns. The SAR levels were examined at different body locations and for patch and dipole antennas applying numerical techniques when an antenna was placed on a digital phantom of the human body. The computed results showed improved SAR levels when using the textile patch antenna, as predicted due to reduced radiation towards the tissue.

6.2.1.2 Wearable Antenna Performance in the 2.4 GHz ISM Band

Body-worn antennas are required to be insensitive to proximity to the body and to have a radiation pattern that minimizes the link loss among wearable antennas and communication units in the WBAN. Two antenna types are chosen as an example of low-profile wearable antennas: microstrip patch and printed monopole. Both antennas are planar and potentially conformal to the human body. They have minimal sensitivity to body proximity, presumably due to the effects of the ground planes, as discussed in the previous section. It is necessary to mention that these antennas are presented as reference designs that could be modified and further miniaturized to suit wearable technologies requirements.

Microstrip patch antennas radiate close to their resonance frequency and the radiator main dimension is usually a half-guided wavelength at the operating frequency. Rectangular patches represent the simplest and most widely used configuration of the microstrip printed antenna. Figure 6.6 shows the design of the patch antenna under investigation that operates at 2.45 GHz. Antenna modelling is performed using the finite integral technique (FIT) utilized within Computer Simulation Technology (CST) Microwave Studio®.

The on-body antenna performance is numerically (in addition to experimentally) investigated by applying a one-layer human tissue slab model (muscle with $\varepsilon_r = 53$ and conductivity $\sigma = 1.7$ S/m at 2.4 GHz, dimensions $120 \times 120 \times 40$ mm) [34]. The detuning experienced

Figure 6.6 Microstrip patch antenna design operating at 2.4 GHz and used for wearable antenna study [28].

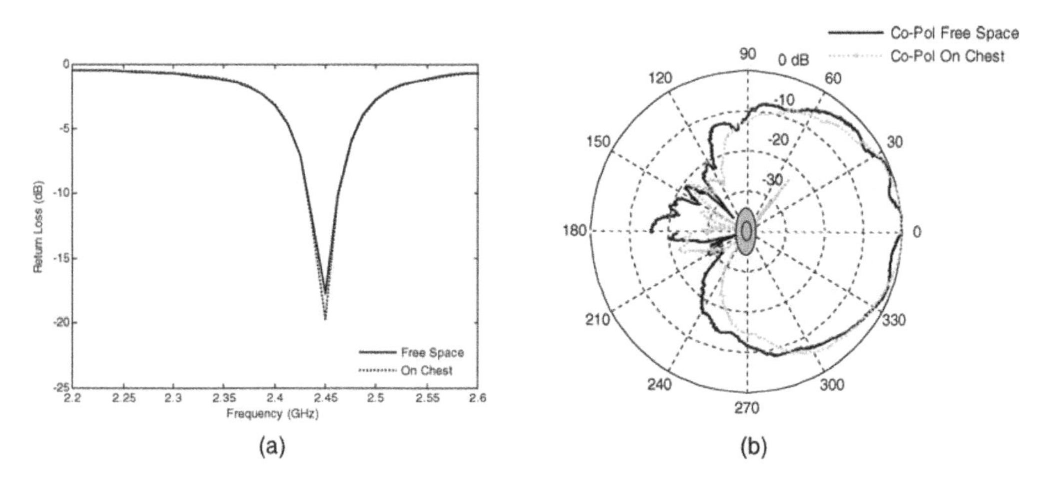

Figure 6.7 Microstrip antenna performance on and off body: (a) return loss and (b) radiation patterns of antenna (azimuth plane patterns). Antenna placed 1 mm away with radiator parallel to the body.

when placed on the body is not significant due to the ground plane size Figure 6.7(a); however, for smaller antennas and ground planes the detuning will be more significant and apparent.

The radiation performance of the antenna when placed on the chest was experimentally measured and compared to the pattern obtained in free space and also to simulated patterns

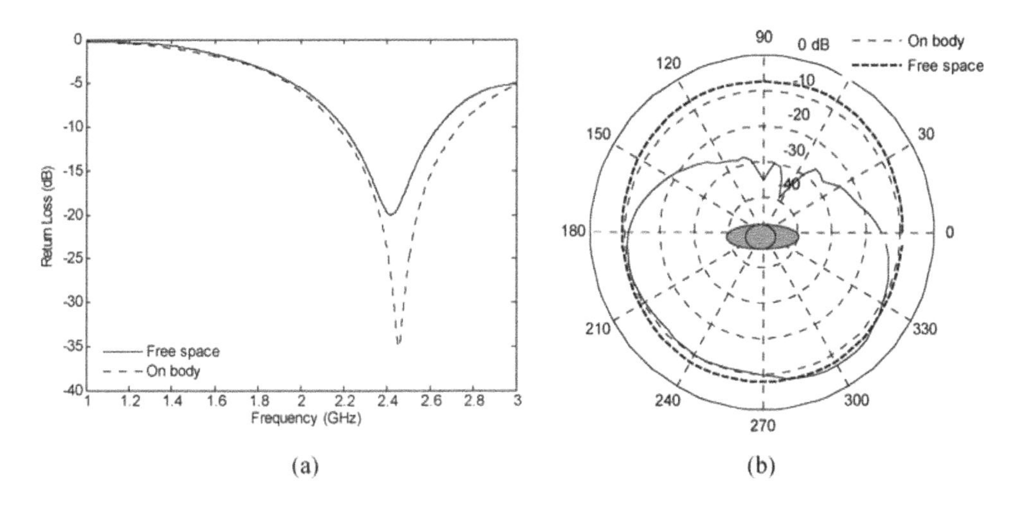

Figure 6.8 Planar monopole performance on and off body: (a) return loss and (b) radiation patterns of antenna (azimuth plane patterns). Antenna placed 1 mm away with radiator parallel to the body.

obtained by numerical modelling techniques. Figure 6.7 presents the measured antenna radiation performance. Antenna gain is increased by 0.5 dB when placed on the body (due to reflections from the highly lossy human tissue) and antenna efficiency reduced by 12 %. The large ground plane minimizes the body presence effect on antenna performance compared with other antenna types.

The second antenna type used in the study is the planar monopole antenna. The vertical and planar monopole is widely used in short-range wireless communications due to its simplicity of fabrication and construction. The antenna consists of a ground plane with optimized dimensions to obtain good matching and a printed copper strip on the opposite side to radiate the electromagnetic waves.

In the same way as for the microstrip patch antenna, the return loss of the monopole is measured on and off the body. The on-body matching response shows a slight frequency detuning, (Figure 6.8(a)). Figure 6.8(b) illustrates the simulated radiation patterns in the antenna azimuth plane on and off the body. The antenna gain is increased by 1 dB when placed on the body due to the high reflections caused by human tissue at this high frequency (2.4 GHz) but radiation efficiency is reduced by a factor of 0.5. One apparent difference between monopole and microstrip antennas is the front–back ratio for radiated power. The difference between front and back radiated power is approximately −30 dB for monopoles, compared to less than 4 dB in the microstrip antenna case, which is associated with the presence of a large ground plane in the microstrip antenna that shields the antenna from the body.

6.2.2 UWB Wearable Antennas

As discussed in the previous section, UWB antennas introduce additional constraints on antenna design. Chen *et al.* [32] presented the proximity effects of the human head on the impedance and radiation performance of two types of planar UWB antennas in terms of impedance matching, gain, radiation patterns and induced electric currents. Due to the high losses in the human tissues at these high frequencies (UWB band 3.1–10.6 GHz), the human body acts as a reflector, in this case the human head, which deforms the antenna radiation patterns and also causes the gain to increase slightly in the more directive spatial angles.

Klemm *et al.* investigated the main differences between deploying directive and omnidirectional UWB antennas as wearable antennas [33]. The study showed the effect of introducing a reflector on the back of the planar stacked patch antenna in reducing backward radiation and hence power absorbed by the body when the antenna is placed on the body. They also demonstrated the transient characteristics of the antennas in free space and how they compare to those obtained in close proximity to the human body. Alomainy *et al.* presented a detailed study of the effect of human presence on UWB antenna parameters (in the frequency and time domain, including spatial variations)[35]. They analysed the dispersion of the channel transfer functions due to electric property variation with frequency and also antenna dispersion. The study proved the suitability of the coplanar waveguide (CPW) fed UWB antenna for BANs in comparison with the microstrip line fed antenna in terms of both frequency and time domain characterization. The CPW-fed planar inverted cone antenna is a compact, low cost and easily fabricated UWB antenna. It is a coplanar version of the conventional radiator above a horizontal ground plane antenna [36, 37]. The antenna consists of two elements, the top element (planar cone) and bottom rectangular-shaped ground plane (incorporating a CPW feed), and is shown in Figure 6.9.

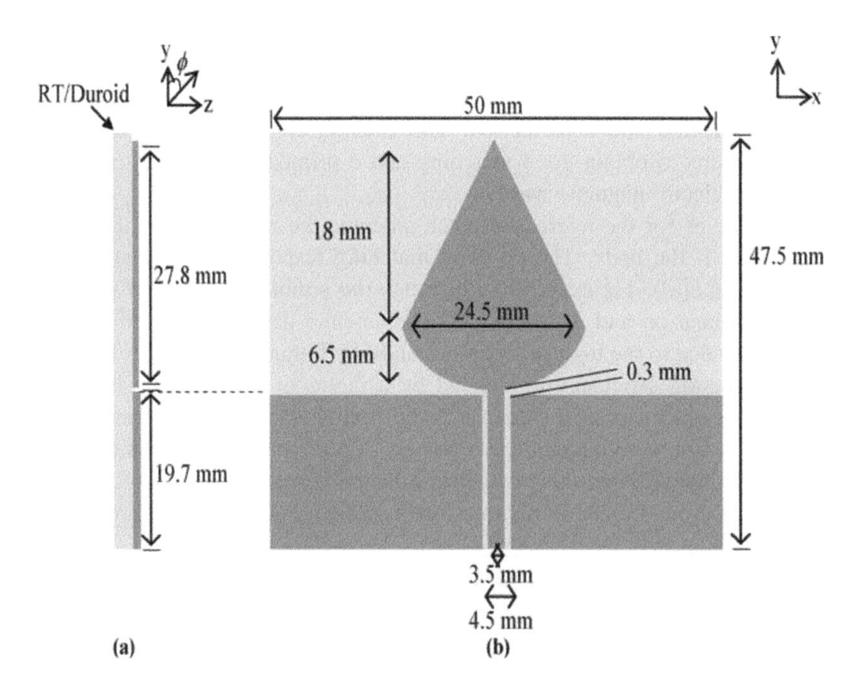

Figure 6.9 Detailed diagram of the CPW-fed planar UWB antenna with dimensions: (a) side view; (b) front view.

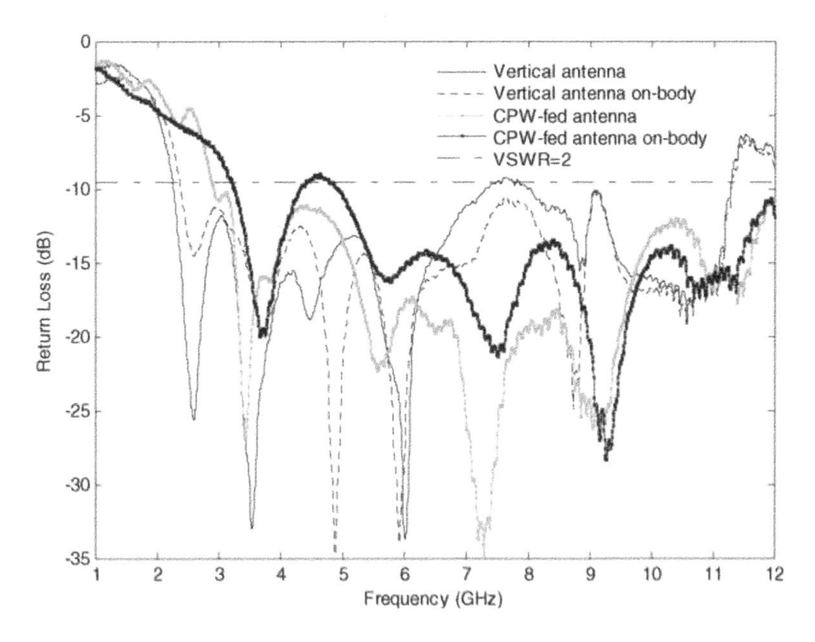

Figure 6.10 CPW-fed antenna return loss on/off the body compared to those of the vertical antenna on horizontal ground.

The return loss responses of the CPW-fed planar antenna are compared to that of the vertical antenna when placed in free space and on the body (Figure 6.10). The bandwidth is 3–12 GHz for a voltage standing wave ratio (VSWR) less than 2. Both antennas seem to preserve good impedance bandwidth even when placed on the human body. The CPW-fed antenna is placed on the centre of the human chest parallel to the body, while the vertical antenna on horizontal ground is placed perpendicular to the body at the same location as for the CPW-fed antenna. The azimuth plane radiation patterns of the CPW-fed antenna are presented in Figure 6.11 for 3 GHz and 10 GHz and compared to those of the vertical antenna over a horizontal ground plane, which has a good cross—polarization performance. Both antennas are omnidirectional across the required band. The measured gain of the antenna ranges from 0 dBi to 2 dBi.

The time domain characteristics of the antenna are obtained by performing free space channel measurements between two identical CPW-fed antennas at a separation distance of 0.5 metres and at different angular positions. The time domain responses are derived by direct application of interpolated fast Fourier transforms on the measured real frequency responses S_{21}. The antenna transient responses are measured for side-by-side and face-to-face configurations. The antenna impulse response at various angular directions with one antenna orientation is changed by 90° around the axis linking both antennas face-to-face. The fidelity of the system impulse response at different directions with reference to the response at 0° (maximum radiated power direction) when the antennas are facing each other is 91.7% and 95.07% for 90° and 180°, respectively Figure 6.12(a). The performance of the antenna is also evaluated by comparing responses obtained for side-by-side and face-to-face configurations Figure 6.12(b). A fidelity of 92.4 % is obtained, which is sufficient to provide a reliable radio link.

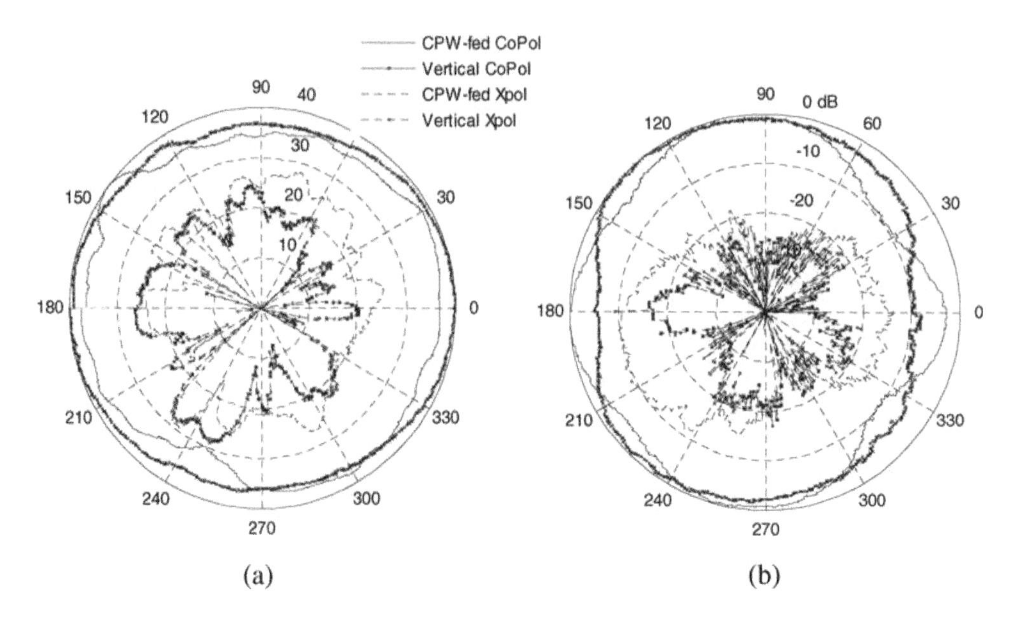

Figure 6.11 Radiation patterns of a CPW-fed antenna (copolar and cross-polar) compared to an antenna on a horizontal ground plane at (a) 3 GHz and (b) 10 GHz (Azimuth plane).

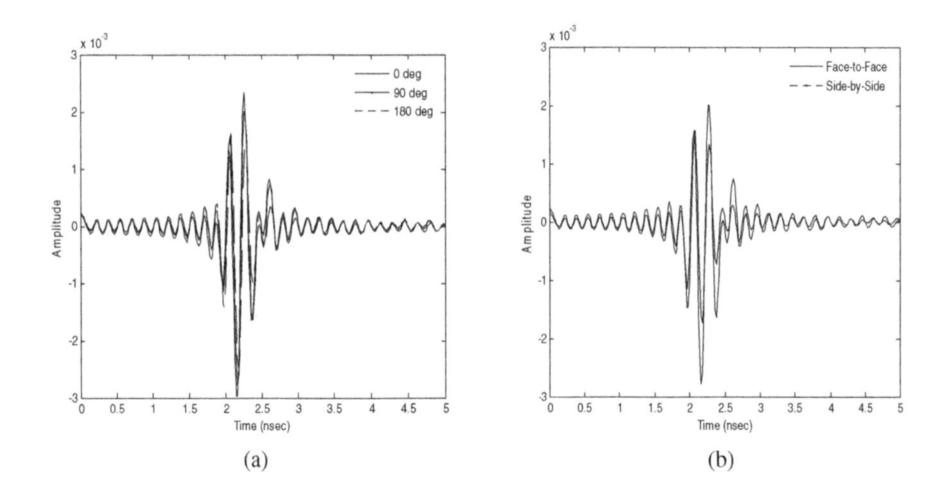

Figure 6.12 Impulse responses of the channel including two identical CPW-fed antennas and free space distance of 0.5 m when antennas set (a) face-to-face at various orientations and (b) face-to-face vs. side-by-side.

The same procedure as outlined above is applied to analyse the transient on-body performance of antennas set side-by-side with variations in human body posture. Figure 6.13 presents the calculated impulse responses and shows the slight changes in pulse template when the body changes posture. The effect of the antenna position is also investigated,

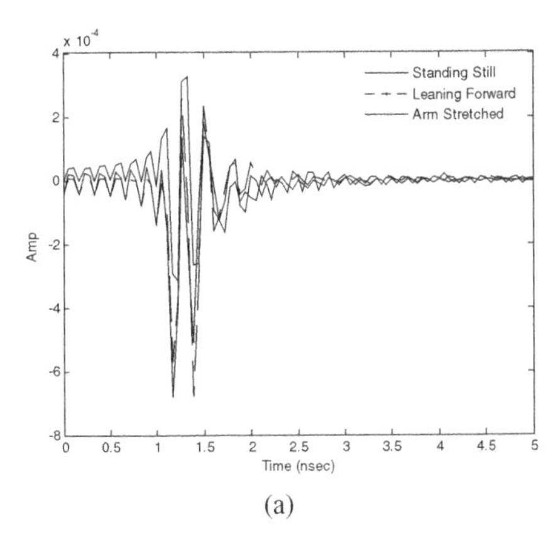

(a)

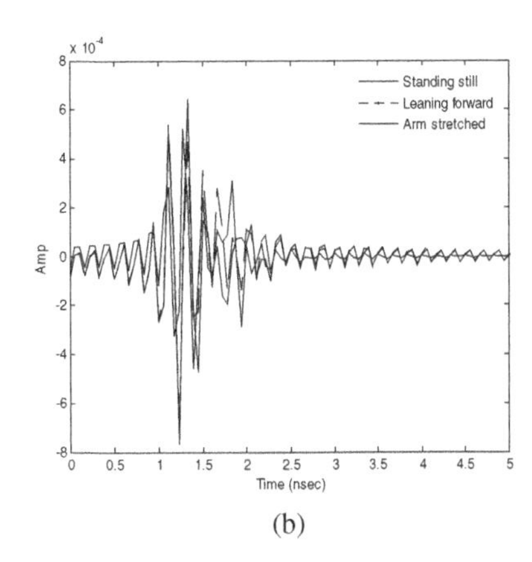

(b)

Figure 6.13 Impulse responses of the channel including two identical CPW-fed antennas placed on the human trunk set side-by-side: (a) lower trunk section (waist); (b) upper trunk section.

showing the distortion introduced when an antenna is placed on the upper part of the human trunk due to reflections from different body parts including shoulders and face, in comparison to lower trunk positions which provide more a stable communication link.

6.3 WBAN Radio Channel Characterization and Effect of Wearable Antennas

Wave propagation around the human body is a complex problem, even though it takes place over short distances. The transmitted signal between two body-worn devices arrives at its destination by means of propagation through the body (high losses vary with frequency), diffraction around the body and reflections from nearby scatterers in wireless communication environment. In BANs, the body is subject to many small movements and, during normal activities, these movements become significant. Thus characterization of radio wave propagation needs to account for both the variable positioning of the communication devices on the body and dramatic changes in the geometry of the local environment. For such small networks, it is quite difficult to separate the antenna's behaviour from the propagation channel.

Characterization and modelling of propagation channels in WBANs have received increasing attention recently due to the demand user-centric communication systems and the lack of radio channel models to provide clear understanding for system designers [34, 36–45]. A channel models for a WBAN at 400 MHz, 900 MHz and 2.4 GHz derived from a numerical modelling simulation of signal propagation around the body is presented in [39]. The electromagnetic waves could also propagate around the body via penetration inside the body and via creeping waves that follow the curvature of the body. UWB wave propagation around the body is numerically and experimentally investigated in [40, 41] to derive a UWB body area network radio channel model for general BAN applications. The radiated signal is diffracted around the torso rather than passing through it, and the distance around the perimeter of the body was used in determining the path loss. However, effects of different antenna parameters and types on the channel are not explored thoroughly, considering that the antenna is an essential part of the whole communication system.

6.3.1 Radio Propagation Measurement for WBANs

UWB antennas impose additional constraints to be accounted for when deriving radio models according to the environment and antenna parameters. UWB antennas play major roles as part of the communication system and, in contrast to conventional antennas, they are non-reciprocal from a transmitting and receiving transfer functions perspective [42]. Modelling of on-body radio channels using both ray tracing and time domain numerical techniques, as presented in [36], led to the inclusion of different antenna directivities and pattern distortion as a function of frequency in deriving reliable and accurate channel models. WBAN propagation channels with planar UWB antennas are characterized and investigated in [37, 43]. The planar antenna performance is compared to that of a reference vertical antenna over a ground plane [37]. The effect of the ground plane's presence on wearable antenna performance and its impact on the radio channel in the ISM band (2.4 GHz) and therefore on the radio system behaviour is presented in [34].

Frequency domain propagation measurement is commonly applied in UWB radio channel characterisation. Post-processing techniques are used to derive and determine the channel parameters in both frequency and time domain by means of fast Fourier transform and digital processing methods. Frequency domain measurements of on-body radio channels using a vector network analyser to collect the channel transfer frequency responses (derived from S_{21}) and hence deducing the channel parameters are presented in [34, 37, 43, 44]. Two identical CPW-fed planar UWB antennas Figure 6.9 are placed on the body to analyse radio channels for various body postures and antenna positions. A matrix of measurement points is adopted to ensure sufficient data collection for reliable and efficient channel characterization and modelling.

Figure 6.14 illustrates the different antenna positions used for measurement and also shows the various angular orientations applied with reference to the transmit antenna (Tx). Those various orientations are studied to emphasize spatial dispersion of the antenna in addition to on-body radio channel dispersion [44]. The minimum distance between Tx and Rx is 10 cm which is the wavelength at the lower frequency (λ_0 at 3 GHz) minimizing antenna coupling effects. The distances between the different receiving points are 5 cm to account for changes with resolution of $\lambda_0/2$ at 3 GHz. In addition to different Tx/Rx orientations, analysis of the effect of body posture changes on the radio channel is also investigated. Six different body postures are adopted: standing still, leaning forward, turning right, turning left, arm stretched aside, and arm folded to form a right angle.

6.3.2 Propagation Channel Characteristics

The impulse responses of the measured frequency transfer functions are shown in Figure 6.15 in comparison with Gaussian derivative pulses. The time domain response at 0.25 m from Tx on the body at the lower torso shows that the impulse is comparable to a high order Gaussian derivative with pulse width 0.65 ns on average and pulse fidelity 75–91% for pulses in Figures 6.15(a)–(c). The impulse response of the on-body channel along the y-axis of Tx is shown in Figure 6.15(c) and presents a distorted response due to the distortion in

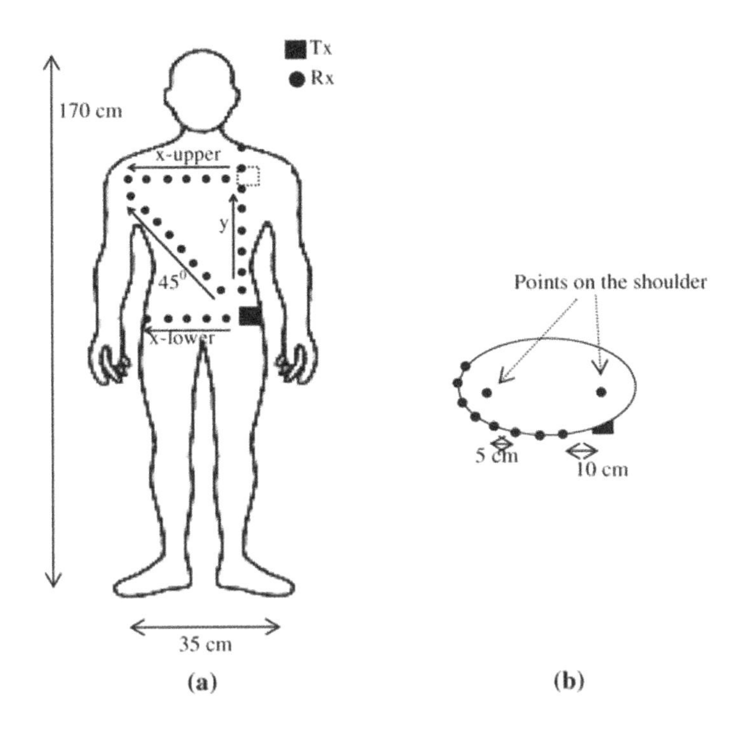

Figure 6.14 Measurement setup for on-body channel characterization with CPW-fed antenna mounted with x–y plane parallel to the body: (a) front view; (b) top view.

the frequency domain at relatively low frequencies. The pulse shape illustrates the resonance behaviour of the antenna specifically in the studied direction. The impulse responses of the on-body channels at different Tx/Rx orientations demonstrate the dispersion of both antenna and radio channel.

To evaluate the performance of the CPW-fed antenna, a comparison with an antenna over a horizontal ground plane is carried out [37, 43]. Figure 6.16 compares the impulse response presented in Figure 6.15(c) of the CPW-fed antenna with that of a vertical antenna over a horizontal ground plane for an on-body channel when Rx is 0.25 m along the y-axis of Tx. The impulse responses are comparable with distortion occurring due to the ringing effect of the monopole-like antennas and also distortion in the frequency transfer function of the antennas at high frequencies. The CPW-fed antenna response has a 30% lower pulse peak than that of the antenna over a horizontal ground plane, caused by better antenna efficiency and gain of the latter antenna.

The received frequency responses at each measurement point are averaged cross the frequency band to derive statistical path loss models for different antenna orientations and body postures. The path loss model is obtained by applying the least squares method to the empirical data. The modelled path loss has an exponent value of $\gamma = 5.1$ and reference path loss of 37.2 dB [44]. The large deviation observed between the measured path losses indicates that the UWB on-body channel is highly mobile and hence highly efficient directive antennas with main radiating beam conformal to the human torso are needed to maximally stabilize the channel (Figure 6.17).

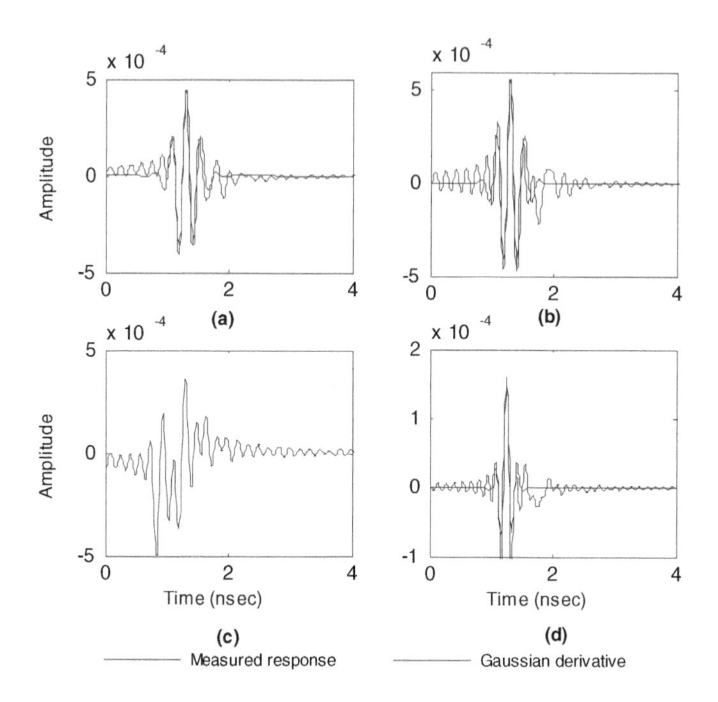

Figure 6.15 Impulse responses of on-body channels at distance 0.25 m for (a) x-lower torso, (b) upper torso, (c) y and (d) 45° with reference to Tx.

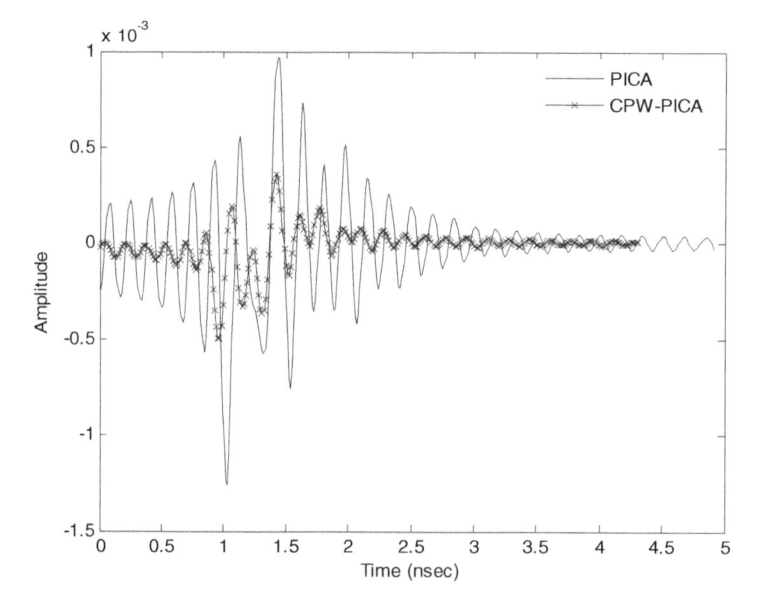

Figure 6.16 Comparing the CPW-fed antenna on-body channel impulse response when Rx is 0.25 m along the y-axis from Tx with the impulse response of the vertical antenna.

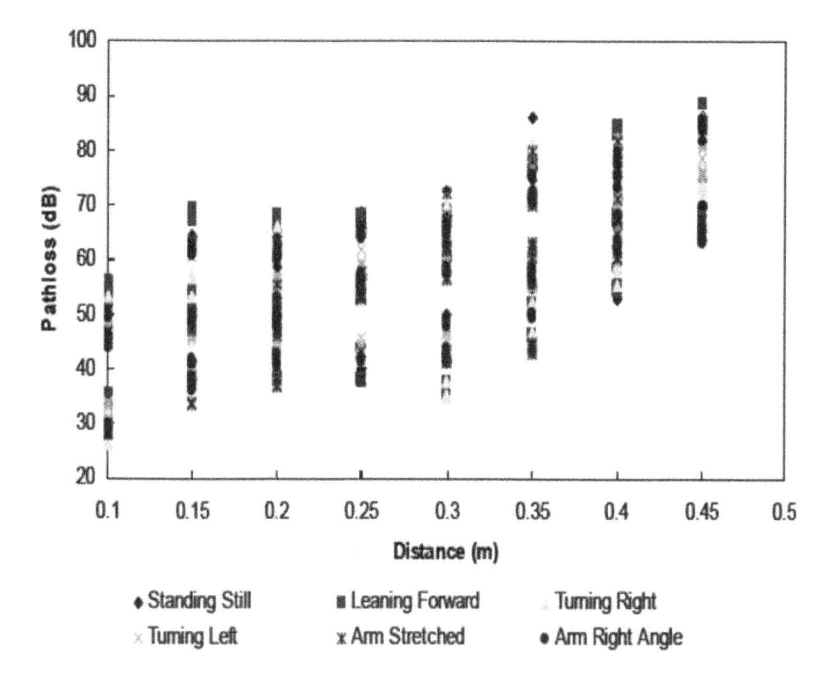

Figure 6.17 Measured on-body channel path loss for all measured data points at different angular positions with various body postures.

Figure 6.18(b) shows the cumulative distribution function (CDF) of the deviation of the measured received power from the calculated average in comparison to that of the vertical antenna over a ground plane. The curve fits a normal distribution fairly well. This shows that applying a CPW-fed antenna for on-body communications will provide better predictability

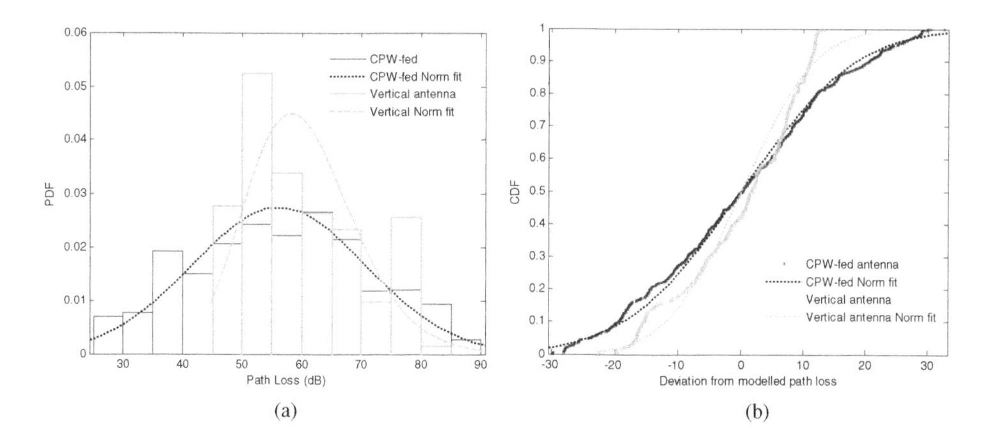

Figure 6.18 Statistical models of the propagation channel path loss for both CPW-fed antenna and vertical antenna: (a) calculated PDF; (b) CDF of deviation from modelled path loss (represents shadowing of the body).

from a system design point of view [34]. The distribution presents shadowing effects of the human body. The probability distribution of the signal strength is obtained from the measured data for the specific body postures and also for the total measured results. The probability distribution (PDF) of measured path losses for CPW-fed and vertical antenna is presented in Figure 6.18(a). The distribution is approximately normal with mean $\mu = 56.12$ dB and standard deviation $\sigma = 14.6$ dB for the CPW-fed antenna. The CPW-fed antenna distribution shows large deviation and spread of calculated value, which indicates the large changes due to body movements and orientations in path losses as predicted earlier.

6.4 Case Study: A Compact Wearable Antenna for Healthcare Sensors

To provide an understanding of practical applications of wearable antennas, a case study of antenna design in healthcare sensors is presented and discussed.

6.4.1 Application Requirements

For wireless sensor applications, antennas need to be efficient and immune from frequency and polarization detuning. Understanding the antenna radiation pattern for wireless sensors (when placed on the human body) is vital in determining sensor performance. It is also important to specify how coupling into the propagation mode, which may be a surface wave or free space wave or a combination of both, occurs. If the sensor is placed too close to the human body, it will have low efficiency due to loss but good coupling to surface wave. On the other hand, if it is placed too far from the body, the antenna efficiency will improve but coupling to surface wave will be poor. This is similar to the classical problem of the VLF antenna over ground but parameters and antenna types will be very different. Hence it is essential to develop special antennas for wireless sensor networks that need monopole patterns for coupling to surface wave and patch like patterns for surface/space wave links (based on preliminary studies).

The case study is based on the sensor designs and modules developed by Healthcare Devices and Instrumentation, Philips Research, Netherlands (Figure 6.19), for operation in the unlicensed ISM band (2.4 GHz with ZigBee specifications). The maximum achievable coverage range is to be delivered by the sensor with respect to the transceiver sensitivity levels and also to the environment. Modelling and characterization of the antenna deployed in the sensor are discussed and investigated with respect to the surrounding components and data connectors. Radio channel characterization of propagation from a body-worn sensor in an anechoic chamber and the typical indoor environment is also presented with numerical simulation results for cross-referencing and validation.

6.4.2 Theoretical Antenna Considerations

The sensor antenna design is restricted by many factors including the sensor size, chip placement, lumped element locations and flexibility of the sensor structure, to be balanced against minimum cost and changes to antenna performance. Figure 6.20 shows the schematic design of the modelled sensor antenna and also an exploded diagram of the proposed sensor.

(a) (b)

(c)

Figure 6.19 Transceiver circuit layer of the medical sensor developed by Philips (applying the Chipcon CC2420 transceiver chip from Texas Instruments [46]) with antenna presented as printed thin wire on the edges of the board: (a) top view; (b) bottom view; (c) prototype fabricated sensor.

The current antenna is a printed quarter-wavelength monopole, etched on the edge of the circular printed circuit board. Hence the antenna is designed with the printed wire wrapping the transceiver chip and other components. The antenna can be compared to the circumference monopole, which is derived from the bent and inverted-L antenna.

The sensor antenna performance is sensitive to the presence of lumped components, pins and copper routings. The surrounding and adjacent components are modelled as a perfect conductor block around which the antenna is printed. The printed circuit board includes a ground plane and supply voltage copper sheets.

The monopole antenna is a progression from the half-wavelength dipole antenna with the lower half replaced by an infinite conduction plate to provide the image of the upper half. Therefore, the monopole has an omnidirectional radiation pattern similar to that of a dipole with radiated power and radiation resistance half of that for a dipole. However, the directivity is doubled due to the halving of the isotropic radiation intensity [8].

When the monopole is placed or printed on a dielectric material with permittivity other than 1, the antenna dimensions have to be modified in order to achieve performance at the frequency of interest. This leads to redefinition of antenna impedance by the approximation

$$Z(\omega, \varepsilon_r) = \frac{1}{\sqrt{\varepsilon_r}} Z\left(\sqrt{\varepsilon_r}\omega, \varepsilon_0\right) \tag{6.15}$$

which leads to the antenna being shortened by a factor $1/\sqrt{\varepsilon_r}$ to ensure the resonance is at the desired frequency. The metal pins and copper lines provide an extension to the antenna which increases the impedance value expected; however, due to the capacitive coupling between the antenna and surrounding components the antenna impedance tends to decrease compared to the general monopole impedance of 40 Ω [8]. Therefore, the radiation characteristics of the antenna are directly affected; hence changes in antenna gain and efficiency are expected.

6.4.3 Sensor Antenna Modelling and Characterization

The antenna design deployed in the proposed sensor is numerically analysed using the finite element method (FEM) utilized in Ansoft's High Frequency Structure Simulator (HFSS). The antenna performance is presented for a sensor in free space and body-worn operations.

6.4.3.1 Free Space Sensor Performance

The printed circular monopole antenna (Figure 6.20) is modelled with an FR4 substrate ($\varepsilon_r = 4.6$ and thickness of 0.3 mm). The printed antenna thickness is 35 μm and the width of the line is 150 μm. The ground and supply voltage layers added have a diameter of

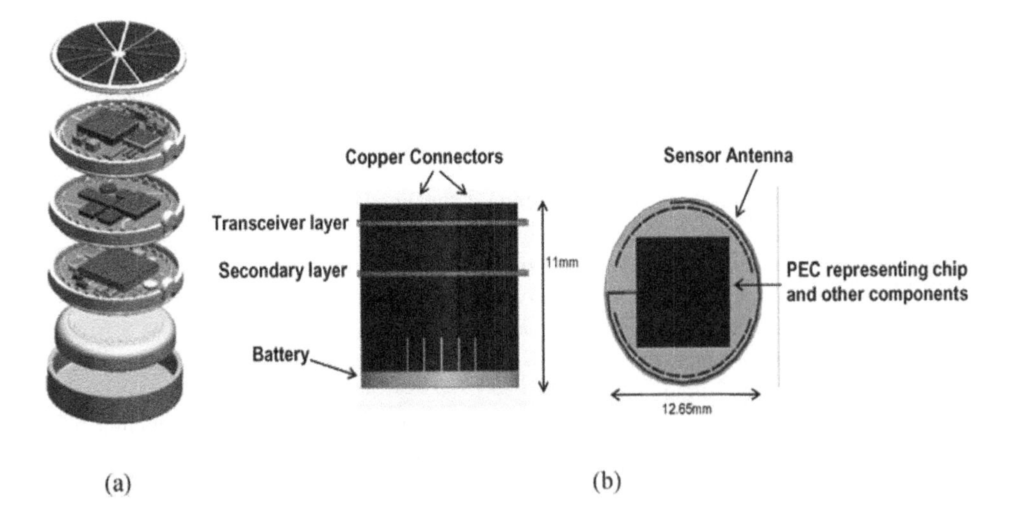

Figure 6.20 (a) Exploded diagram of the proposed healthcare sensor and (b) dimensions and detailed structure of the modelled sensor including the antenna.

5.5 mm, thickness of 17.5 μm each and separation between the layers of 80 μm. The actual antenna length is 31.5 mm. The antenna is a circumference quarter-wavelength monopole. The complex impedance at the RF transceiver differential output is 115 + j180 Ω, therefore a matching circuit is applied in order to match the output to the single-ended monopole (matching to 50 Ω) [46].

Initial analysis of the monopole impedance at 2.4 GHz demonstrated the effect of adjacent components on antenna performance and hence antenna impedance, in addition to the bend introduced in the antenna design. The antenna has a complex impedance of 35 + j320 Ω at 2.4 GHz. A capacitor of value 0.2 pF is introduced is to match the antenna to the 50 Ω impedance seen at the matching circuit output.

Figure 6.21 presents the sensor antenna return loss and the resulting narrowband caused by capacitor addition. The 3D antenna radiation pattern at 2.4 GHz is shown in Figure 6.22. The pattern is different than that of a conventional vertical monopole due to the introduced bend and also the effect of surrounding elements. The antenna gain calculated an numerically is around 2 dB with an efficiency of 77%. This demonstrates the potential reliable application of the sensor in free space with sufficient coverage area obtainable.

6.4.3.2 Wearable Sensor Performance

Body-worn sensor antenna performance is numerically investigated when the sensor is placed on the human chest. A simplified human model which consists of a slab of lossy

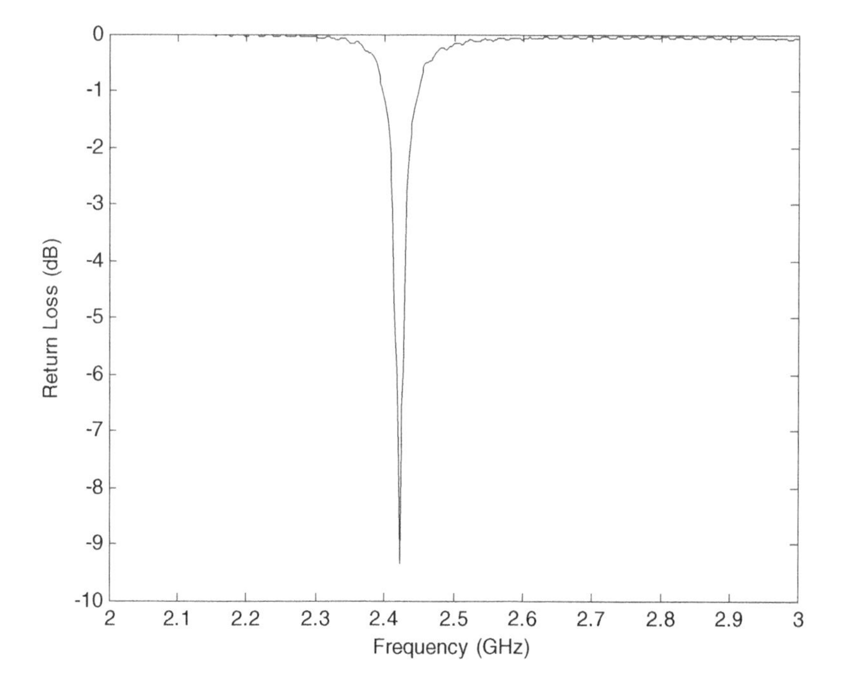

Figure 6.21 Simulated return loss of sensor antenna in free space with resonance around the desired frequency.

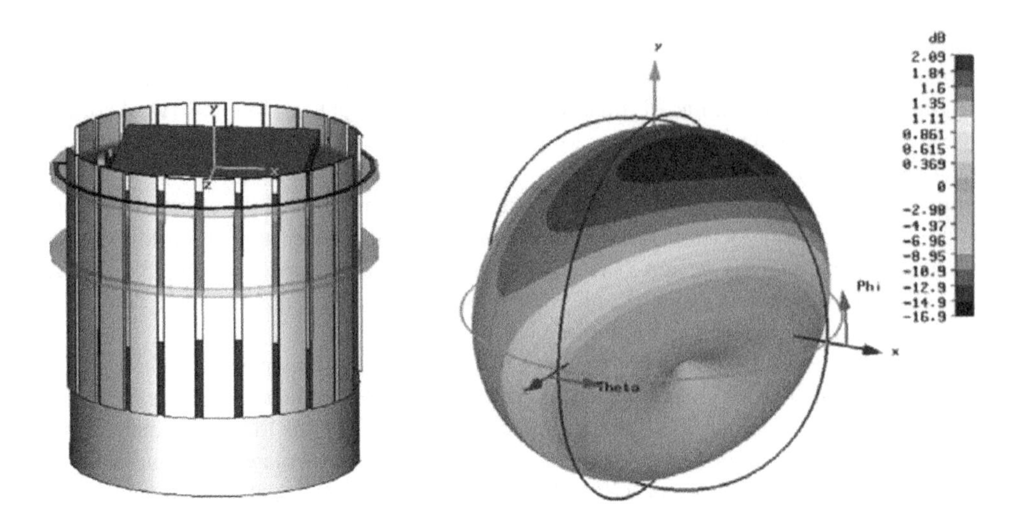

Figure 6.22 Sensor antenna radiation pattern numerically calculated at 2.4 GHz.

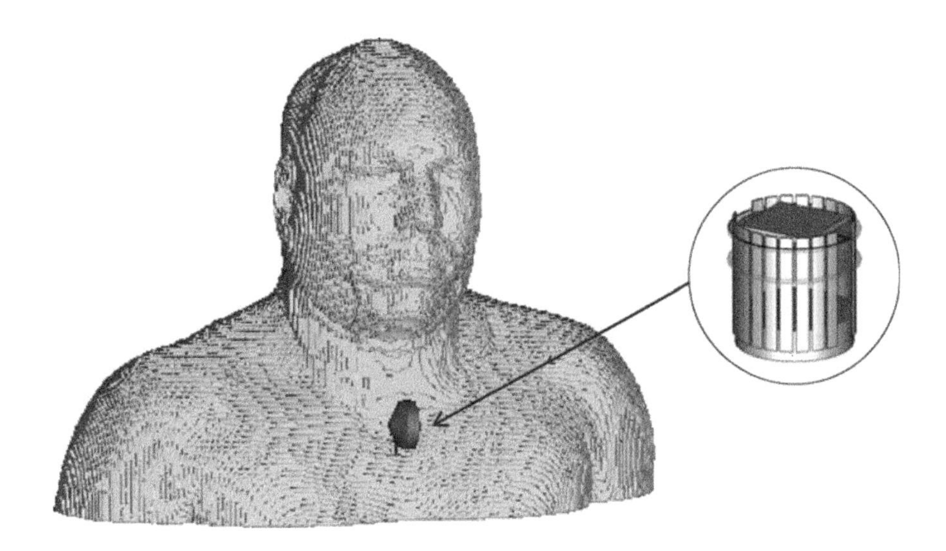

Figure 6.23 Sensor placed on male model.

human is initially applied for cross-referencing and evaluation. The tissue slab measures 120 × 110 × 44 mm with muscle electric properties ($\varepsilon = 52.8$ and $\sigma = 1.7$ S/m at 2.4 GHz). Once satisfactory initial results are obtained, the sensor is placed on the chest of the detailed multi-layer human model, namely the visible male model developed by the US Air Force (http://www.brooks.af.mil/AFRL/HED/hedr/); see Figure 6.23.

The simulated return loss of the sensor shows the slight detuning due to presence of lossy tissue with sensor placed 2 mm away from the body (Figure 6.24). Although the detuning

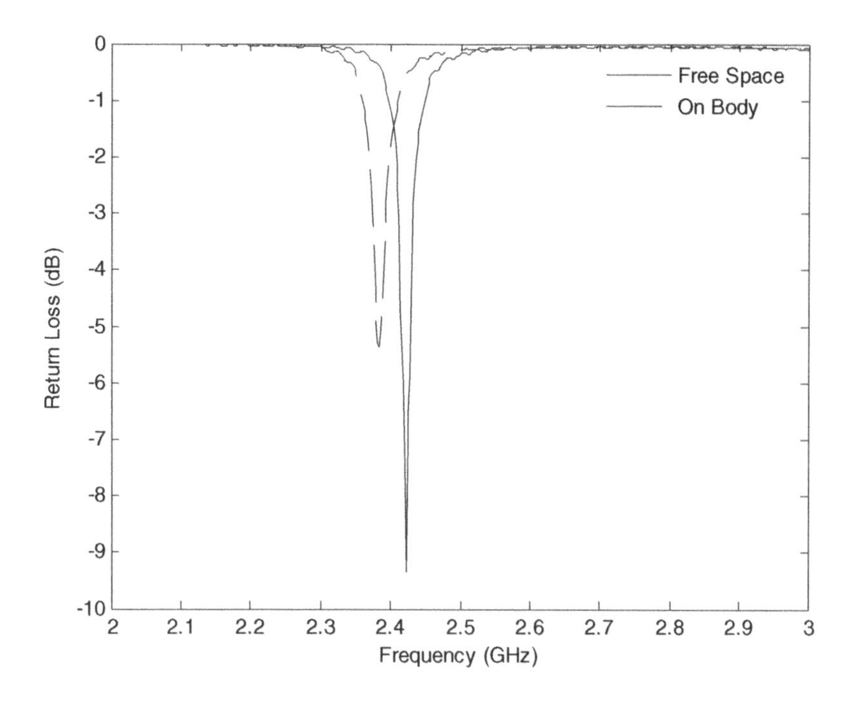

Figure 6.24 Return loss of sensor antenna when placed in free space compared to on-body return loss.

effect is minimal, the narrowband nature of the antenna (with capacitor used for matching) signifies such changes and its direct influence on radiated energy.

The antenna gain when placed on the body is increased to 2.4 dB caused by reflections from the human body, which at this high frequency is considered as a large reflector due to high losses and also very small penetration depth. The azimuth plane radiation patterns shown in Figure 6.25 illustrate the effect of the human body on the antenna performance and the reduced radiated power in the backward direction with front–back ratio of around 2–30 dB.

6.4.4 Propagation Channel Characterization

The angular radiation of the sensor is measured in an anechoic chamber to evaluate the spatial performance of the antenna at various angular directions. The sensor is placed on a turntable with angular capabilities of 0°–360°. The microstrip patch antenna introduced in Figure 6.6 is used as the receiver in the measurement setting shown in Figure 6.26. The patch antenna is placed 88 cm away from the transmitting sensor. The sensor is placed on the turntable in horizontal and vertical orientations. A similar setting is used when the sensor is worn by the user (Figure 6.27). The receiving antenna was connected through a 5 meter long cable to a spectrum analyser set with centre frequency of 2.4 GHz and frequency span of 200 MHz. The received spectrum showed a received signal at 2.405 GHz, which is covered by both sensor and Rx patch antennas.

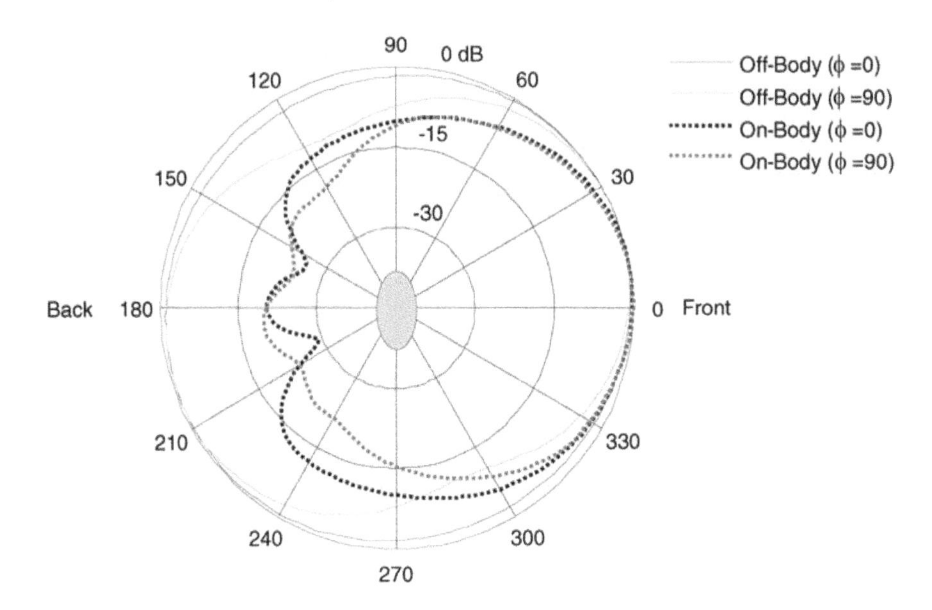

Figure 6.25 Azimuth plane radiation pattern of sensor antenna when placed in free space and on the body.

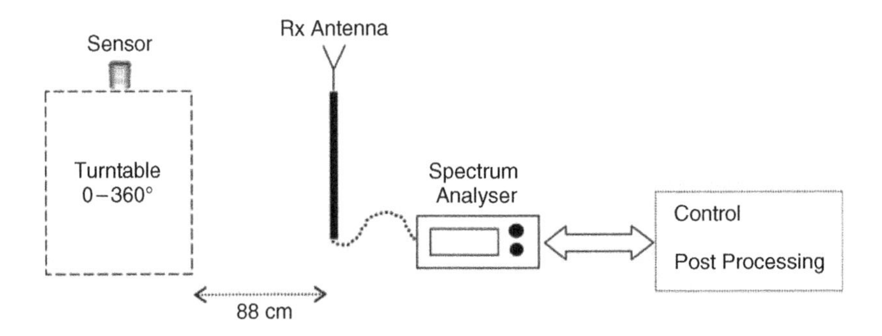

Figure 6.26 Measurement setup for sensor angular pattern performance using a patch antenna as the receiving antenna.

Figure 6.28 shows the results obtained for measurement in both copolar and cross-polar positions of the sensor antenna in free space and also when placed on the body with the antenna parallel to the body. When the body shadows the communication link between Tx and Rx at 180° the loss due to the shadowing is around 18–20 dB.

The angular patterns (Figure 6.28) present reasonable omnidirectional behaviour of the sensor antenna with maximum variation of 8–10 dB for free space cases (off-body). Following the set-up described above, path loss analysis of the radio channel between the Tx sensor and a receiving antenna for cases where the sensor placed is in free space and on the body in the anechoic chamber and in the indoor environment is performed. Figure 6.29 shows the

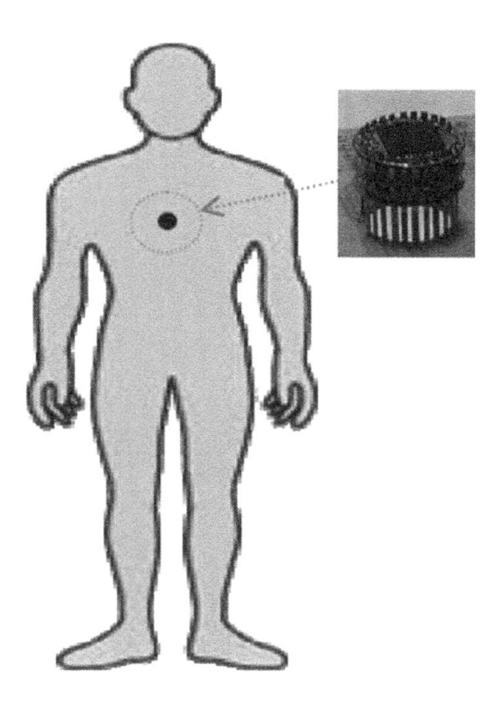

Figure 6.27 Philips test module sensor placed on the body for radio channel characterization measurement.

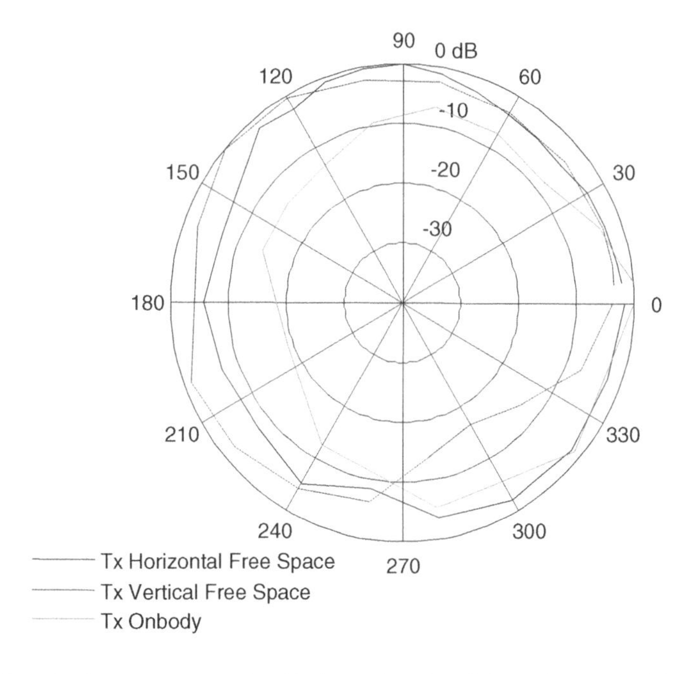

Figure 6.28 Received power pattern when Tx (sensor) is placed 88 cm from a receiving patch antenna for horizontal and vertical sensor placements.

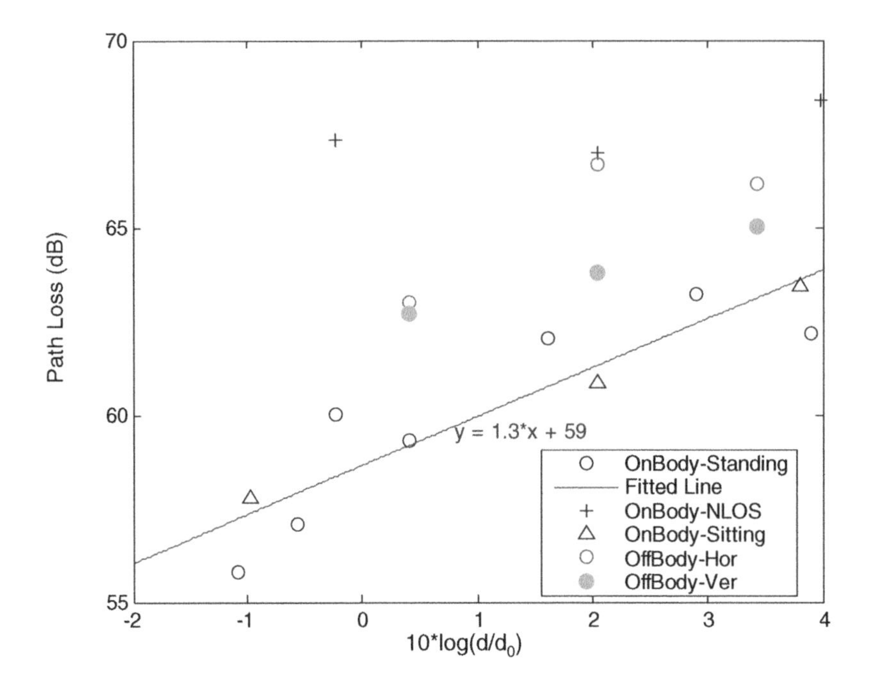

Figure 6.29 Indoor measured path loss when sensor is placed off and on body with modelled path loss using the least fit square technique.

path loss measured in the indoor environment. As predicted, the exponent is lower than that of free space with a value of 1.3 when the sensor is placed on the body due to multipath components from the different scatterers. For similar distances the loss is higher for non-line-of-sight (NLOS) cases. The directivity of the antenna increases when it is placed on the body, as discussed earlier, due to high losses at 2.4 GHz of the human tissue which leads to greater received power for the same distances as applied in the standalone sensor case.

6.5 Summary

Wireless body area networks have been made possible by the emergence of small and lightweight wireless systems such as Bluetooth™ enabled devices and PDAs. Antennas are an essential part of any WBAN system and, due to varying requirements and constraints, careful consideration of their design and deployment is needed.

This chapter introduced wireless body area networks and their progression from WLAN and WPAN to satisfy the demand for more personal systems. The main requirements and features of wearable antennas were presented with regard to design and implementation issues. A review of the latest developments in body-worn antennas and devices provided a clearer picture of the current state of the art and the potential areas for additional investigations and applications. As an inseparable part of the whole communication system, specifically in WBAN, the influence of different antenna parameters and types on the radio propagation channel is of great significance, especially when designing antennas for wearable personal technologies.

A case study on a compact wearable antenna used in sensors designed for healthcare applications was presented. Antenna performance was investigated numerically with regard to impedance matching, radiation patterns, gain and efficiency. The small size of the sensor made it susceptible to variable changes caused by the human body and movements, specifically radiated power, efficiency and the front–back ratio of radiated energy. The antenna performance evaluation and radio propagation characterization provided indications of potential developments in designing optimum performance sensors. Improvements are necessary in antenna design, matching circuitry and also sensor layout for better coverage area and also to achieve the maximum range with respect to the transceiver module.

References

[1] http://grouper.ieee.org/groups/802/15/.

[2] E. Jovanov, A. Milenkovic, C. Otto and P.C de Groen, A wireless body area network of intelligent motion sensors for computer assisted physical rehabilitation. *Journal of NeuroEngineering and Rehabilitation*, March 2005.

[3] S. Park and S. Jayaraman, Enhancing the quality of life through wearable technology. *IEEE Engineering in Medicine and Biology Magazine*, 22 (2003), 41–48.

[4] J. Bernard, P. Nagel, J. Hupp, W. Strauss, and T. von der Grün, BAN – Body area network for wearable computing. Paper presented at 9th Wireless World Research Forum Meeting, Zurich, July 2003.

[5] S. Matsushita, A headset-based minimized wearable computer. *IEEE Intelligent Systems*, 16 (2001), 28–32.

[6] P. Lukowicz, U. Anliker, J. Ward, G. Troster, E. Hirt, C. Neufelt, AMON: a wearable medical computer for high risk patients. *Proceedings of the Sixth International Symposium on Wearable Computers 2002*, Seattle, WA, October 2002, pp. 133–134.

[7] C. Kunze, U. Grossmann, W. Stork, and K. Müller-Glaser, Application of ubiquitous computing in personal health monitoring systems. *Biomedizinische Technik: 36th Annual Meeting of the German Society for Biomedical Engineering*, 2002, pp. 360–362.

[8] C. Balanis, *Antenna Theory Analysis and Design*. New York: John Wiley & Sons, Inc., 1997.

[9] http://niremf.ifac.cnr.it/tissprop/

[10] C. Gabriel and S. Gabriel, Compilation of the dielectric properties of body tissues at RF and microwave frequencies, 1999. http://www.brooks.af.mil/AFRL/HED/hedr/reports/dielectric/Title/Title.html

[11] Federal Communications Commission, First Report and Order, Revision of the Part 15 Commission's Rules Regarding Ultra-Wideband Transmission Systems, ET-Docket 98–153, April 22, 2002.

[12] D. Lamensdorf and L. Susman, Baseband-pulse-antenna techniques. *IEEE Antennas and Propagation Magazine*, 36 (1994), 20–30.

[13] X. Qing and Z.N. Chen, Transfer functions measurement for UWB antenna. *Proceedings of the IEEE Antennas and Propagation Society International Symposium and USNC/URSI National Radio Science Meeting*, Monterey, CA, June 2004.

[14] J.S. McLean, H. Foltz and R. Sutton, Pattern descriptors for UWB antennas. *IEEE Transactions on Antennas and Propagation*, 53 (2005).

[15] Internet resources, Smart textiles offer wearable solutions using Nanotechnology, URL: http://www.fibre2fashion.com/news/

[16] Internet resource, Ubiquitous Communication Through Natural Human Actions, URL: http://www.redtacton.com/en/

[17] B. Sinha, Numerical modelling of absorption and scattering of EM energy radiated by cellular phones by human arms. *IEEE Region 10 International Conference on Global Connectivity in Energy, Computer, Communication and Control*, New Delhi, December 1998, Vol. 2, pp. 261–264.

[18] J. Wang, O. Fujiwara, S. Watanabe, Y. Yamanaka, Computation with a parallel FDTD system of human-body effect on electromagnetic absorption for portable telephone. *IEEE Transactions on Microwave Theory and Techniques*, 52 (2004), 53–58.

[19] H. Adel, R. Wansch and C. Schmidt, Antennas for a body area network. *Proceedings of the IEEE Antennas and Propagation Society International Symposium*, Columbus, OH, June 2003, Vol. 1, pp. 471–474.

[20] Body worn squad level antennas. http://www.natick.army.mil /soldier/media/fact/individual/Antenna_BodyWorn.PDF
[21] Wearable antennas: integration of antenna technologies with textiles for future warrior systems. http://www.natick.army.mil/soldier/media/fact/individual/Antenna_Wearable.html
[22] Harris Broadband Body-Worn Dipole Antenna (30–108 MHz). http://www.rfcomm.harris.com/products/antennas-accessories/
[23] Wearable Antenna Designs LBE Integrated Shoulder Antenna (LISA). http://www.megawave.com/wearable.htm
[24] P. Salonen, L. Sydänheimo, M. Keskilammi, and M. Kivikoski, A small planar inverted-F antenna for wearable applications. *Third International Symposium on Wearable Computers*, 18–19 October 1999, pp. 95–100.
[25] P. Salonen, M. Keskilammi, and L. Sydänheimo, Antenna design for wearable applications. Tampere University of Technology, Finland.
[26] P. Salonen, Y. Rahmat-Samii, H. Hurme and M. Kivikoski, Dual-band wearable textile antenna. *Proceedings of the IEEE Antennas and Propagation Society International Symposium*, Monterey, CA, 20–25 June 2004, Vol. 1, pp. 463–466.
[27] P. Salonen and L. Hurme, A novel fabric WLAN antenna for wearable applications. *Proceedings of the IEEE Antennas and Propagation Society International Symposium*, Columbus, OH, 22–27 June 2003, Vol. 2, pp. 700–703.
[28] C. Cibin, P. Leuchtmann, M. Gimersky, R. Vahldieck and S. Moscibroda, A flexible wearable antenna. *Proceedings of the IEEE Antennas and Propagation Society International Symposium*, Monterey, CA, 20–25 June 2004, Vol. 4, pp. 3589–3592.
[29] A. Tronquo, H. Rogier, C. Hertleer and L. Van Langenhove, Robust planar textile antenna for wireless body LANs operating in 2.45 GHz ISM band. *IEE Electronics Letters*, 42 (2006), 142–143.
[30] M. Klemm, I. Locher and G. Troster, A novel circularly polarized textile antenna for wearable applications. *7th European Conference on Wireless Technology*, 2004, pp. 285–288.
[31] P. Salonen, Y. Rahmat-Samii and M. Kivikoski, Wearable antennas in the vicinity of human body, *Proceedings of the IEEE Antennas and Propagation Society International Symposium*, Monterey, CA, 20–25 June 2004, Vol. 1, pp. 467–470.
[32] Z.N. Chen, A. Cai, T.S.P. See, X. Qing and M.Y.W. Chia, Small planar UWB antennas in proximity of the human head. *IEEE Transactions on Microwave Theory and Techniques*, 54 (2006), 1846–1857.
[33] M. Klemm, I.Z. Kovacs, G.F. Pedersen and G. Troster, Novel small-size directional antenna for UWB WBAN/WPAN applications. *IEEE Transactions on Antennas and Propagation*, 53 (2005), 3884–3896.
[34] A. Alomainy, Y. Hao, A. Owadally, C.G. Parini, Y. Nechayev, P.S. Hall and C.C. Constantinou, Statistical analysis and performance evaluation for on-body radio propagation with microstrip patch antennas. *IEEE Transactions on Antennas and Propagation.*
[35] A. Alomainy, Y. Hao, C. G. Parini and P.S. Hall, Characterisation of printed UWB antennas for on-body communications. *IEE Wideband and Multi-band Antennas and Arrays*, Birmingham, UK, 7 September 2005.
[36] Y. Zhao, Y. Hao, A. Alomainy and C.G. Parini, UWB on-body radio channel modelling using ray theory and sub-band FDTD method. *IEEE Transactions on Microwave Theory and Techniques*, Special Issue on Ultra-Wideband, 54 (2006), 1827–1835.
[37] A. Alomainy, Y. Hao, X. Hu, C.G. Parini and P.S. Hall, UWB on-body radio propagation and system modelling for wireless body-centric networks. *IEE Proceedings Communications*, Special Issue on Ultra Wideband Systems, Technologies and Applications, 153 (2006).
[38] T. Zasowski, F. Althaus, M. Stager, A. Wittneben, and G. Troster, UWB for noninvasive wireless body area networks: channel measurements and results. *Proceedings of the IEEE Conference on Ultra Wideband Systems and Technologies*, Reston, VA, November 2003, pp. 285–289.
[39] J. Ryckaert, P. De Doncker, R. Meys, A. de Le Hoye and S. Donnay, Channel model for wireless communication around human body. *Electronics Letters*, 40 (2004), 543–544.
[40] A. Fort, C. Desset, J. Ryckaert, P. De Doncker, L. Van Biesen and S. Donnay, Ultra wideband body area channel model. *International Conference on Communications*, Seoul, May 2005.
[41] A. Fort, C. Desset, J. Ryckaert, P. De Doncker, L. Van Biesen and P. Wambacq, Characterization of the ultra wideband body area propagation channel. *International Conference on Ultra-WideBand*, Zurich, September 2005.
[42] X. Qing and Z.N. Chen, Transfer functions measurement for UWB antenna. *IEEE Antennas and Propagation Society International Symposium and USNC/URSI National Radio Science Meeting*, Monterey, CA, June 2004.

[43] A. Alomainy, Y. Hao, C.G. Parini and P.S. Hall, Comparison between two different antennas for UWB on-body propagation measurements. *IEEE Antennas and Wireless Propagation Letters*, 4 (2005), 31–34.
[44] A. Alomainy and Y. Hao, Radio channel models for UWB body-centric networks with compact planar antenna. *Proceedings of the IEEE Antennas and Propagation Society International Symposium*, Albuquerque, NM, 9–14 July 2006.
[45] P.S. Hall and Y. Hao, *Antennas and Propagation for Body-Centric Wireless Networks*. Boston: Artech House, 2006.
[46] Chipcon CC2420 transceiver chip, 2.4 GHz IEEE 802.15.4 / ZigBee-ready RF Transceiver, URL: http://www.chipcon.com/files/CC2420_Data_Sheet_1_4.pdf

7

Antennas for UWB Applications

Zhi Ning Chen and Terence S.P. See
Institute for Infocomm Research, Singapore

Ultra-wideband (UWB) is one of the most promising technologies for future high-data-rate wireless communications, high-accuracy radars, and imaging systems. Compared with conventional broadband wireless communication systems, the UWB system operates within an extremely wide bandwidth in the microwave band and at a very low emission limit. Due to the system features and unique applications, antenna design is facing a variety of challenging issues such as broadband response in terms of impedance, phase, gain, radiation patterns as well as small or compact size. This chapter will address the antenna design issues in UWB systems. First, the UWB technology and regulatory environment is briefly introduced; general information on UWB systems is provided. Next, the challenges in UWB antenna design are described. The special design considerations for UWB antennas are summarized. State-of-the-art UWB antennas are also reviewed. UWB antennas for fixed and mobile devices are presented. Finally, a new concept for the design of a small UWB antenna with reduced ground-plane effect is introduced and applied to a practical scenario where a small printed UWB antenna is installed on a laptop computer.

7.1 UWB Wireless Systems

The term 'ultra-wideband' (UWB) usually refers to a technology for the transmission of information spread over an extremely large operating bandwidth where the electronic systems should be able to coexist with other electronic users. UWB technology has been around for decades. Its original applications were mostly in military systems. However, the first Report and Order by the Federal Communications Commission (FCC) authorizing the unlicensed use of UWB on February 14, 2002, gave a huge boost to the research and development efforts of both industry and academia [1]. The intention is to provide an efficient use of scarce frequency spectra, while enabling short-range but high-data-rate wireless personal area network (WPAN) and long-range but low-data-rate wireless connectivity applications, as well as radar and imaging systems, as shown in Table 7.1.

Antennas for Portable Devices Zhi Ning Chen

Table 7.1 Frequency ranges for various types of UWB systems under −41.3 dBm EIRP emission limits [1]

Applications	Frequency range (GHz)
Indoor communication systems	3.1–10.6
Ground-penetrating radar, wall imaging	3.1–10.6
Through-wall imaging systems	1.61–10.6
Surveillance systems	1.99–10.6
Medical imaging systems	3.1–10.6
Vehicular radar systems	22–29

According to Part 15.503 of the FCC rules, the following technical terms can be defined for UWB operation.

- *UWB bandwidth* is the frequency range bounded by the points that are 10 dB below the highest power emission with the upper edge f_h and the lower edge f_l. Thus, the *center frequency* f_c of the UWB bandwidth is designated as

$$f_c = \frac{f_h + f_l}{2}. \tag{7.1}$$

Accordingly, the *fractional bandwidth BW* is defined as

$$\begin{aligned} BW &= 2\frac{f_h - f_l}{f_h + f_l} \times 100\% \\ &= \frac{f_h - f_l}{f_c} \times 100\%. \end{aligned} \tag{7.2}$$

- *A UWB transmitter* is an intentional radiator that, at any point in time, has a fractional bandwidth BW of at least 20 % or has a UWB bandwidth of at least 500 MHz, regardless of the fractional bandwidth.
- *Effective isotropically radiated power* (EIRP) represents the total effective transmit power of the radio, i.e. the product of the power supplied to the antenna with possible losses due to an RF cable and the antenna gain in a given direction relative to an isotropic antenna. The EIRP, in terms of dBm, can be converted to the field strength, in dBμV/m at 3 meters, by adding 95.2. With regard to this part of the rules, EIRP refers to the highest signal strength measured in any direction and at any frequency from the UWB device, as tested in accordance with the procedures specified in Part 15.31(a) and 15.523 of the FCC rules.

The emission limit masks are regulated by the regulators such as the FCC as shown in Figure 7.1. The emission power limits are lower than the noise floor in order to avoid possible interference between UWB devices and existing electronic systems. The masks vary in different regions, but the maximum emission levels are always kept lower than −41.3 dBm/MHz.

Furthermore, according to the FCC, any transmitting system which emits signals having a bandwidth greater than 500 MHz or 20 % fractional bandwidth can gain access to the UWB

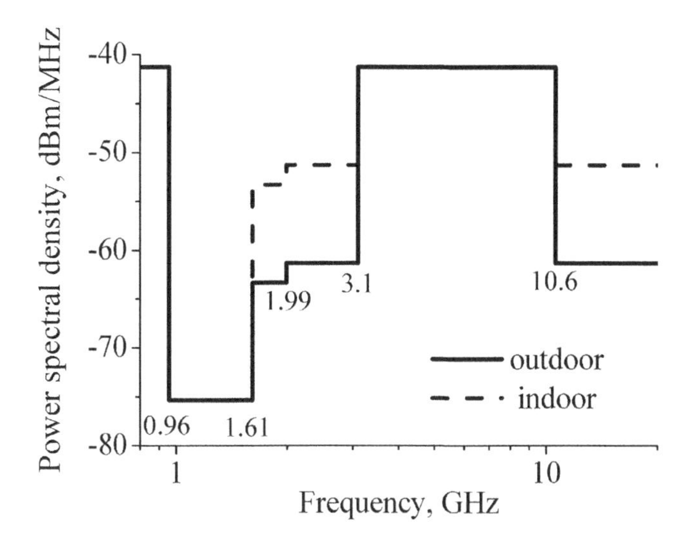

Figure 7.1 Emission limit masks for indoor and outdoor UWB applications.

spectrum. Thus, both the traditional pulse-based systems transmitting each pulse which entirely or partially occupies the UWB bandwidth, and the carrier systems based on, for instance, the orthogonal frequency-division multiplexing (OFDM) method with a collection of narrowband carriers of at least 500 MHz can utilize the UWB spectrum under the FCC's rules.

The extremely large spectrum provides the room to use extremely short pulses in the order of picoseconds. Thus, the pulse repetition or data rates can be low or very high, typically several gigapulses per second. The pulse rates are dependent on the applications. For instance, radar and imaging systems prefer low pulse rates in the range of a few megapulses per second. Pulsed or OFDM communication systems tend to use high data rates, typically in the range of 1–2 gigapulses per second, to achieve gigabit-per-second wireless connection, although the communication range may be very short, typically a few meters. However, the use of high data rates can enable the efficient transfer of data from digital camcorders, wireless printing of digital pictures from a camera without the need for an intervening personal computer, as well as the transfer of files among cellphones and other handheld devices such as personal digital audio, video players, and laptops.

7.2 Challenges in UWB Antenna Design

One of the challenges for the implementation of UWB systems is the development of a suitable or optimal antenna. From a systems point of view, the response of the antenna should cover the entire operating bandwidth. The response or specifications of an antenna will vary according to system requirements. Therefore, it is important for an antenna engineer to be familiar with the requirements of the system before designing the antenna.

Generally, in UWB antenna design, both the frequency and time-domain responses should be taken into account. The frequency-domain response includes impedance, radiation, and transmission. The impedance bandwidth is measured in terms of return loss or voltage standing wave ratio (VSWR). Usually, the return loss should be less than $-10\,\mathrm{dB}$ or

VSWR < 2:1. An antenna with an impedance bandwidth narrower than the operating bandwidth tailors the spectrum of transmitted and received signals, acting as a bandpass filter in the frequency domain, and reshapes the radiated or received pulses in the time domain. The radiation performance includes radiation efficiency, radiation patterns, polarization, and gain. The radiation efficiency is an important parameter especially for small antenna design, where it is difficult to achieve impedance matching due to small radiation resistance and large reactance. For a small antenna with weak radiation directivity, the radiation efficiency is of greater practical interest than the gain. The radiation patterns show the directions where the signals will be transmitted.

Different from narrowband and conventional broadband systems, the requirements of the antennas are dependant on modulation schemes. So far, two modulation schemes, namely the multiple-carrier OFDM and pulsed direct sequence code division multiple access (DS-CDMA) have been proposed for high-data-rate wireless communications. In these schemes, the UWB band can be occupied in different ways. Figure 7.2 illustrates the spectra of the OFDM and pulse-based UWB systems, which are compliant with the FCC's emission limit masks for indoor and outdoor applications. For instance, the emission mask can be divided into 15 sub-bands, with each band having a bandwidth of 500 MHz as shown in Figure 7.2(a). Alternatively, the entire UWB band can be occupied by a single pulse or several pulses, as shown in Figure 7.2(b).

In order to coexist with the devices based on IEEE 802.11a (UNII) within the operating frequency range of 5.150-5.825 GHz, some methods have been applied in such UWB systems. In an OFDM-based UWB system, the sub-bands falling in the UNII range, namely the fourth, fifth and sixth lower sub-bands in Figure 7.2(a), can be suspended. In a pulse-based UWB system, by modulating the pulses with carriers, the spectrum can be notched to solve the possible interference problem as depicted in Figure 7.2(b). In the figure, the spectrum can be notched at 5-6 GHz by modulating the pulses at the carrier frequencies of 4 GHz and 8.5 GHz.

Due to the different occupancy of the UWB band in the two types of UWB system shown in Figure 7.2, the considerations for selection of the source pulses and templates, and design of antennas are distinct, as discussed by Chen *et al.* [2]. Chen *et al.* concluded that the response of an antenna to UWB pulses can be described in terms of its temporal characteristics, while it may be more intuitive for antenna engineers to consider the antenna performance in the frequency domain [2]. In the frequency domain, an ideal UWB antenna is required to work well across the entire UWB band with acceptable radiation efficiency, gain, return loss, radiation pattern and polarization.

In an OFDM-based system, each sub-band having a few hundred megahertz (larger than 500 MHz) can be considered as broadband. Within the sub-bands, the effect of non-linearity of the phase shift on the reception performance can be ignored because the phase varies very slowly with frequency. Therefore, the design of the antenna is more focused on achieving constant frequency response in terms of the radiation efficiency, gain, return loss, radiation patterns, and polarization over the operating band, which may fully or partially cover the UWB bandwidth of 7.5 GHz.

For pulse-based systems, in order to prevent the distortion of the received pulses, an ideal UWB antenna should produce radiation fields of constant magnitude and a phase shift that varies linearly with frequency.

By way of comparison, four types of antenna are shown in Figure 7.3: a thin strip dipole antenna operating with a narrow bandwidth (which we will refer to as antenna A); a diamond

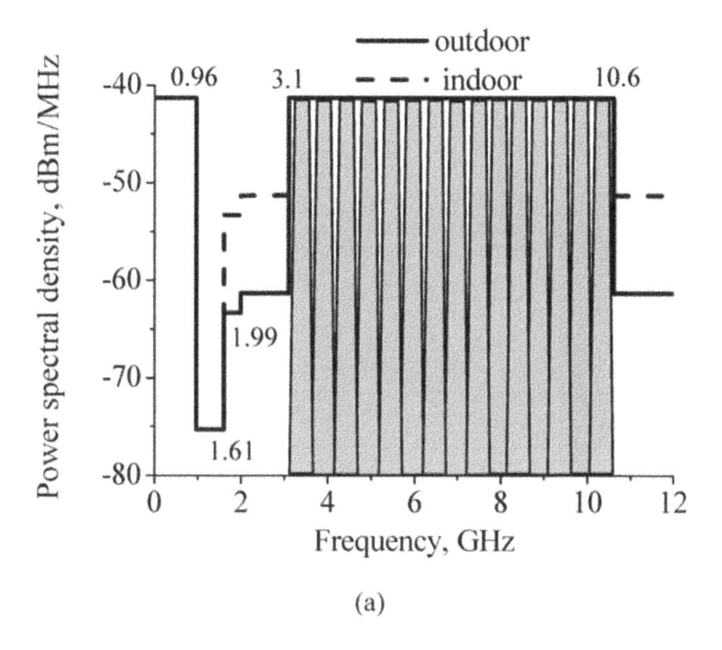

(a)

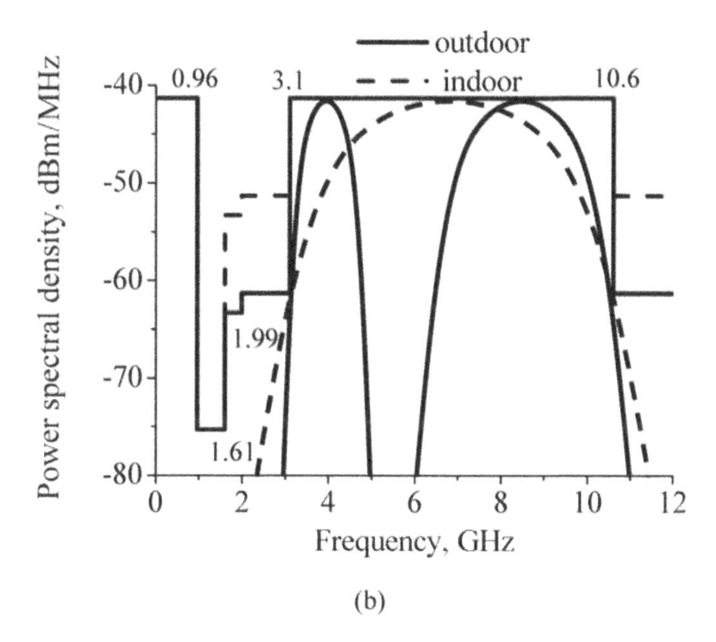

(b)

Figure 7.2 Spectra of OFDM and pulse-based UWB systems compliant with the FCC's emission limit masks for indoor and outdoor UWB applications.

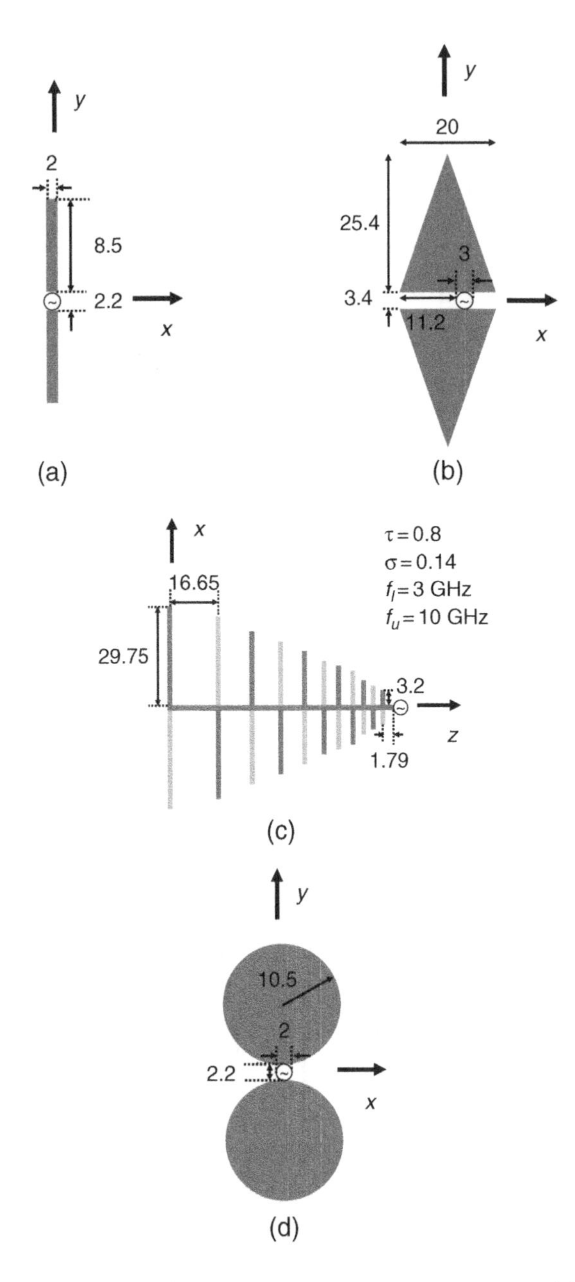

Figure 7.3 Four types of antennas. (a) Antenna A: thin strip dipole antenna; (b) Antenna B: diamond dipole antenna; (c) Antenna C: log-periodic antenna; and (d) Antenna D: circular dipole antenna. Dimensions in millimetres.

dipole antenna having a broad operating bandwidth (B); a typical log-periodic antenna with high gain and a broad operating bandwidth (C); and a circular dipole antenna with a very broad operating bandwidth (D). The spectral and temporal characteristics of these antennas are compared using Zeland IE3D, an electromagnetic simulator based on the method of

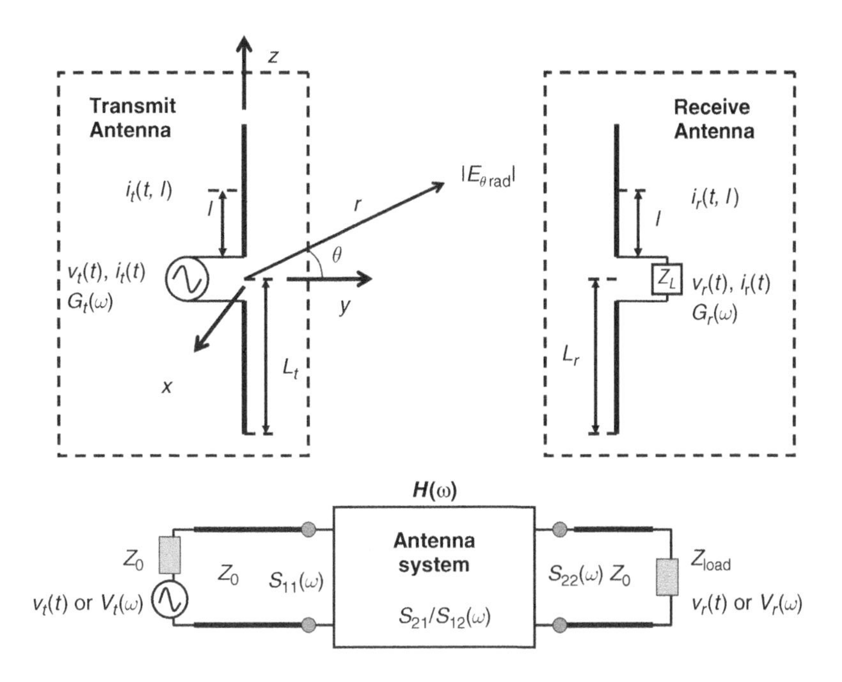

Figure 7.4 A transmit–receive antenna system.

moments. In the comparison, the transfer function can be defined using the system shown in Figure 7.4. As mentioned in [2], it is clear that the UWB system response between the transmit and receive antennas is frequency-dependent. The conventional Friis transmission formula is modified as follows:

$$\frac{P_r(\omega)}{P_t(\omega)} = \left(1 - |\Gamma_t(\omega)|^2\right)\left(1 - |\Gamma_r(\omega)|^2\right) G_r(\omega) G_t(\omega) |\hat{\rho}_t(\omega).\hat{\rho}_r(\omega)|^2 \left(\frac{\lambda}{4\pi r}\right)^2. \tag{7.3}$$

The relationship between the source and output signal (voltage) can be written as:

$$\begin{aligned} &[V_t(\omega)/2]^2/2[= P_t(\omega)Z_0], \\ &V_r^2(\omega)/2[= P_r(\omega)Z_{\text{load}}]. \end{aligned} \tag{7.4}$$

Thus, the transfer function in (7.3) can be simplified to give

$$\begin{aligned} H(\omega) &= \frac{V_r(\omega)}{V_t(\omega)} = \left|\sqrt{\frac{P_r(\omega)}{P_t(\omega)} \frac{Z_{\text{load}}}{4Z_0}}\right| e^{-j\phi(\omega)} = |H(\omega)|\, e^{-j\phi(\omega)} \\ \phi(\omega) &= \phi_t(\omega) + \phi_r(\omega) + \omega r/c \end{aligned} \tag{7.5}$$

where c denotes velocity of light, and $\phi_t(\omega)$ and $\phi_r(\omega)$ are the phase shift due to the transmit and receive antennas, respectively. As a result, if the effect of the RF channel is not taken into account, the transfer function $H(\omega)$ is determined by the characteristics of both transmit and receive antennas, such as impedance matching, gain, polarization matching, the distance between the antennas, and the orientation of the antennas. Therefore, the transfer function $H(\omega)$ can be used to describe general antenna systems, which may be dispersive.

Furthermore, the transmit-receive antenna system can be considered as a two-port network. The transfer function $H(\omega)$ can be measured in terms of S_{21} when the source impedance and load are matched to the antenna input and output, respectively. This implies that the measurable parameter S_{21} or $H(\omega)$ is able to integrate all the important system parameters in terms of gain, impedance matching, polarization matching, path loss, and phase delay. Therefore, they can be used to assess the performance of UWB antenna systems and other antenna systems whose performance is frequency-dependent.

In the measurement of $H(\omega)$, the orientations of the transmit and receive antennas are shown in Figure 7.5. Identical antennas are used as transmit and receive antennas in the test setup shown in the figure. Figure 7.5(a) shows a pair of antennas B with a separation

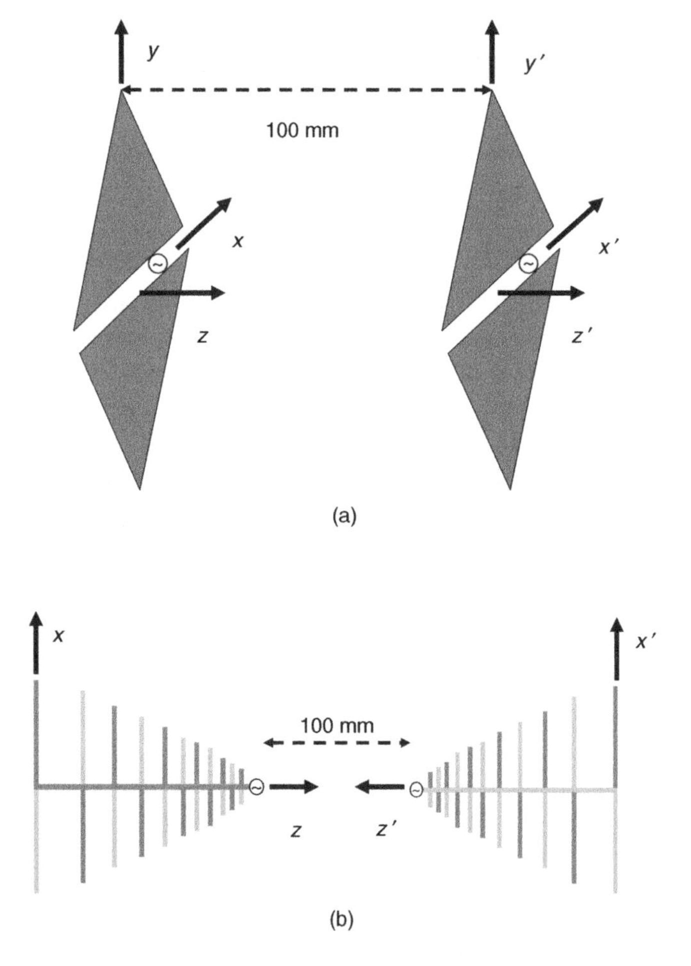

Figure 7.5 Orientation of antennas: (a) antenna B (antennas A and D placed in the same position); (b) antenna C.

of 100 mm and positioned in parallel and face-to-face. Similarly, antennas A and D are positioned face-to-face in the same orientation with a separation of 100 mm, while a pair of antennas C is placed tip-to-tip with a separation of 100 mm as shown in Figure 7.5(b).

Figure 7.6 demonstrates the return losses $|S_{11}|$ and the magnitude of the transfer function $|S_{21}|$ for antennas A – D. It is clear from Figure 7.6(a) that antenna A has a narrow-band impedance and transmission response. The antenna is well matched around the center frequency of the UWB band (7 GHz). Around 6–7 GHz, the transmission reaches its peak and subsequently decreases gradually.

From Figure 7.6(b), it is evident that the impedance bandwidth of antenna B, for 10 dB return loss, covers the whole UWB band very well. However, the 10 dB bandwidth for the transmission only ranges from 2 to 6 GHz, partially covering the lower range of the UWB band.

Antenna C displays broadband characteristics for both impedance and transmission, as shown in Figure 7.6(c). Compared with the other three antenna designs, antenna C is much larger in size and more directional in radiation with a high system gain of −18 to −15 dB.

Antenna D shows broad impedance bandwidth covering the entire UWB band and broad transmission coverage from 2 to 8.5 GHz within a 10 dB variation (Figure 7.6(d)).

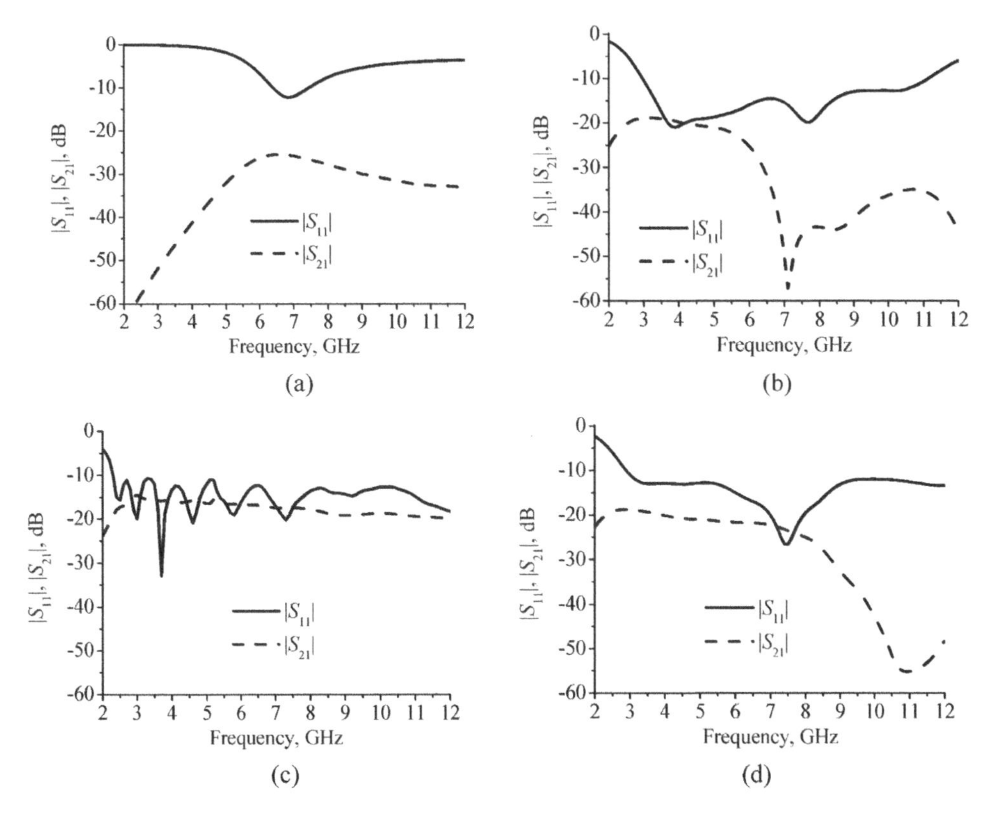

Figure 7.6 Comparison of the return loss $|S_{11}|$ and of the transfer function $|S_{21}|$ for (a) antenna A; (b) antenna B; (c) antenna C; (d) antenna D.

The comparison of the system gain for transmission shows that antenna C has the highest peak gain of −15 dB, whereas antenna A has the lowest peak gain of −25 dB. Antennas B and D have the same peak gain of −18 dB.

Due to the ultra-wide operating bandwidth of UWB systems, the phase response of the transmission may not be linear. This feature differentiates the design considerations for the pulse-based UWB antennas from those in narrowband systems and OFDM-based UWB systems [2]. The non-linear phase response may severely distort the waveforms of short pulses in the form of ringings. The phase responses of $|S_{21}|$ for antennas A – D are shown in Figure 7.7. The non-linear phase response of antenna C over the UWB band is demonstrated in Figure 7.7(c). The phase centers shift with frequency because the main radiation at a frequency f_r always occurs at the dipole with a length of around a half-wavelength at f_r. Therefore, the phase centers at lower operating frequencies are located around longer dipoles, and conversely around shorter dipoles at higher operating frequencies.

Figure 7.8 shows the radiation patterns for antennas A – D at 3, 5, 7, and 9 GHz. Antennas A, B, and D are basically dipole antennas and show typical radiation characteristics especially at the lower operating frequencies, namely an omnidirectional radiation in the horizontal plane

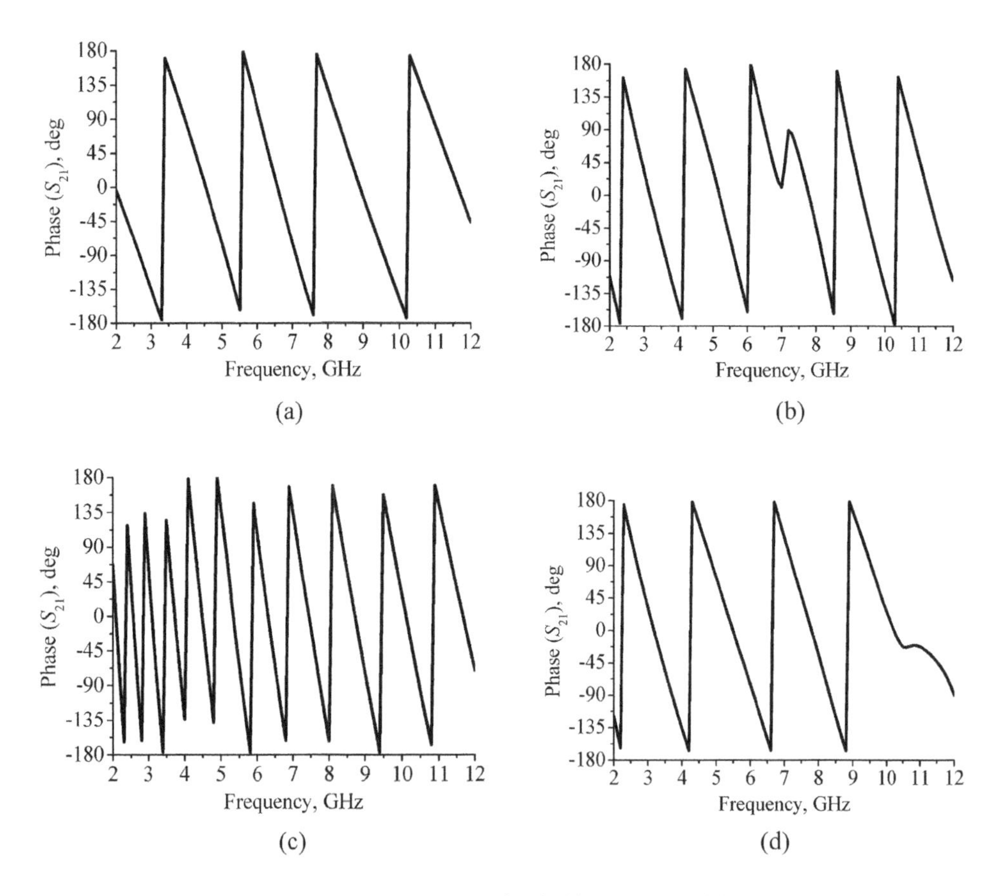

Figure 7.7 Comparison of the phase response of $|S_{21}|$: (a) antenna A; (b) antenna B; (c) antenna C; and (d) antenna D.

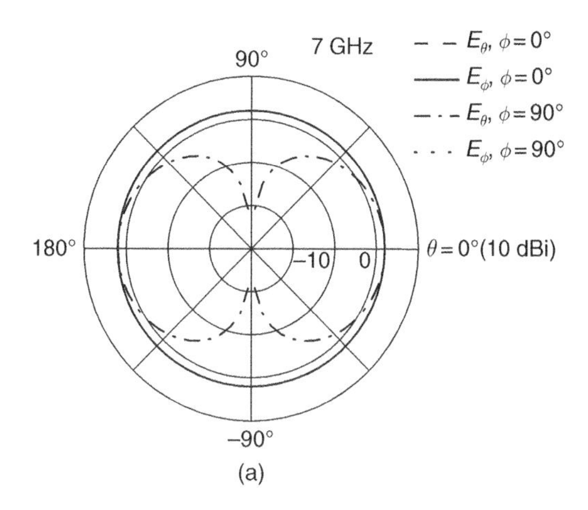

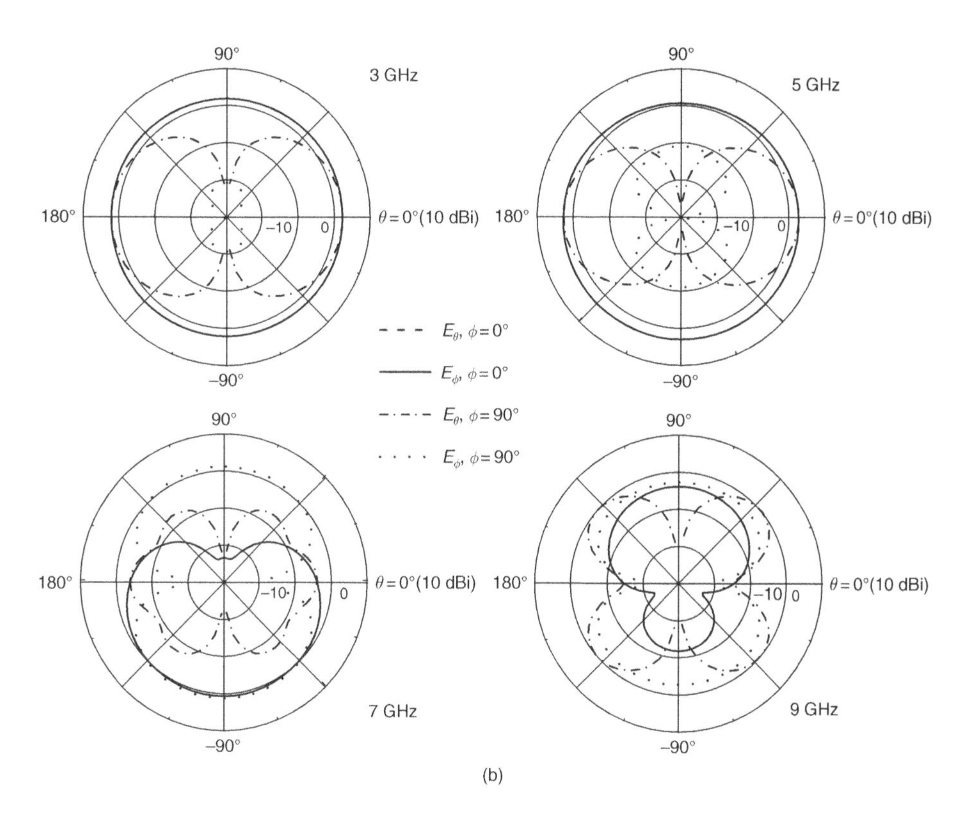

Figure 7.8 Comparison of the radiation patterns at 3, 5, 7, and 9 GHz: (a) antenna A; (b) antenna B; (c) antenna C; (d) antenna D.

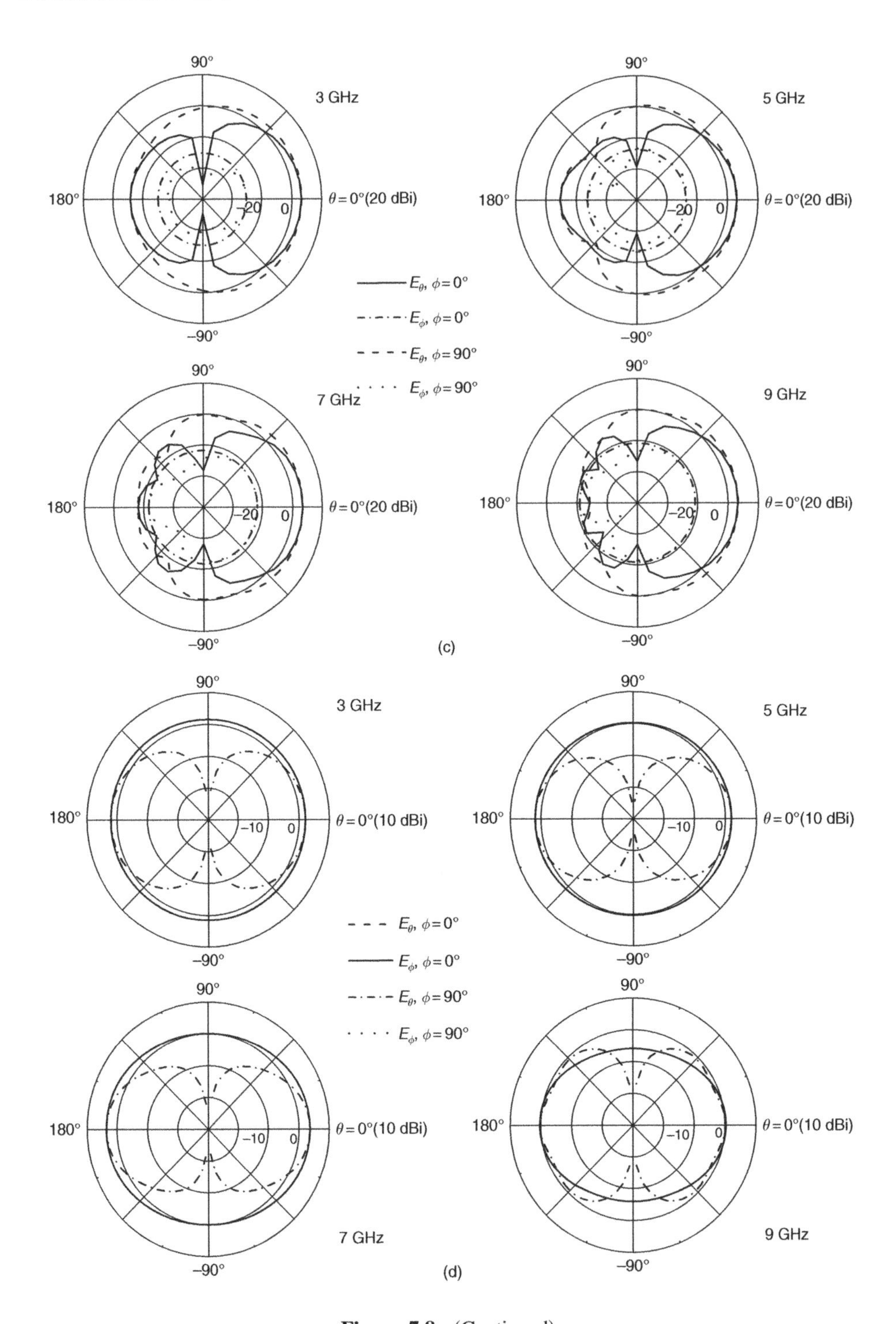

Figure 7.8 (Continued).

and inverted figure-of-eight radiation in the vertical planes. However, at higher frequencies the radiation patterns of the broadband antennas B and D vary significantly because the antenna size has become electrically large. In the horizontal plane, the radiation becomes directional. The radiation patterns of antenna B have nulls in the horizontal plane at 7 and 9 GHz. This has narrowed the transmission response of antenna B to a great extent, as shown in Figure 7.6(b) along the z-axis direction. As mentioned above, antenna C is a directional antenna with a high and stable gain along its tip direction.

In order to examine the effects of performance of the transmit and receive antennas on the received signals in a pulsed system, the impulse responses of the four antenna systems are illustrated in Figure 7.9. Because antenna A has the narrowest bandwidth and lowest gain, the pulse received has the lowest magnitude and experiences more ringing. For antenna C, with the broadest bandwidth and highest gain but a non-linear phase response, the received pulse has a higher magnitude but experiences the most severe distortion with significant ringing. Comparing the waveforms of the pulses received by antennas B and D with acceptable gain, it is evident that the magnitude of the received pulse

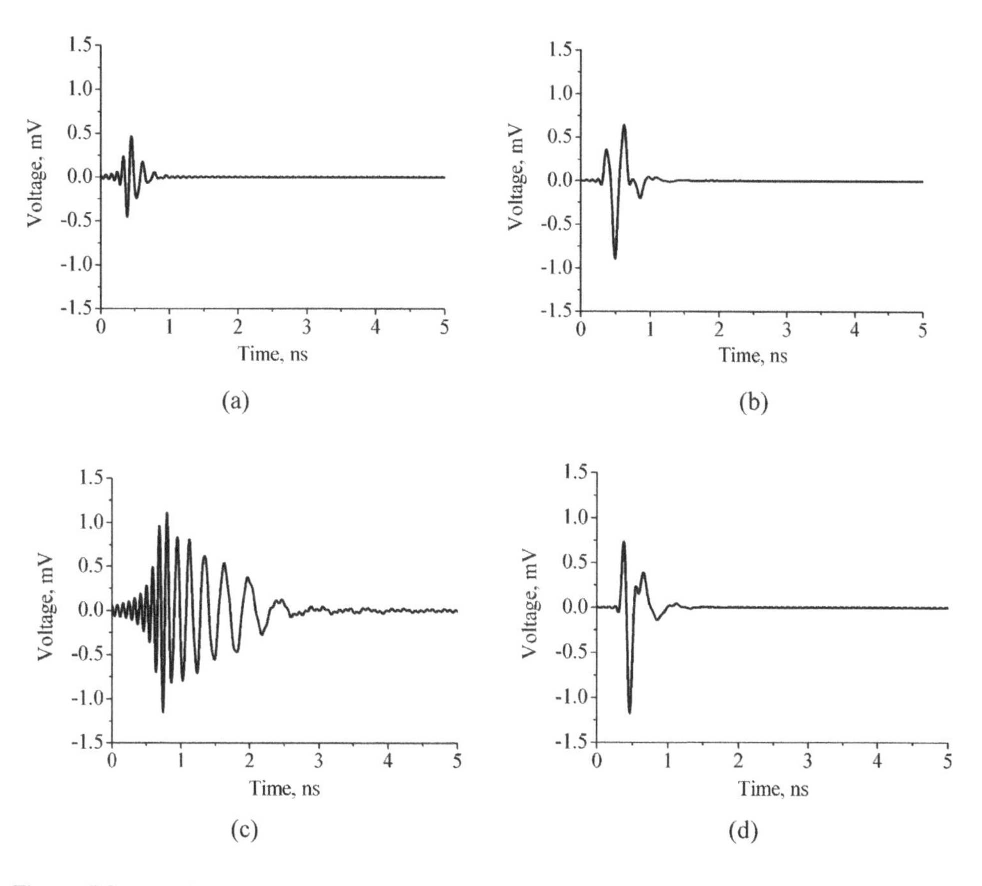

Figure 7.9 Impulse response at the load of the receive antenna: (a) antenna A; (b) antenna B; (c) antenna C; (d) antenna D.

Table 7.2 Summary of antenna performance.

Antenna	Impedance bandwidth	System gain/bandwidth	Phase response	Suitable for systems OFDM/pulsed
A	Narrow	Lowest/narrowest	Linear	No/No
B	UWB band	Acceptable/narrower	Linear	No[a]/Yes
C	UWB band	Highest/widest	Non-linear	Yes/No
D	UWB band	Acceptable/acceptable	Linear	Yes/Yes

[a] The bandwidth of the system gain is still broad and covers part of the UWB band, for example, the lower portion of the UWB band of 3.1–5 GHz, which is widely used in high-speed/short-range mobile devices.

of antenna D is 30 % higher than that of antenna B, and they both have less ringing than antenna C. A summary of the performance of antennas A–D is given in Table 7.2.

The assessment of the antennas should be performed from an overall systems point of view and not just that of an antenna element. As such, there are three parameters, namely the fidelity, system gain, and EIRP bandwidth, which can be used to analyze the performance of the antenna in transmit – receive antenna systems. In order to illustrate the concept of the three parameters more clearly, we will take the circular dipole antenna (D) as an example.

In this study, the sine modulated Gaussian pulse has been chosen as the source pulse. The monocycle pulses are modulated at frequencies of $f_s = 4, 7,$ and 8.5 GHz, which corresponds closely to the center frequency of the lower UWB band (3.1–5 GHz), the entire UWB band (3.1–10.6 GHz), and the upper UWB band (6–10.6 GHz), respectively. With optimization, the pulse parameter σ for the template pulse as well as the modulation frequency of the template pulse can be obtained for a fixed modulation frequency of the source pulse f_s.

First, the fidelity parameter is used to measure the performance of the pulsed UWB systems in [2, 3]. The fidelity of the pulses of an antenna system can be calculated to assess the quality of a received pulse and select a proper detection template. The fidelity can be defined as

$$F = \max_{\tau} \int_{-\infty}^{\infty} L[p_{\text{source}}(t)] p_{\text{output}}(t - \tau) dt \tag{7.6}$$

where the source pulse $p_{\text{source}}(t)$ and output pulse $p_{\text{output}}(t)$ are normalized by their respective energies. The fidelity F is the maximum value of the integral by varying time delay τ with respect to the source pulse. The linear operator $L[\cdot]$ operates on the input pulse $p_{\text{source}}(t)$. Evidently, the template at the output of a receiving antenna may most probably be $L[p_{\text{source}}(t)]$ and not $p_{\text{source}}(t)$ in order to achieve maximum fidelity.

Figure 7.10 plots the fidelity obtained for different modulation frequencies of the source pulse f_s. With a proper selection of the source pulse parameter σ and the modulation frequency f_t for the template pulse, maximum fidelity can be achieved. It should be noted that in this case the antenna has a broad operating bandwidth so that the distortion of the waveforms of the received pulses is slight. As a result, the frequency f_t is very close to the

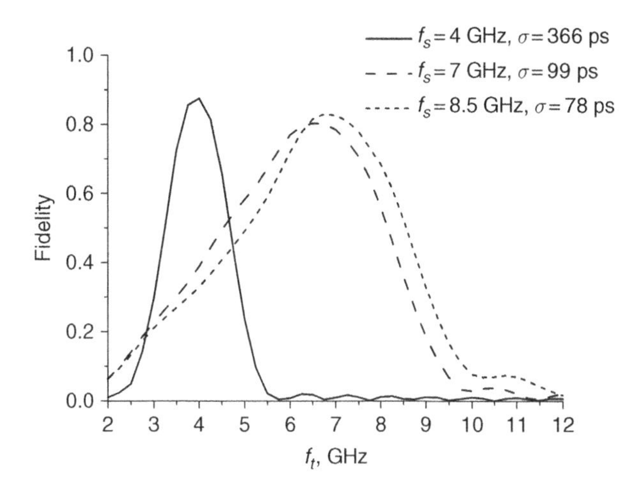

Figure 7.10 Comparison of fidelity against template pulse.

frequency f_s. Otherwise, the difference between the frequencies f_s and f_t will be large due to the severe distortion of the waveforms of the received pulses [2].

Also, the source pulse has been properly chosen such that the radiated spectra are able to comply with the FCC's emission limits as shown in Figure 7.11. Therefore, in the selection and optimization of the source and template pulses, the compliance with the FCC's emission limit masks and maximum fidelity are two key factors which should be taken into consideration.

The system gain parameter can be used to evaluate the efficiency of a transmit-receive antenna system and can be calculated from

$$G_{\text{sys}} = 16\pi r^2 \frac{R_0 \int_0^\infty V_r^2(t, \theta, \phi, \theta', \phi')dt}{R_{\text{load}} \int_0^\infty V_t^2(t)dt} \tag{7.7}$$

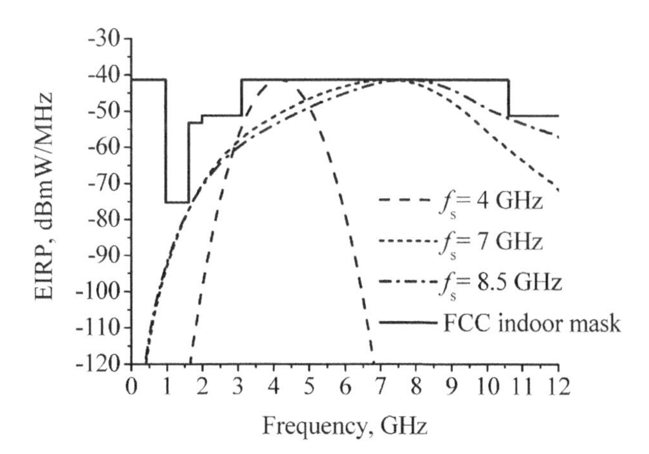

Figure 7.11 Spectra of radiated pulses for different source pulses.

where V_t and V_r are the source and output voltages, respectively; R_0 and R_{load} are the source and load resistances, respectively; θ, ϕ indicate the elevation and azimuth angles of the transmitting antenna; θ', ϕ' denote the elevation and azimuth angles of the receiving antenna; and r is the distance between the antennas. Clearly, G_{sys} not only depends on the impedance matching of the antennas but also on their directivities, which are angle-dependent.

The EIRP bandwidth is defined as the bandwidth within which the EIRP of the radiated spectrum is 10 dB below the maximum. This is the parameter used to measure the efficiency of occupation of the operating bandwidth. In this case, since the radiated spectrum is to be within the indoor emission limits from 3.1 to 10.6 GHz, the EIRP bandwidth will not exceed 7.5 GHz.

Figure 7.12 plots the variation of the system gain and EIRP bandwidth against the source pulse parameter σ. The lower bound for σ is set at 366, 99, and 78 ps for $f_s = 4, 7$, and 8.5 GHz, respectively. It should be noted that any value less than the lower bound will result in the failure of the radiated spectrum to conform to the FCC's emission limits. Due to the variation of the antenna gain along the z-axis as shown in Figure 7.6(d), the optimum system gain varies from $-33\,\text{dB*m}^2$ (for $f_s = 8.5$ GHz) to $-28.5\,\text{dB*m}^2$ (for $f_s = 4$ GHz).

Figure 7.13 plots the variation of the fidelity against different σ when the template is fixed. In order to achieve a broad EIRP bandwidth performance, σ should be chosen such that good fidelity, high system gain, and broad EIRP bandwidth performance can be obtained. From Figure 7.13, it is evident that the fidelity can exceed 0.8 around the optimal σ.

In conclusion, the UWB antenna design plays a unique role in UWB wireless communication systems. UWB systems based on distinct modulation schemes have different requirements in antenna design. In an OFDM-based UWB system, the requirements are almost the same as those in a broadband system but with an extremely broad bandwidth which usually varies from 50 % (for the lower UWB band of 3.1–5 GHz) to 100 % (for the entire UWB band of 3.1–10.6 GHz). Special attention must be given to pulse-based UWB systems, where a UWB antenna functions as a bandpass filter and reshapes the spectra of the radiated/received pulses. In order to avoid undesirable distortions of the radiated and received pulses, the critical requirements in antenna design include, as far as the electrical parameters are concerned,

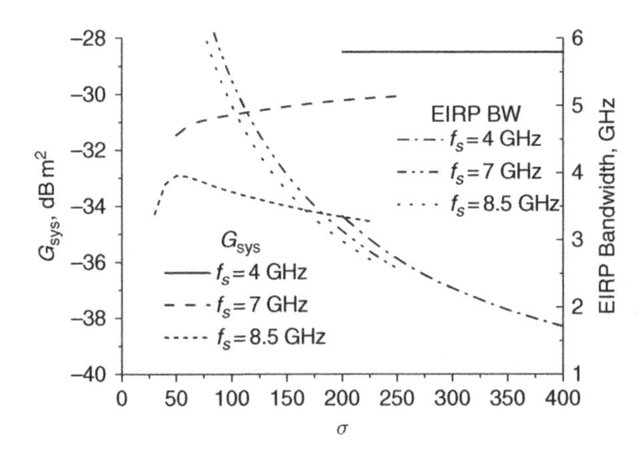

Figure 7.12 System gain and EIRP bandwidth vs. σ.

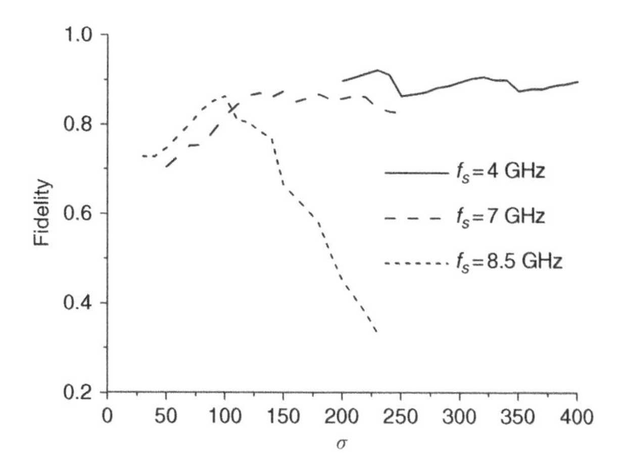

Figure 7.13 Fidelity vs. σ.

- ultra-wide impedance bandwidth covering the bandwidth where most of the energy of the source pulse is concentrated;
- steady directional or omnidirectional radiation patterns;
- constant gain;
- constant group delay or linear phase;
- constant polarization; and
- high radiation efficiency.

All of the electrical performance should be achieved over the entire operating band. The mechanical requirements in antenna design – small size, embeddability, low profile and low cost – stem from the applications directly because the most promising application of UWB technology is in short-range wireless connections between mobile or handheld consumer devices. Therefore, the size and cost issues have become critical.

In particular, OFDM-based multiband UWB systems require flat impedance and gain response, which can cover the entire operating bandwidth. Pulse-based UWB systems require linear phase, impedance, and gain response which can entirely cover the operating bandwidth or partially cover the bandwidth where the majority of the pulse energy is distributed.

7.3 State-of-the-Art Solutions

7.3.1 Frequency-Independent Designs

The UWB antennas under discussion can actually be considered as frequency-independent with an extremely broad frequency response, typically 50–100 %. For example, transverse electromagnetic (TEM) horns feature very broad well-matched bandwidths and have been widely studied and applied [4–6]. TEM horn antennas in their basic forms are illustrated in Figure 7.14. A pair of triangular metal flares forms a TEM horn antenna and a feed excites the end of the horn, as shown in Figure 7.14(a). To enhance the gain of the horn antenna,

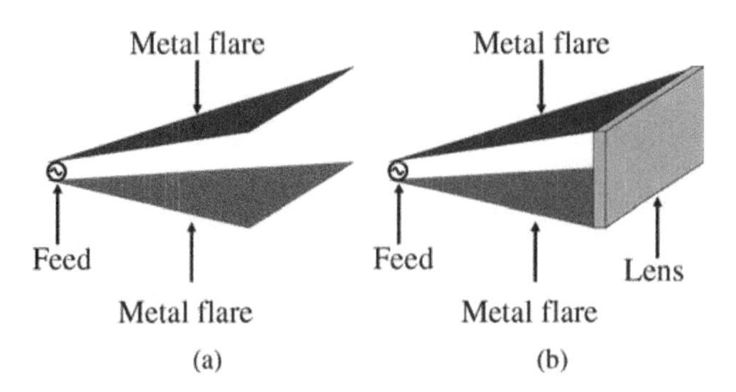

Figure 7.14 TEM horn antennas in their basic forms.

a lens is used to cover the aperture of the horn, as shown in Figure 7.14(b). The antenna radiates linearly polarized TEM waves.

Theoretically, frequency-independent antennas, which have a constant performance at all frequencies, can also be applied to broadband antenna design. The self-complementary log-periodic structures, such as planar log-periodic slot antennas, bidirectional log-periodic antennas, log-periodic dipole arrays, two-or four-arm log-spiral antennas, and conical log-spiral antennas, are a typical design [7]. The geometry of a balanced conical log-spiral antenna is shown in Figures 7.15(a) and 7.15(b). A planar two-arm spiral antenna is shown in Figure 7.15(c). However, for log-periodic antennas, the frequency-dependent phase centers severely distort the waveforms of radiated pulses [8]. They radiate circularly polarized wave at the boresight. Spiral antennas can be completely specified by angles, and they feature constant impedance and radiation pattern performance with frequency. In practice, spiral antennas have a finite size so the frequency-independent behavior is only exhibited over a certain frequency range which is determined by its inner and outer radius. The planar spiral antenna usually requires a backed cavity which is typically lossy to improve the low-frequency impedance behavior and axial ratio by reducing the reflections from the end of the spiral arm. The lossy cavity also absorbs the back radiation from the spiral to enhance pattern bandwidth and achieve unidirectional, radiation usually at the boresight with an undesirable reduction in gain. Alternatively, the spiral antennas backed by conducting cavities have been widely used in applications which require the radiation to be directional.

Biconical antennas, constructed by Sir Oliver Lodge in 1897, are the earliest antennas used in wireless systems, as mentioned by John D. Kraus [9]. They have relatively stable phase centers with broad impedance bandwidths due to the excitation of TEM modes. Many diverse variations of biconical antennas, such as finite biconical antennas, discone antennas, single-cone with resistive loadings have since been constructed and optimized for broad impedance bandwidth [10–12]. Figure 7.16 shows a typical biconical antenna and its variations.

Cylindrical antennas with resistive loading also feature broadband impedance characteristics by forming traveling waves along the dipole arms [13–15].

7.3.2 Planar Broadband Designs

However, the antennas mentioned in Section 7.3.1 are seldom used in portable/mobile devices due to size and cost constraints. Antennas with bulky size are usually employed

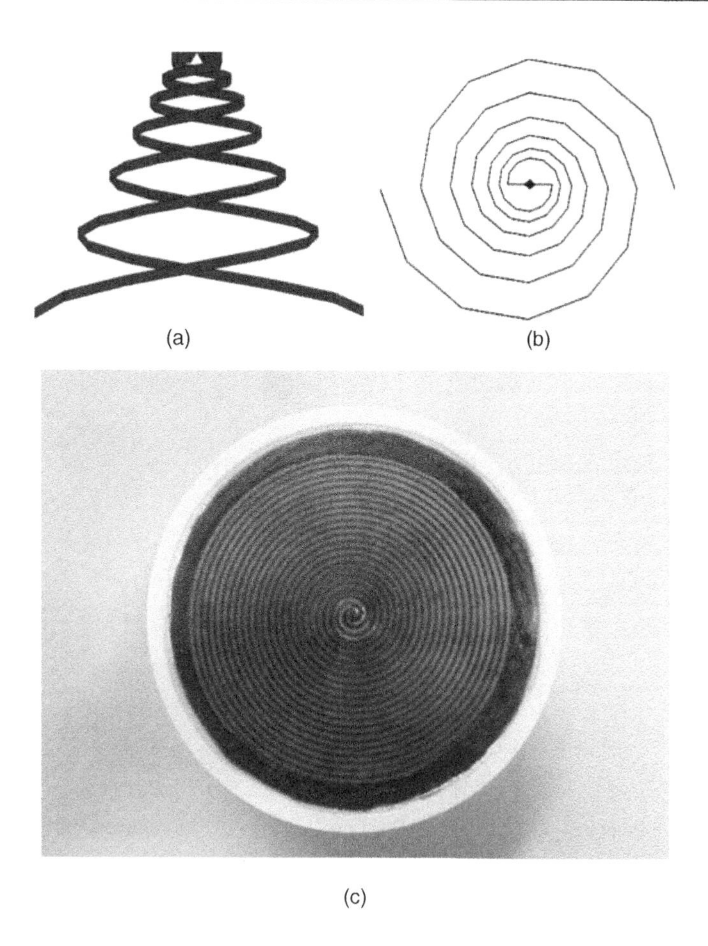

Figure 7.15 Frequency-independent antennas: (a) conical log-spiral antenna, side view; (b) conical log-spiral antenna, top view; (c) planar two-arm spiral antenna backed by a conducting cavity.

in measurement or fixed stations with a certain coverage. Very often, the cost of manufacture and materials are expensive for portable devices. In addition, the directional radiation of the antennas at the elevation and/or azimuth angles may result in pulse distortion.

Instead, planar monopoles (dipoles) or disk antennas have been proposed because they have shown excellent performance in impedance and radiation as well as the significant advantage of small size/volume [16–19]. The earliest planar dipole may be the Brown–Woodward bow-tie antenna, which is a simple and planar version of the conical antenna [9, 20]. The planar antennas for broadband and multiband applications were reviewed in [21, 22].

Figure 7.17 depicts three typical trapezoidal planar designs [23–26]. The trapezoidal radiators with dimensions of W_u, W_b, H, and h form a dipole. A bow-tie antenna is formed if $W_b = 0$, whereas a diamond antenna is formed if $W_u = 0$. The dipoles are fed by balanced feeding structures. However, it is difficult to engineer the balanced feeding structures in practical applications. Also, the effects of the feeding structure on the performance (especially the impedance performance) of the dipole should be taken into account.

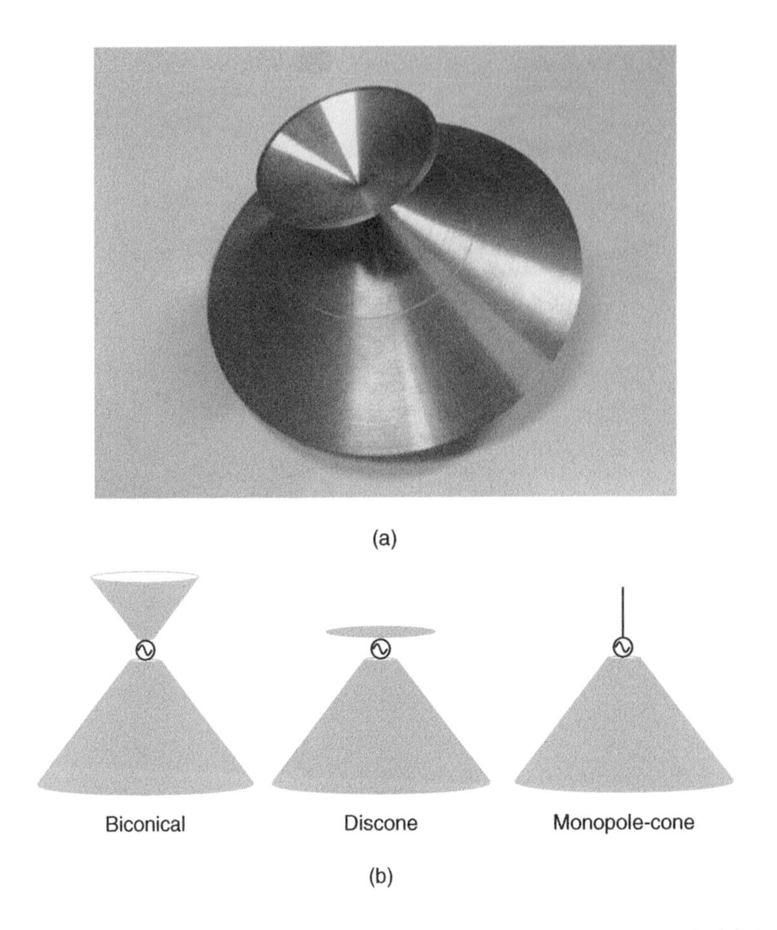

Figure 7.16 A biconical antenna developed by the Institute for Infocomm Research; (b) three typical biconicals.

Besides, the antennas can be excited by a coaxial cable through surface-mounted adapter (SMA) connectors where they are in their monopole forms with large ground planes. The frequency (in gigahertz) corresponding to the lower edge of impedance bandwidth can be estimated by

$$F = \frac{0.25 \times 300}{H + h + (W_u + W_b)/4\pi} \tag{7.8}$$

or

$$F = \frac{0.25 \times 300}{\sqrt{H^2 + \left((2\max(W_u, W_b) - W_u - W_b)/2\right)^2} + h + \dfrac{W_u}{2\pi}} \tag{7.9}$$

where

$$\max(W_u, W_b) = \begin{cases} W_u & W_u \geq W_b, \\ W_b & W_b \geq W_u. \end{cases}$$

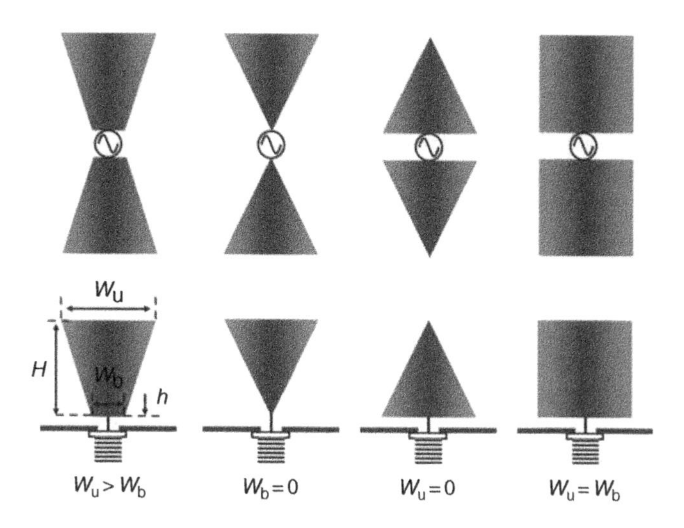

Figure 7.17 Trapezoidal planar antennas.

All right-hand-side dimensions are in millimetres. Equation (7.9) is the modified version of (7.8) to estimate the frequency corresponding to the lower edge of the impedance bandwidth. Equation (7.8) has been used to estimate the frequency corresponding to the lower edge of the impedance bandwidth of a rectangular radiator, which is a special case of the polygonal radiator.

Table 7.3 compares the frequencies corresponding to the lower edge of the impedance bandwidth of a trapezoidal antenna with dimensions $H = 40$ mm, $h = 1$ mm, W_u/W_b varying from 0 to 20 as well as $\max(W_u)$ and $\max(W_b) = 40$ mm by measurement and by calculation using (7.8) and (7.9). From the table, several points can be observed. First, the estimated frequencies using both the formula (7.8) and the modified version (7.9) agree well with the measurements, with less than 6 % difference between them when the ratio W_u/W_b is around 1.0 (i.e. the antenna is nearly a rectangular planar antenna). Second, the difference in frequency increases when W_u/W_b is larger than 1.0. The evaluated frequency using (7.8) increases, whereas, using (7.9), it decreases when W_u/W_b increases. The latter case is in much better agreement with the measured results. Finally, the impedance bandwidth can be more than 80 % for VSWR = 2:1 when the ratio W_u/W_b varies between 1 and 2. It should

Table 7.3 Comparison of impedance bandwidth for VSWR = 2:1 and estimated frequency corresponding to the lower edge of impedance bandwidth by measurement and equations (7.8) and (7.9).

W_u/W_b	0	0.25	0.5	0.75	1	1.33	2	4	20
Bandwidth, %	36	42	51	62	63	>80	76	36	5
F_1, GHz	2.08	1.86	1.67	1.52	1.49	1.36	1.32	1.30	1.25
F_2, GHz	1.69	1.66	1.64	1.61	1.58	1.61	1.64	1.64	1.70
F_3, GHz	1.64	1.66	1.65	1.62	1.58	1.65	1.54	1.48	1.44

F_1: measured; F_2: equation (7.8), F_3: equation (7.9).

be noted that the location of feed point and feed gap also affect the impedance matching significantly. By optimizing all the parameters, the antenna with $W_u/W_b = 0$ can also achieve a very broad impedance bandwidth, as shown in Figure 7.6(b).

The polygonal planar antenna is derived from the trapezoidal planar antenna. A rectangular planar monopole comprising a planar radiator is fed by a coaxial probe as shown in Figure 7.18(a) [27–31]. The antenna's performance is determined by the shape and size of the planar radiator as well as the feeding section. As discussed above, the size and shape of the radiator mainly determine the frequency corresponding to the lower edge of the impedance bandwidth. The feed gap, location of the feed point, and the shape of the bottom of the radiator determine the impedance matching as shown in Figures 7.18(b)–(d) [32–35]. The shape of the bottom of the radiator is modified for impedance matching. The impedance matching is determined by the impedance transition between the coaxial probe and the radiator. A broadband impedance transition will ensure impedance matching across a broad bandwidth. A comparison of the square planar antenna and its variations was carried out in [34]. The impedance bandwidth of a basic square antenna (Figure 7.18(a)) was enhanced by offsetting the feed point. Also, a bevel can be employed in the square antenna to improve the impedance bandwidth, as shown Figures 7.18(d) and 7.18(e) [34, 35]. A shorting pin can also be used to make the planar antenna more compact [35, 36].

The implementation of modified feeding structures can enhance the impedance performance of planar antennas. For instance, the coaxial probe excites the planar radiator via a U-shaped or inverted E-shaped transition which forms an impedance transformer for broadband impedance matching, as depicted in Figures 7.18(f) and 7.18(g) [37]. Also, the planar radiator can be fed by two probes simultaneously, which are fed by a microstrip through a power divider printed on a printed circuit board (PCB) [38]. In order to reduce the height of the planar monopole, a strip can be attached to the top side of the radiator as shown in Figure 7.18(h) [39].

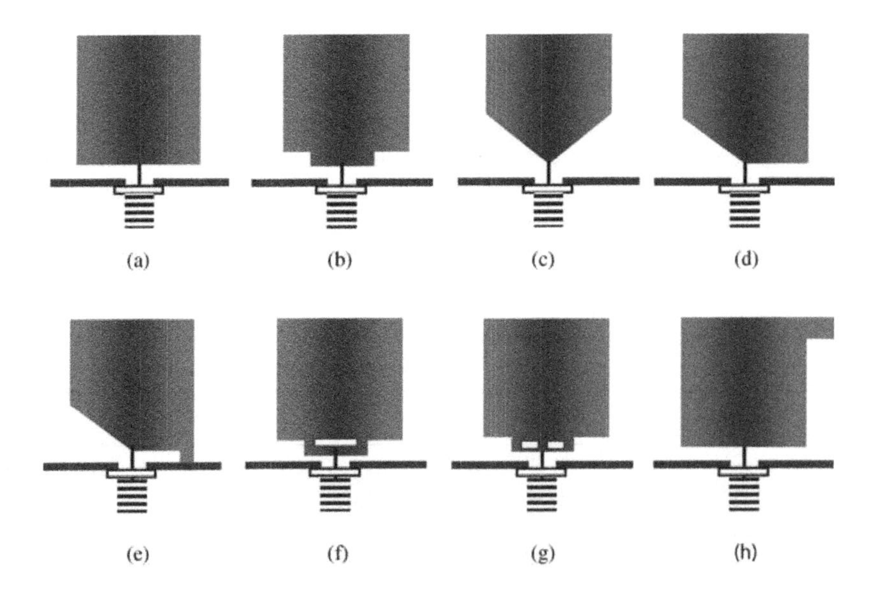

Figure 7.18 Modified versions of rectangular planar antennas.

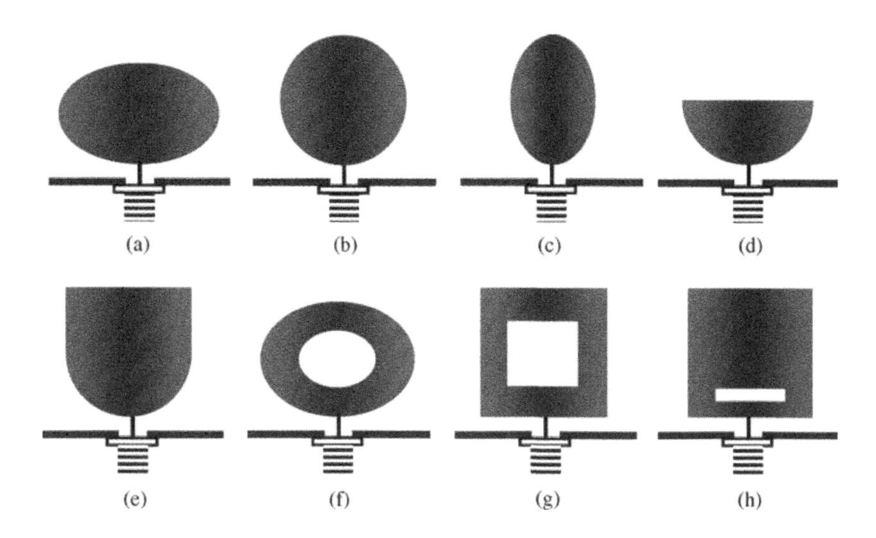

Figure 7.19 Elliptical planar antennas and their variations, and slotted planar antennas.

In addition, the radiator can theoretically be of any shape. Figure 7.19 shows a variety of shapes which have been used in planar antenna design. Among these, elliptical planar antennas, shown in Figures 7.19(a)–(e), are of importance to planar antenna design due to their broadband and high-pass impedance performance [18, 40–44]. The slots or apertures in annular and slotted planar antennas, shown in Figures 7.19(f)–(h), are often employed to improve the impedance bandwidth by changing the current distributions on the radiators [45–48].

Besides the probe feeding structure, the planar radiator may also be fed by electromagnetic coupling [47, 49]. Such a feeding scheme can be used in printed planar monopole antenna designs.

7.3.3 Crossed and Rolled Planar Broadband Designs

Studies have demonstrated that the asymmetrical planar structure severely distorts the omnidirectional radiation characteristics of the planar monopole antenna. This occurs especially at higher operating frequencies, where the planar antenna has become electrically larger than at lower operating frequencies so that the radiation from different parts of the planar radiator has a larger phase difference [34, 50, 51]. The problem becomes more severe if the UWB band covers a broad bandwidth of 50 % (3.1–5 GHz) or 105 % (3.1–10.6 GHz). Moreover, omnidirectional radiation is desirable for mobile UWB applications. In order to alleviate this problem, two planar radiators with a crossed configuration have been used to form a monopole as shown in Figure 7.20(a) [51]. Table 7.4 compares the largest deviation from the maximum gain in the H plane for a planar and a crossed monopole [51].

However, the crossed monopole has a larger volume than the planar monopole, although it is lighter than the thick solid cylindrical monopole. As an alternative, the concept of a roll monopole has been presented which features both broadband well-matched impedance response, as in the case of a planar monopole, and broadband omnidirectional radiation

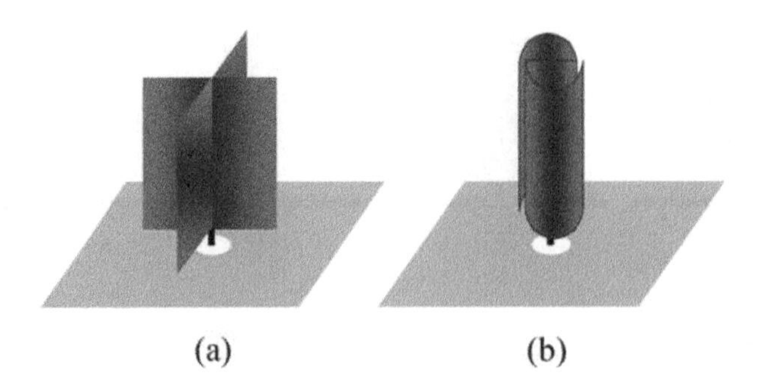

Figure 7.20 (a) A crossed monopole. (b) A two-arm roll monopole.

Table 7.4 Comparison of the largest deviation from the maximum gain in the H plane for a planar and a crossed monopole [51].

f, GHz	Largest deviation from maximum gain in H plane, dB	
	Planar monopole	Crossed monopole
4.8	3.0	1.1
6.8	5.7	2.2
9.0	9.3	2.8

response, as in the case of a circular cylindrical monopole [52, 53]. Figure 7.20(b) shows a two-arm roll monopole which has been applied to UWB applications, where the performance of a pair of roll antennas has been examined in both time and frequency domains [53]. The roll antenna can be constructed with multiple layers. The coupling between adjacent layers introduces capacitance, while the spiral cross-section provides the increase in inductance. Moreover, the roll monopole is much more compact and has a lighter weight than the planar monopole and the thick cylindrical monopole. Figure 7.21 illustrates the example discussed in [50], where a $75 \times 50\,\mathrm{mm}$ copper sheet fed at its left-hand corner was rolled into a cylindrical monopole with a radius of $r = r_o + \alpha\phi$, with $r_o = 4\,\mathrm{mm}$ and $\alpha = 0.5/360°$.

7.3.4 Planar Printed PCB Designs

In practical mobile UWB applications, antennas which can be embedded or integrated into UWB devices are attractive to system designers. Antennas that can be directly printed onto PCBs are the most promising candidates for such applications. The antenna is usually constructed by etching the radiator onto the dielectric substrate of a piece of the PCB and a ground plane near the radiator. The antenna can be fed by a microstrip transmission line or a coplanar waveguide (CPW) structure, as shown in Figure 7.22.

Figure 7.22(a) shows a typical PCB monopole fed by a microstrip line. The radiator is etched onto a dielectric substrate. The ground plane is etched onto the opposite side of the PCB. Such a PCB antenna can easily be integrated into the system circuits for a compact design and fabricated at low manufacturing cost.

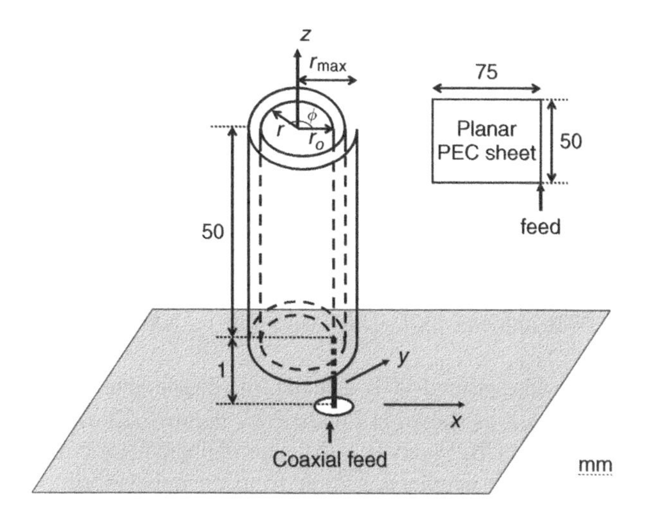

Figure 7.21 Multilayered roll monopole with broad bandwidth for impedance and omnidirectional radiation. Measurements in millimetres.

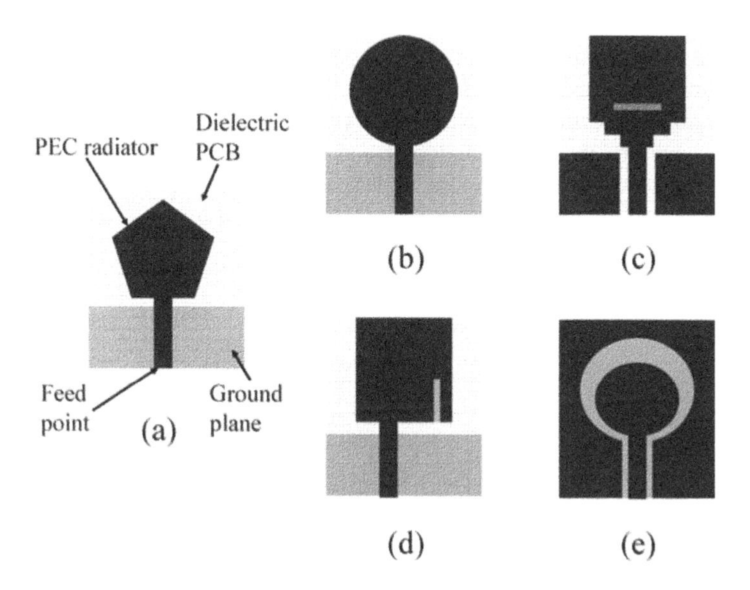

Figure 7.22 Monopoles printed onto a PCB: (a) PCB antenna; (b) microstrip-fed circular PCB antenna; (c) CPW-fed slotted rectangular PCB antenna; (d) microstrip-fed rectangular notched PCB antenna; (e) CPW-fed elliptical slot antenna.

The radiator of the PCB antenna, which may be of any shape, is optimized to cover the UWB bandwidth and to miniaturize the antenna. Its shape may be elliptical, rectangular, triangular, or some combination or variation thereof, as shown in Figure 7.22(b)–(d) [54–56].

Furthermore, the radiator can be slotted for good impedance matching and size reduction, as shown in Figure 7.22(c) [45]. The impedance matching can also be enhanced by notching the radiator as shown in Figure 7.22(d) [57]. The CPW-fed antenna is another important type of planar antenna, as shown in Figure 7.22(e) [58]. This antenna is also known as the planar volcano-smoke slot antenna [59].

The printed PCB antenna is essentially an unbalanced antenna different from a balanced dipole and also a monopole with a large ground plane. The effect of the ground plane on the impedance and radiation performance of the PCB antenna is usually significant. The change in shape, size, and/or orientation of the ground plane may affect the impedance and radiation performance of the PCB antenna [60]. This issue will be addressed in the case study in section 7.4.

Besides the monopole-like printed PCB antenna, the dipole antenna printed onto a PCB is also used in UWB devices, as shown in Figure 7.23. Figure 7.23(a) shows the concept of printed dipole antenna on a PCB. Figure 7.23(b) is an implementation of the dipole antenna printed onto a PCB, where a simple transition from an unbalanced-to-balanced feeding structure is formed by two microstrip lines which are etched on the opposite surfaces of the dielectric substrate. One of the microstrip lines is fed at its end and the other directly connected to the system ground plane [61]. In such applications, the important design issues include the design of balanced feeding structure or transition between an unbalanced to a balanced feeding structure, and the effect of the system ground plane on the performance of the printed PCB dipoles [61, 62]. Similar to the monopole-like PCB antenna, the radiator of a printed dipole antenna can be of any shape, chosen so as to optimize the impedance matching and radiation performance within the operating UWB range.

It should be noted that the planar monopole and dipole antennas feature broad impedance bandwidth but suffer from high cross-polarization radiation levels. The large lateral size and/or asymmetric geometry of the planar radiator have resulted in the cross-polarized

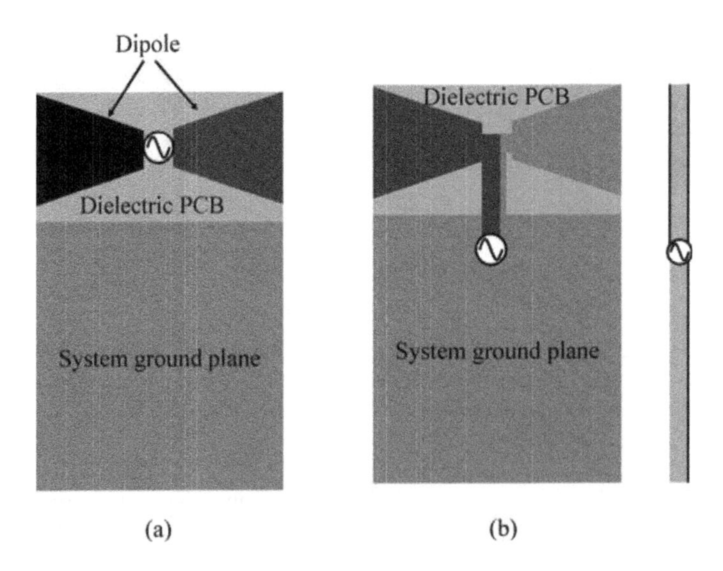

Figure 7.23 Dipoles printed onto a PCB: (a) dipole antenna; (b) microstrip-fed dipole antenna.

radiation. Fortunately, the purity of the polarization issue is not critical, particularly for the antennas used in portable devices.

7.3.5 Planar Antipodal Vivaldi Designs

Omnidirectional radiation performance is important for portable UWB devices, but the antennas with stable directional radiation may also be of interest, for instance, in portable radar apparatus. However, it is difficult to design an antenna with stable radiation performance across the UWB bandwidth due to the change in the magnitude and phase of the current induced on the radiators. As a type of endfire traveling-wave antenna, tapered slot antennas (TSAs) are capable of providing consistent radiation performance across the UWB bandwidth.

Linear TSAs and Vivaldi antennas are the simplest version of TSAs but with broadband impedance and radiation performance [63–67]. In order to enhance the performance of the Vivaldi antenna, a modified version of it, the antipodal Vivaldi antenna, has been proposed [67–70], as shown in Figure 7.24(a). In order to make the design more compact and

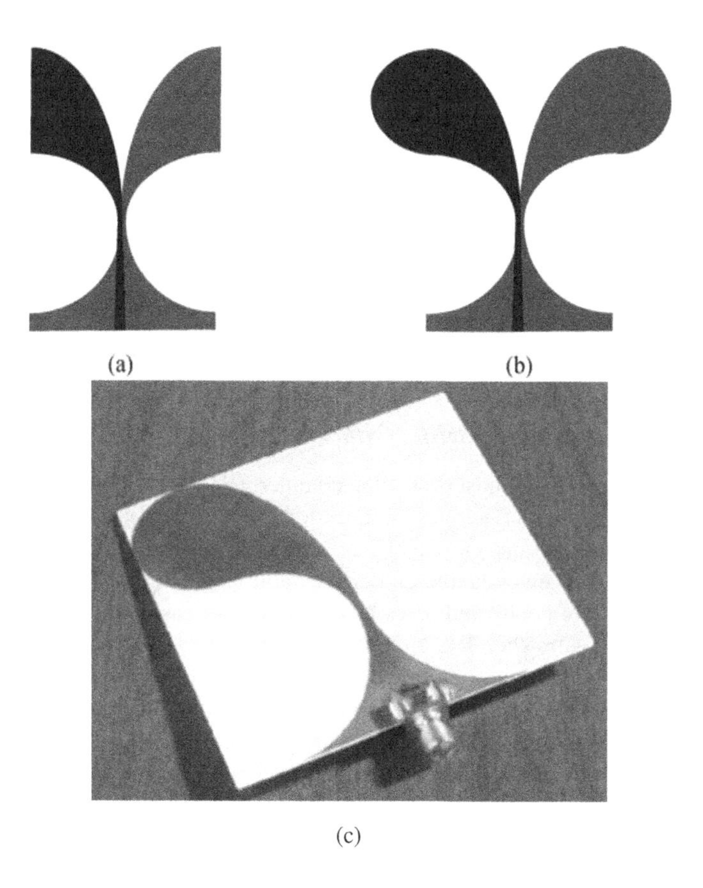

Figure 7.24 The antipodal Vivaldi antennas: (a) conventional antipodal Vivaldi antenna; (b) modified antipodal Vivaldi antenna; (c) photo of the modified antipodal Vivaldi antenna.

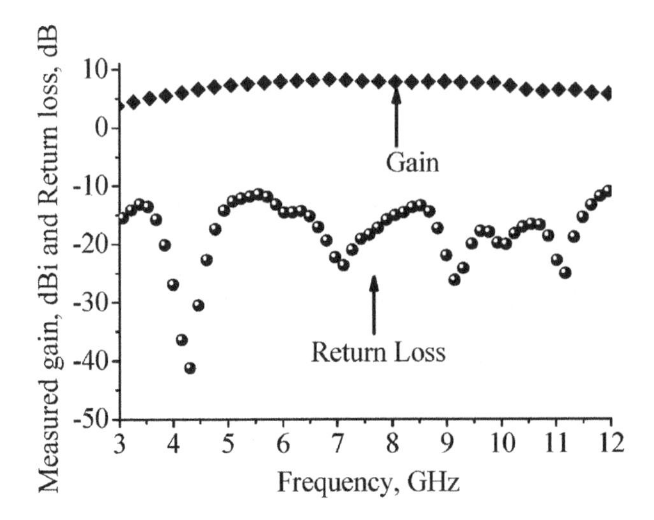

Figure 7.25 Measured return loss and gain at boresight across the UWB bandwidth.

improve the impedance matching, the antipodal Vivaldi antenna is modified by attaching two semi-circles to the ends of the arms as shown in Figures 7.24(b) and 7.24(c), where a broad-band impedance and unbalanced-to-balanced transition is achieved by a simple microstrip structure instead of a conventional microstrip line-slot feeding structure [71]. Figure 7.25 shows the measured impedance and gain response of the antenna shown in Figure 7.24 within the UWB bandwidth. The broadband impedance and radiation characteristics have been observed.

7.4 Case Study

7.4.1 Small Printed Antenna with Reduced Ground-Plane Effect

As mentioned above, one of the most promising commercial applications of UWB technology is in short-range high-data-rate wireless connections. The devices used in such wireless connections will be portable and mobile. Therefore, the UWB antennas should be small in size and light for possible embeddable and/or wearable applications. In such applications, small printed antennas are good candidates because they are easily embedded into wireless devices or integrated with other RF circuits. The printed UWB antenna can achieve a broad impedance bandwidth by optimizing the radiator, ground plane, and feeding structure [72–76]. However, such UWB antennas usually suffer from the need for an additional impedance matching network and/or large system ground planes. In addition, due to the unbalanced structure of the printed UWB antenna, consisting of a planar radiator and system ground plane, the shape and size of the ground plane will inevitably have significant effects on the performance of the printed UWB antenna in terms of the operating frequency, impedance bandwidth, and radiation patterns [62, 77]. Such ground-plane effects cause severe practical antenna engineering problems such as complexity of design and difficulties with deployment.

7.4.1.1 Antenna Design

A small printed UWB antenna is presented to alleviate the ground-plane effects. The printed rectangular antenna shown in Figure 7.26 is designed to cover the UWB band of 3.1–10.6 GHz. A rectangular slot was notched onto the upper radiator etched on a piece of PCB (RO4003, $\varepsilon_r = 3.38$ and 1.52 mm in thickness). The notch of $w_s \times l_s$ is cut close to the attached strip of $w_{rs} \times l_{rs}$ at a distance d_s. Two bevels are cut to improve the impedance matching, especially at higher frequencies. Both the feed gap g and the position of feed point d affect the impedance matching. The length of the ground plane, l_g, has been optimized for good impedance matching to achieve a miniature design.

The optimized dimensions are $w_s \times l_s = 4\,\text{mm} \times 12\,\text{mm}$, $w_{rs} \times l_{rs} = 2\,\text{mm} \times 6\,\text{mm}$, $d = 6\,\text{mm}$, $d_s = 4\,\text{mm}$, $g = 1\,\text{mm}$, and $l_g = 9\,\text{mm}$. The feeding strip is 3.5 mm in width. A Cartesian coordinate system (x, y, z) is oriented such that the bottom plane of the PCB in Figure 7.26 lies in the x–y plane.

7.4.1.2 Antenna Performance

Figure 7.27 shows good agreement between the simulated and measured return losses. The measured bandwidth for −10 dB return loss covers the range of 2.95–11.6 GHz with multiple resonances. It should be noted that in simulations, the antenna is fed by a delta-type source at the end of the feeding strip and close to the edge of the PCB. The excitation source with a 50 Ω internal resistance is between the end of the feeding strip and ground plane right beneath, namely a vertical excitation in the Zeland IE3D software without any RF feeding cables. In the measurements, a 50 Ω SMA is connected to the end of the feeding strip and grounded to the edge of the ground plane. An RF cable from the vector network analyzer is connected to the SMA to excite the antenna. In small-antenna measurements, the RF cable usually affects the performance of the antenna under test (AUT) greatly. From the comparison in

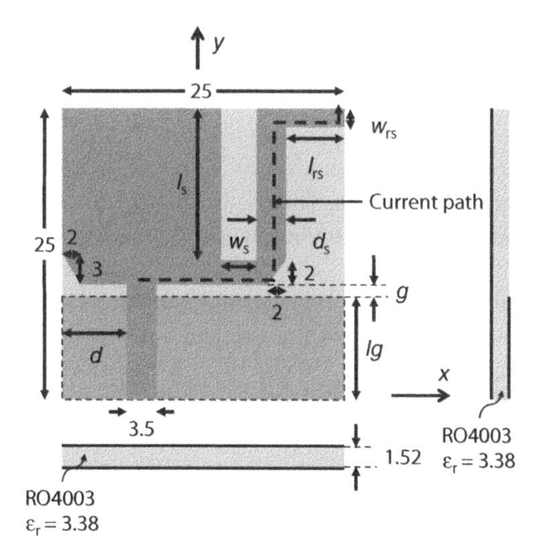

Figure 7.26 Geometry of the small printed antenna. Dimensions in millimetres.

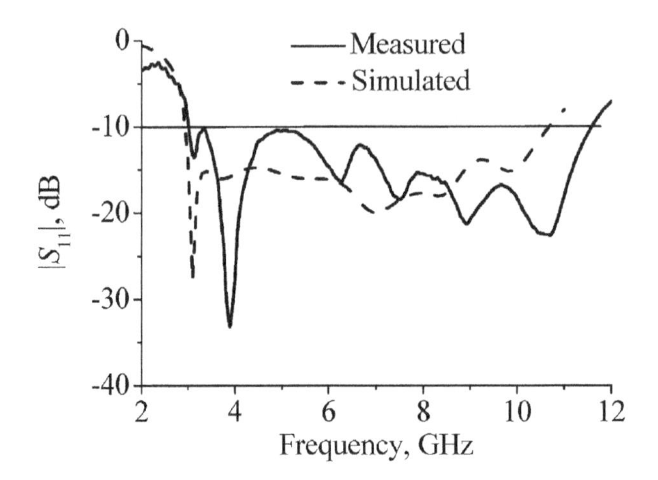

Figure 7.27 Comparison of simulated and measured return loss.

Figure 7.27, it is evident that the presence of the RF cable hardly affects the lower edge frequencies around 3 GHz. This implies that the design is less dependent on the ground plane in terms of impedance matching. This feature makes the printed antenna design flexible and suitable for practical applications where the antenna is to be integrated into various circuits or devices.

Figure 7.28 compares the simulated current distributions on antennas with and without the notche at 3 GHz. The majority of the electric current is concentrated around the notch at the right-hand part of the radiator. The currents on the left-hand part of the radiator and the ground plane are very weak. This suggests that the notch has a significant effect on the antenna performance at the lower operating frequencies. As a result, the impedance matching at 3 GHz is more sensitive to the notch dimensions than the shape and size of the ground plane. As a result, the effects of the ground plane and RF cable on the antenna performance at the lower frequencies can be greatly suppressed. By way of comparison, the electric current for the antenna without notch is mainly concentrated around the feeding strip portion at 3 GHz such that the ground plane significantly affects the impedance and radiation performance of the antenna without notch. Therefore, the performance of the notched antenna has the advantage of the suppressed ground-plane effects over the conventional designs without notch.

The lowest resonant frequency, f_l, of a planar monopole antenna in its symmetrical and basic form can be estimated [26]. That of the notched antenna design can be estimated by the longest effective current path $L = \lambda_l/2$, where λ_l is the wavelength at f_l, although the antenna is an unbalanced asymmetrical dipole with an irregular shape. From the electric current distribution on the antenna at the lowest frequency of 3 GHz, it can be seen that most of the electric current is concentrated on the right-hand part of the upper radiator. Thus, the path length L can be determined by the edge length of the right-hand part of the upper radiator, namely 12 mm (the horizontal path from feed point) +13 mm (the vertical path from the bottom of the upper radiator) +6 mm (the length of the horizontal strip) +2 mm (the width of the horizontal strip) = 33 mm, as depicted in Figure 7.26 [26]. Thus, f_l ($= c/\lambda_l$

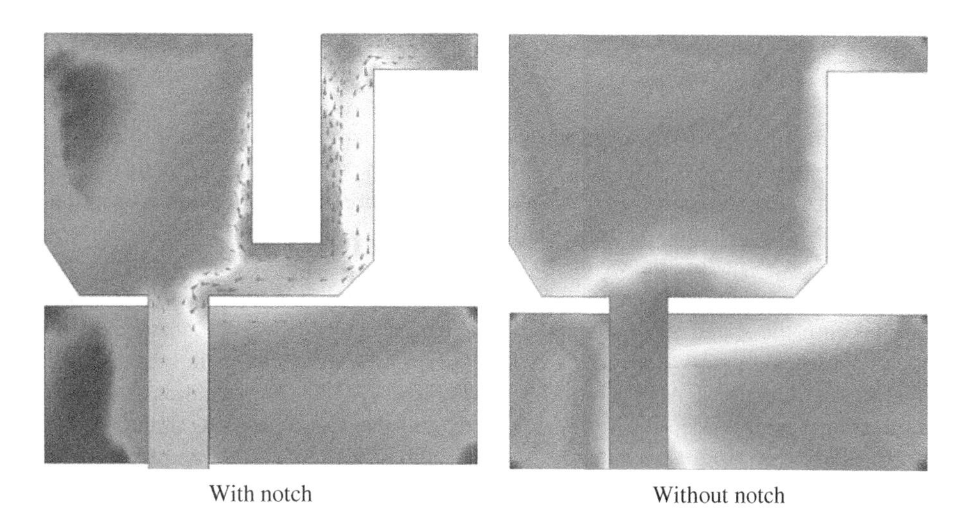

Figure 7.28 Simulated current distributions on the antenna with and without the notch.

where $\lambda_l = 2L\sqrt{(\varepsilon_r + 1)/2}$ and c is speed of light) is 3.07 GHz. This has been validated by simulated and measured results of 3.10 GHz as shown in Figure 7.27.

With the estimation of the lowest resonant frequency f_l, it is found that the path length of the electric current at the right-hand part of the radiator is around a half-wavelength at f_l. In order to explain the effect of the notch cut from the radiator, Figure 7.29 illustrates the current distributions on the upper portions (stems), where the path length of the electric current at the stems is around a quarter- and a half-wavelength, respectively. The current at the junction between the bottom RF cable and the quarter-wavelength stem is strong, whereas the current is relatively weak at the junction between the RF cable and half-wavelength

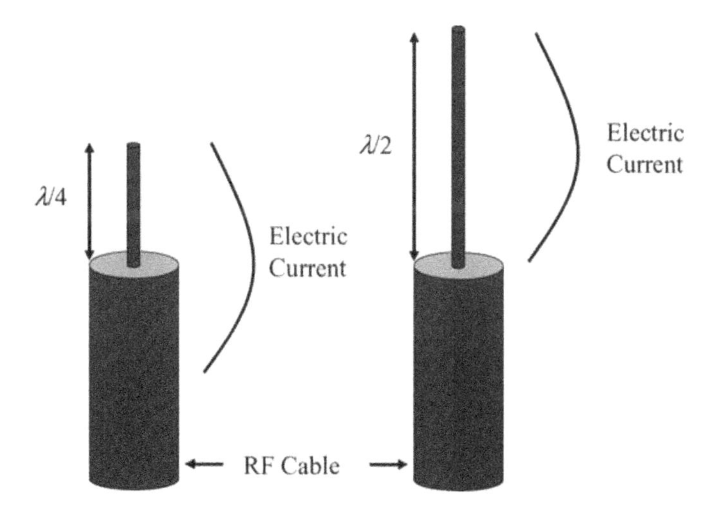

Figure 7.29 Illustration of electric current on unbalanced antennas.

stem, as shown in Figure 7.29. Therefore, very little current will flow into the RF cable so that the effects of the ground plane (RF cable) on the antenna performance are significantly reduced.

The three-dimensional (3D) radiation patterns for total radiated electric fields were measured at frequencies of 3, 5, 6, and 10 GHz by using the Orbit-MiDAS system, as shown in Figure 7.30. Antennas designed for mobile devices require 3D radiation and high radiation efficiency. In the 3D radiation patterns, the lighter shading indicates the stronger radiated E-fields and the darker shading the weaker ones. It is evident from the figure that the radiation at 3, 5, and 6 GHz is almost 3D omnidirectional, which is unlike a typical monopole/dipole antenna because the x and y-components of the electric currents on the antenna are both strong, as shown in Figure 7.28. The radiation is slightly weak along the negative y and negative x-axis directions. At the higher frequency of 10 GHz, the radiation has become more directional with a deep dip in the x–z plane and the negative y-axis direction due to the electrically larger size of the antenna. Such 3D omnidirectional radiation performance is conducive to the application of these antennas in mobile devices.

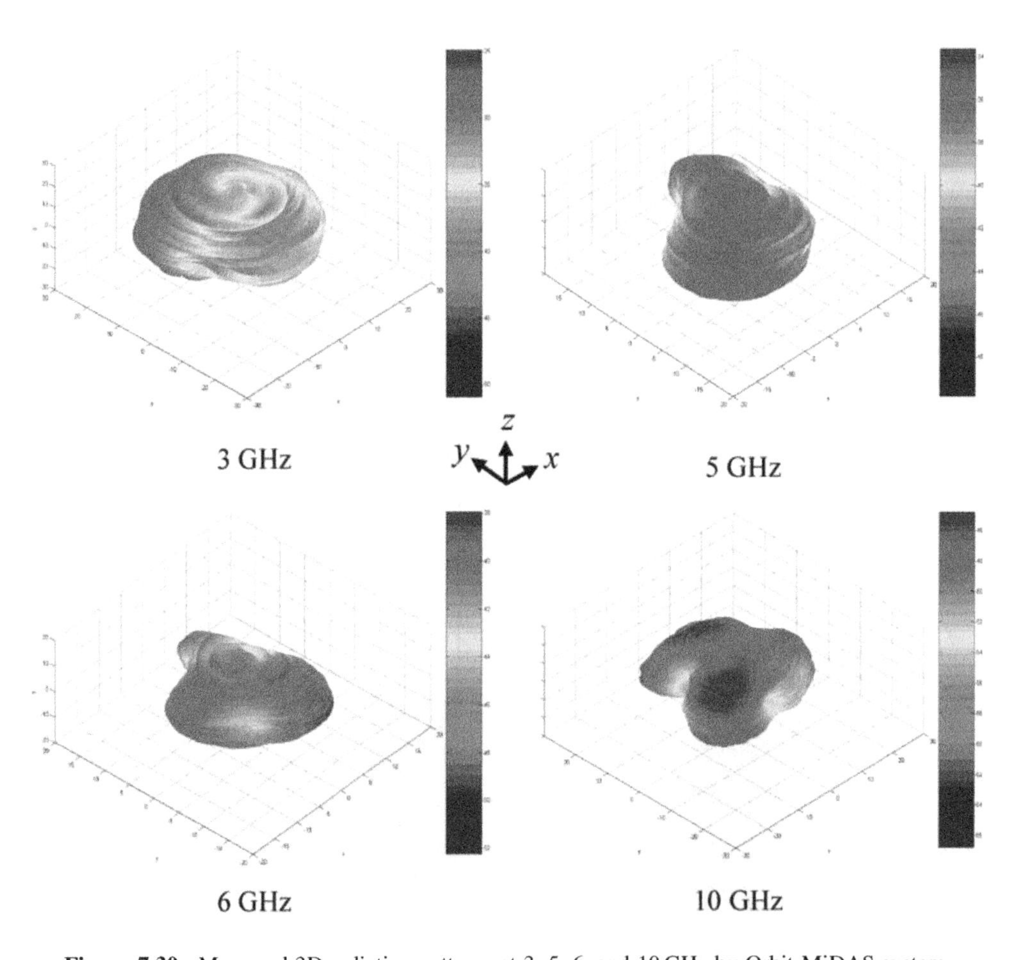

Figure 7.30 Measured 3D radiation patterns at 3, 5, 6, and 10 GHz by Orbit-MiDAS system.

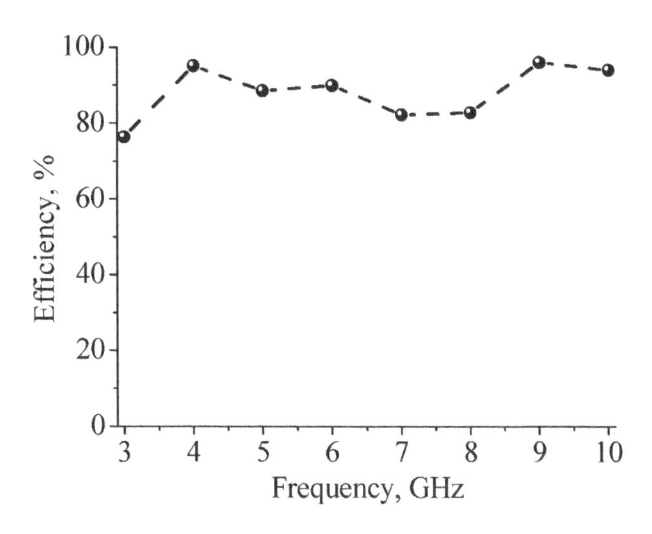

Figure 7.31 Measured radiation efficiency by Orbit-MiDAS system.

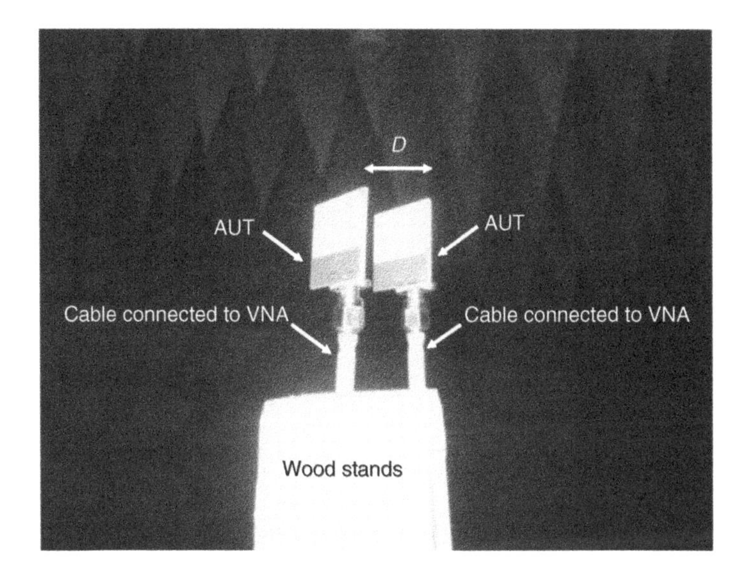

Figure 7.32 Transfer function measurement setup.

Furthermore, Figure 7.31 shows that the measured radiation efficiency varies from 79 % at 3.1 GHz to 95 % at 4 GHz across the bandwidth of 3.1–10.6 GHz.

In addition, the transmission between the two identical proposed antennas is examined in an electromagnetic anechoic chamber. The setup is shown in Figure 7.32. The antennas under test (AUT) are placed face-to-face at a separation of D. The antennas are connected to the RF cables through the SMAs. The RF cables are connected to the HP8510C vector network analyzer.

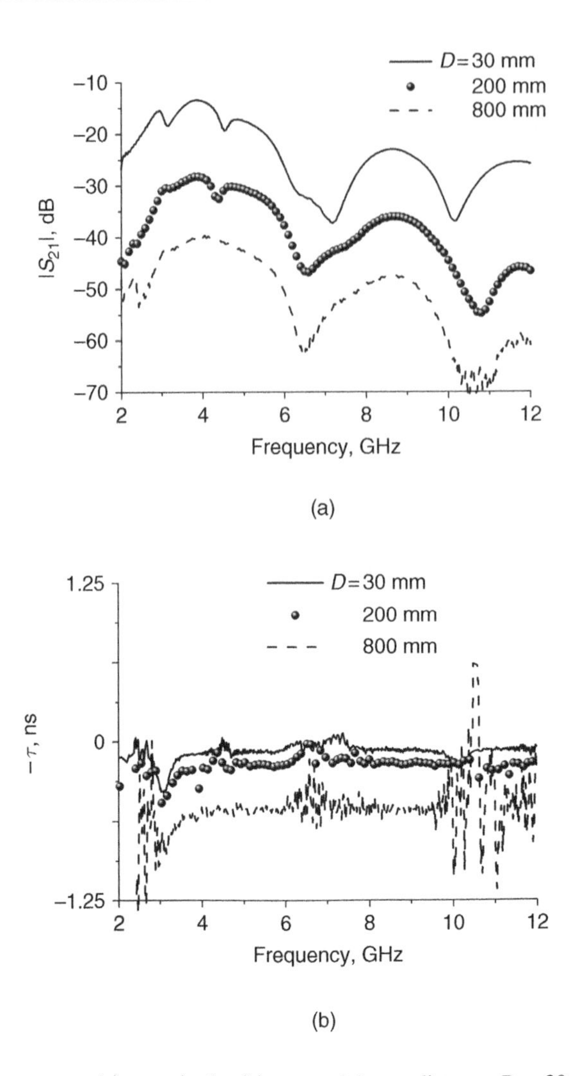

Figure 7.33 Measured S_{21}: (a) magnitude; (b) group delay at distance $D = 30, 200$, and 800 mm.

Figure 7.33 shows the measured $|S_{21}|$ for $D = 30$ mm, 200 mm, and 800 mm. Figure 7.33(a) plots the magnitude of S_{21}. At different distances D, the measured $|S_{21}|$ varies. At lower frequencies, the ripples due to the effect of the mutual coupling between the two antennas can be observed when D is 30 mm (0.3λ at 3 GHz). When the antennas are placed in each other's far-field zone, the measured $|S_{21}|$ changes gradually against frequency, as shown in the case of $D = 800$ mm, because of the gain variation against frequency.

Moreover, the phase response of the UWB antenna has a significant effect on the waveforms of the transmitted and received pulses, in particular, in pulsed UWB systems, where an extremely broad operating bandwidth is occupied by the pulsed signals. The group delay (in seconds) is given by:

$$\tau_{\text{group delay}} = -\frac{d\phi(\text{rad})}{d\omega(\text{rad} \cdot \text{Hz})}, \qquad (7.10)$$

where ϕ is the phase of measured S_{21} and ω indicates the angular frequency. Figure 7.33(b) shows that the measured group delay is about -0.5 ns for $D = 30$ mm and 200 mm, and -2 ns for $D = 800$ mm. At around 7 GHz, the noise when $D = 800$ mm increases due to the weaker $|S_{21}|$. From the results, it is recommended that the AUT be separated at a distance of 1–3 times the largest operating wavelength for S_{21} measurements, since the measurement in a far-field zone is of more practical interest.

It should be noted that, from the measured $|S_{21}|$ shown in Figure 7.33, the transmission gain along a specific direction experiences large variation for operating frequencies higher than 5.5 GHz due to a change in the radiation patterns. Therefore, this antenna is able to meet the demands of the UWB systems operating in the lower band of 3.1–5 GHz very well in terms of 10 dB bandwidth, which is widely used for wireless UWB devices such as wireless universal serial bus (WUSB) dongles.

Moreover, the characteristics in the time domain are examined by displaying the waveforms of the received impulses. The source impulses applied to the transmit antenna shown in Figure 7.32 are selected to be the Rayleigh pulses given by

$$v(t) = \frac{e^{-(t/\sigma)^2}}{t}. \tag{7.11}$$

As examples, the waveforms in the time-domain and spectra in the frequency domain of the impulses with parameters $\sigma = 35, 50$, and 100 ps are shown in Figure 7.34. The time-domain responses of the antennas are depicted in Figure 7.35 for $D = 30$, 200, and 800 mm and $\sigma = 35, 50$, and 100 ps. Previous studies have shown that the time-domain response is determined mainly by both the transfer functions of the antenna systems and source pulses [2].

Here, the aim is to examine time response of the proposed antenna system. The results show that the distance between the antennas has a slight effect on the waveforms of the received impulses. The main impact is on the increased tail-end ringing of the impulses. The waveforms of the impulses with $\sigma = 50$ ps experience less distortion than the other two impulses with $\sigma = 35$ and 100 ps. However, it should be noted that the impulses with $\sigma = 100$ ps have the highest amplitude because the majority of energy of the Rayleigh impulse is concentrated within the lower frequency range at around 2 GHz, where the $|S_{21}|$ is higher, as shown in Figure 7.33(a). Furthermore, the impulses can be optimized or modulated to comply with the emission limits and maximize the output signals, as suggested in [2].

7.4.1.3 Antenna Parametric Study

Parametric studies are carried out to provide antenna engineers with more design information. The performance of the antenna is mainly affected by geometrical and electrical parameters, such as the dimensions related to the notch, top strip, feeding strip, feed gap, ground plane, and the dielectric constant of the substrate.

The parameters related to the *notch* include its dimensions (w_s, l_s) and location (d_s). Figure 7.36 shows the effect of varying the parameters on the impedance matching. It is clear from Figure 7.36(a) that the length of the notch has a significant effect on the impedance matching, especially at lower operating frequencies. Increasing the length l_s lowers the lower edge frequency of the bandwidth due to the extended current path so that the size of the antenna can be reduced. The width and location of the notch (w_s, d_s) have a slight effect on the lower edge frequency. In general, all the notch-related parameters influence the

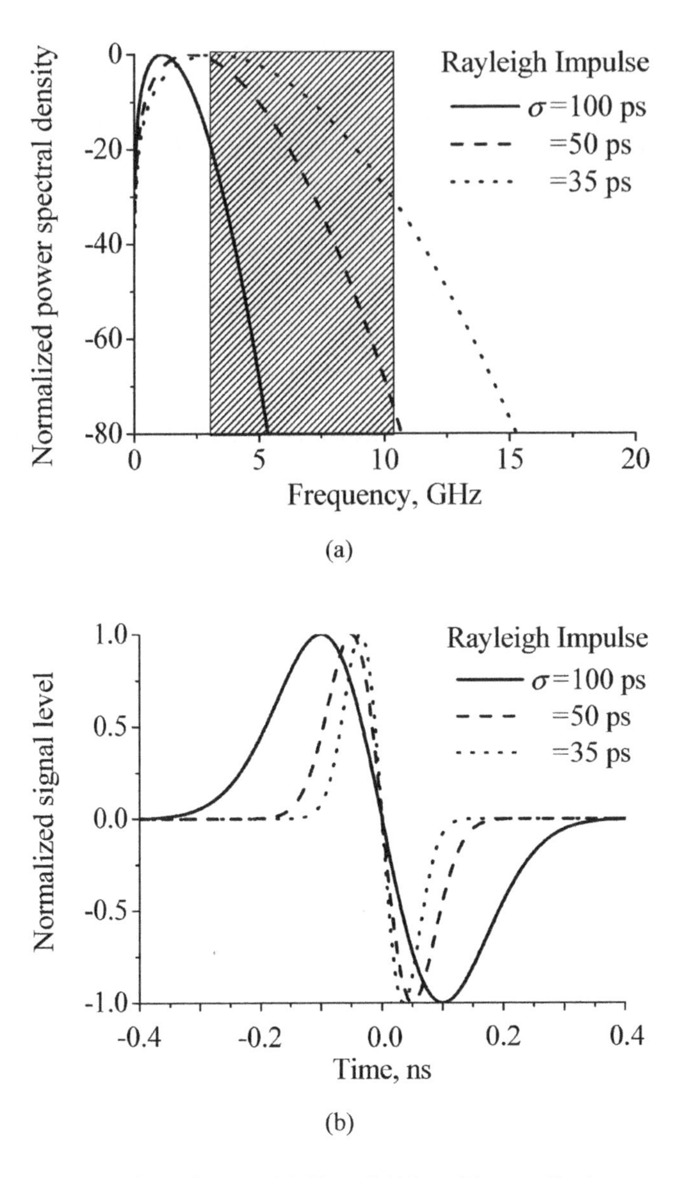

Figure 7.34 Rayleigh impulses with $\sigma = 35, 50$, and 100 ps: (a) normalized power spectral density; (b) normalized signal levels.

impedance matching to a certain extent. This conclusion accords with the findings from the current distribution at 3 GHz.

The *top strip* has dimensions $w_{rs} \times l_{rs}$. Figure 7.37(a) shows that increasing the top strip length can reduce the lower edge frequency by increasing the overall size of the antenna. In antenna design, this technique has been used to reduce the antenna height, for example in inverted-L or inverted-F antennas. Figure 7.37(b) demonstrates that the effect of the top strip width on the impedance matching can be ignored for widths between 1 mm and 3 mm.

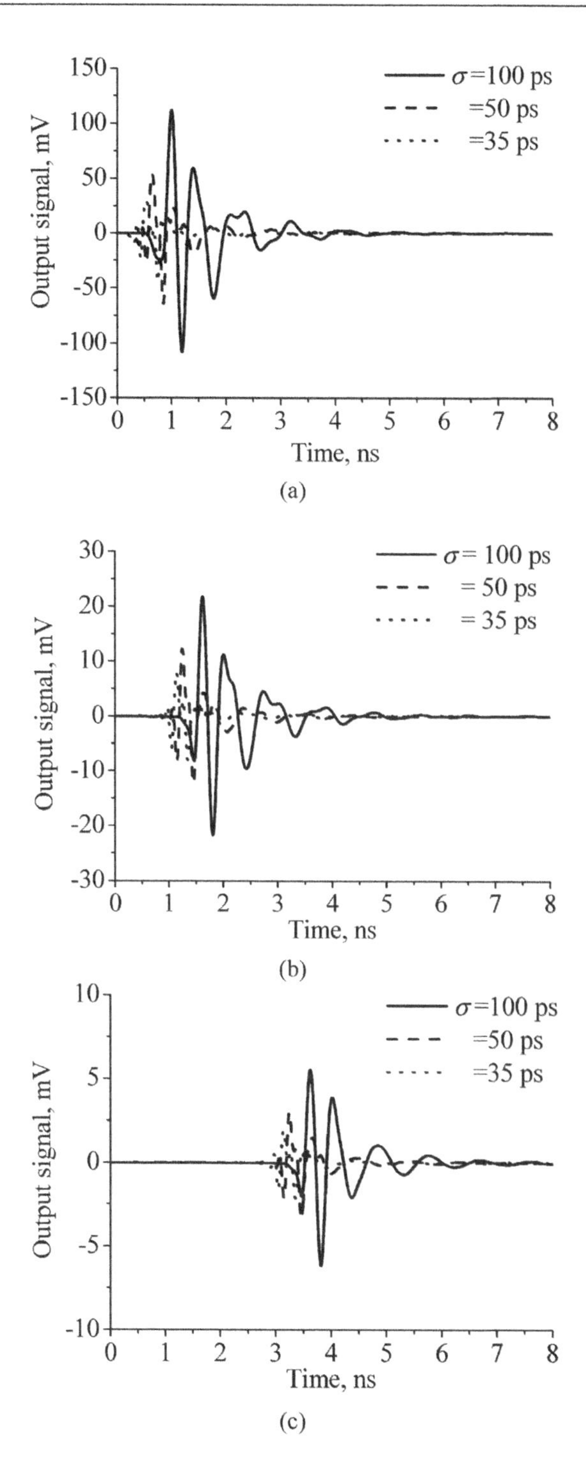

Figure 7.35 Time-domain responses of Rayleigh impulses with $\sigma = 35, 50$, and 100 ps at distance (a) $D = 30$ mm, (b) $D = 200$ mm, and (c) $D = 800$ mm.

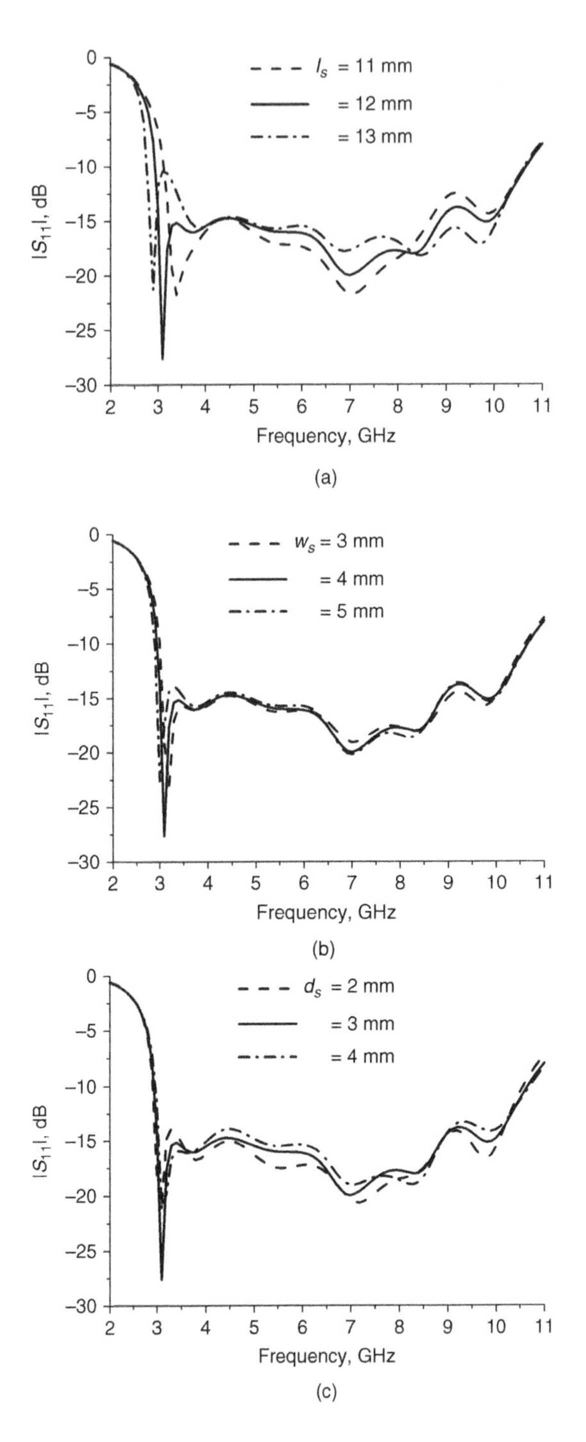

Figure 7.36 Effects of varying notch-related parameters: w_s, l_s, d_s.

The lower edge frequency is slightly increased when $w_{rs} = 1$ mm. Furthermore, the size of the top strip does not have a significant effect on the impedance response of the antenna at higher frequencies.

Figure 7.37(c) demonstrates the effect of varying the location of the *feeding strip, d*, on the impedance matching. It is clear that optimizing the location of the feeding strip can significantly improve the impedance matching, especially at the higher frequencies. This technique has been used in planar broadband antenna designs [30].

Ground-plane/Substrate-related: From Figure 7.38 three important points can be observed. First, Figure 7.38(a) shows that the impedance matching is very sensitive to

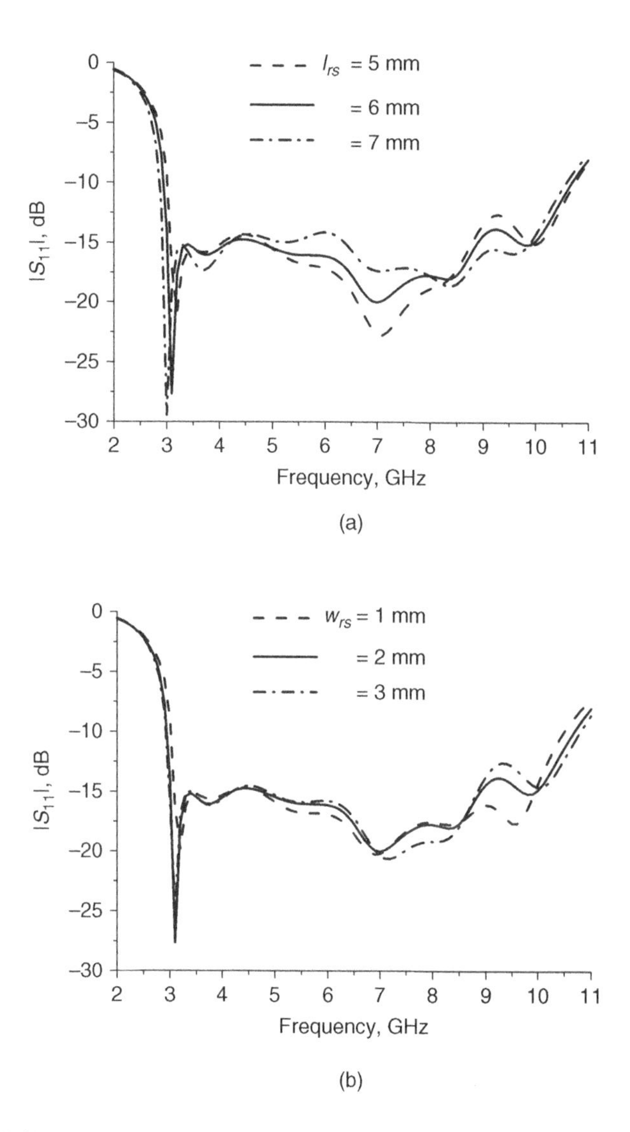

Figure 7.37 Effects of varying top and feed strip-related parameters: $w_{rs} \times l_{rs}$, d.

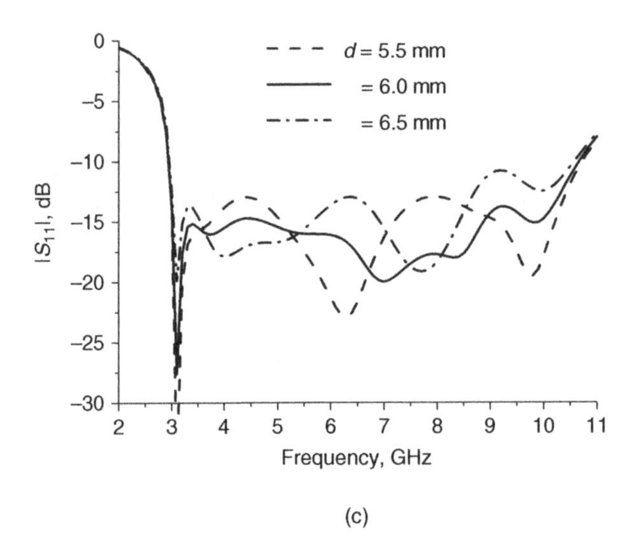

Figure 7.37 (Continued).

the *feed gap*, *g*, especially at higher frequencies. Second, the length of the *ground plane*, l_g, affects the impedance matching more significantly at higher frequencies than at lower frequencies, as shown in Figure 7.38(b). This finding is consistent with the current distributions in Figure 7.28, where more current is concentrated on the ground plane at the higher frequencies than at lower frequencies. Finally, the impedance response is also affected by the *dielectric constant*, ε_r, as shown in Figure 7.38(c). In this study, a change in the dielectric constant leads to a shift in the characteristic impedance of the feeding strip from 50 Ω.

The key to this design is to notch the radiator. The characteristics of two designs, namely with and without notch, are compared to understand the function of the notch in the design. In order to examine the effect of the notch on the impedance matching, the notch in the design shown in Figure 7.26 has been removed while keeping all the other dimensions the same. Also, the antenna without notch has been optimized in order to compare it with the notched antenna, while maintaining the overall size at 25 mm × 25 mm. It can be seen that the antenna without notch is only able to achieve a lower edge frequency of 3.7 GHz. The return losses are simulated and compared in Figure 7.39. It is clear that the notch not only reduces the effect of the ground plane on the antenna's performance but also miniaturizes the antenna size, as mentioned in the previous section.

In short, the notion of designing small antennas with reduced ground-plane effect has been proposed for UWB antenna designs to be applied in promising ultra-wideband mobile applications. In the following section, this concept is applied to UWB antennas designed for WUSB devices installed on a laptop computer.

7.4.2 Wireless USB

As mentioned previously, one of the most promising applications of UWB technology is in short-range and high-speed wireless interfaces. The wireless USB (WUSB) may be the

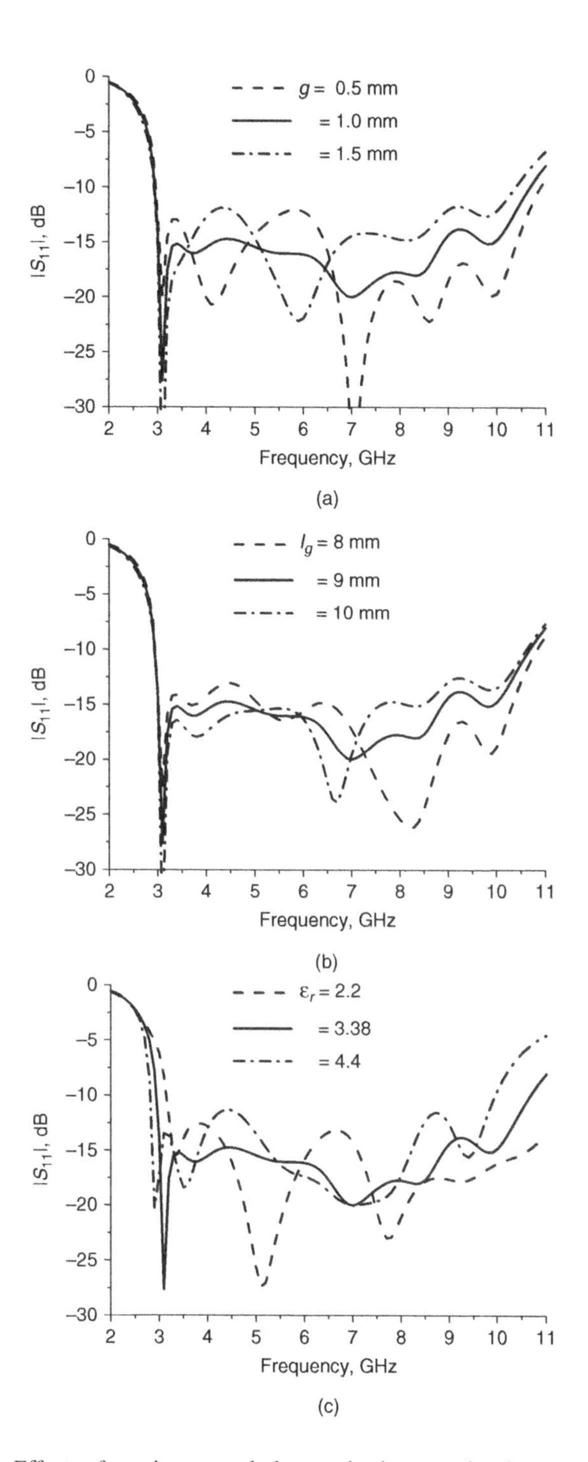

Figure 7.38 Effects of varying ground plane and substrate-related parameters: g, l_g, ε_r.

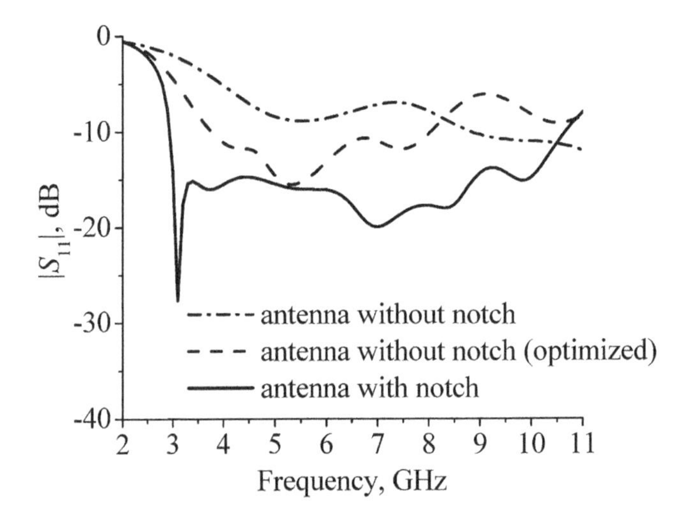

Figure 7.39 Effects of the notch on impedance response.

first commercial UWB product in market. The WUSB will be a replacement for wired USB and will match the USB 2.0 data rate of 480 Mbps. The technology is a hub-and-spoke connection that supports dual-role devices in which a product such as a camera can either act as a device to a host laptop/desktop or as a host to a device such as a printer.

This section will address the issues related to the UWB antennas applied in a WUSB dongle which is used in a laptop environment.

7.4.2.1 Planar Antenna Design

Figure 7.40 shows the geometry of the planar antenna and the Cartesian coordinate system. The radiator and ground plane are etched on opposite sides of the PCB (RO4003, $\varepsilon_r = 3.38$, 1.52 mm in thickness). The radiator consists of a rectangular section measuring 15.5 mm × 15 mm and a horizontal strip which has been designed based on the ideas proposed in the last section. A rectangular notch of $w_s \times l_s = 1\,\text{mm} \times 14\,\text{mm}$ is cut close to the horizontal strip of $w_{rs} \times l_{rs} = 1\,\text{mm} \times 5\,\text{mm}$ at a distance of $d = 2\,\text{mm}$. The radiator is fed by a microstrip line of 3.5 mm width located at $d_s = 2\,\text{mm}$ from the left-hand side of the radiator with a feed gap $g = 1\,\text{mm}$. The excitation is located at the edge of a microstrip line of length 21 mm. The ground plane has size $l_g \times w_g = 48.5\,\text{mm} \times 20\,\text{mm}$, which is the dimension of a typical USB dongle.

7.4.2.2 Antenna Performance

The impedance response was examined by the Agilent N5230A vector network analyzer. From Figure 7.41 it can be seen that the antenna covers a well-matched bandwidth of 3–5 GHz for $|S_{11}| < -10\,\text{dB}$. This lower UWB band has been widely used in WUSB designs.

Figure 7.42 shows the measured radiation patterns for the total field of the antenna in the horizontal (x–y) plane at 3.5 GHz, 4 GHz, and 4.5 GHz in free space. The maximum

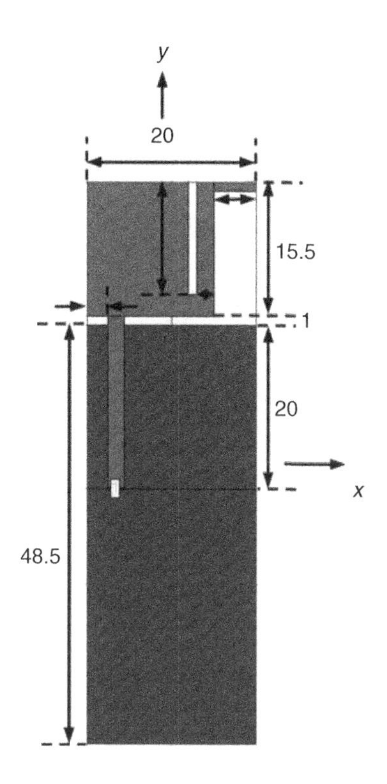

Figure 7.40 Geometry of the planar antenna. Dimensions in millimetres.

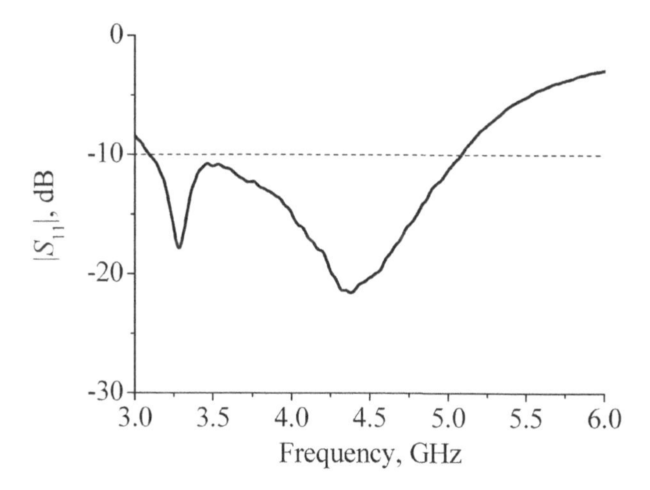

Figure 7.41 Measured return loss of the planar antenna.

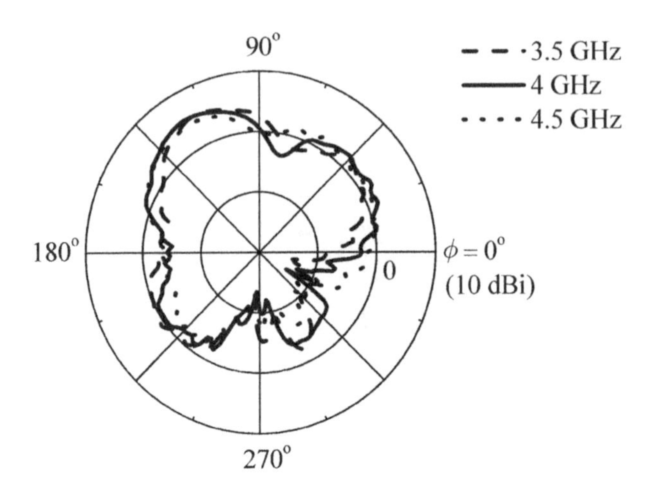

Figure 7.42 Measured total field radiation patterns of the planar antenna in free space at 3.5, 4, and 4.5 GHz in the x–y plane.

Table 7.5 Average gain for the total field in the x–y plane.

f, GHz	3	3.5	4	4.5	5
Average gain, dBi	−3.68	−2.84	−2.32	−2.58	−3.15

radiation is along the $\phi = 120°$ direction with gain of 5 dBi due to the asymmetric structure of the antenna. Table 7.5 gives the average gain of the total fields of the antenna. It can be seen that the average gain varies from around −2 dBi to −4 dBi across the impedance bandwidth, which is acceptable for wireless interfaces.

For a small antenna to be used in mobile devices such as laptops, printers, and DVD players, in an indoor environment where the polarizations are random, the gain or radiation efficiency is a vital performance indicator. Compared to the peak gain, the average gain of the total field is of greater interest for mobile devices. The average gain (dBi) is defined as

$$G_{\text{average}} = \frac{\sum_{n=1}^{N} G_n(\theta_n, \phi_n)}{N}, \tag{7.12}$$

where G_{average} stands for the average gain of the total field along a specific cut or orientation; G_n (θ_n, ϕ_n) is the gain measured at a particular orientation (θ_n, ϕ_n) or along a specific cut; and N is the total number of measured G_n. In this study, the gain for both the E_θ and E_ϕ components at a certain point in the x–y plane is first measured, and the gain for the total field is calculated. With this process repeated at intervals of $\phi = 2°$ $(N = 181)$, the average gain for the total field can be found using (7.12).

In mobile applications, 3D radiation patterns are of greater interest than 2D radiation patterns. Figure 7.43 shows the simulated 3D radiation patterns for the total field at 3.5 GHz, 4 GHz, and 4.5 GHz. The radiation performance is quite stable across the operating bandwidth, and at the higher frequencies the radiation becomes more directional due to the increased electrical size of the antenna.

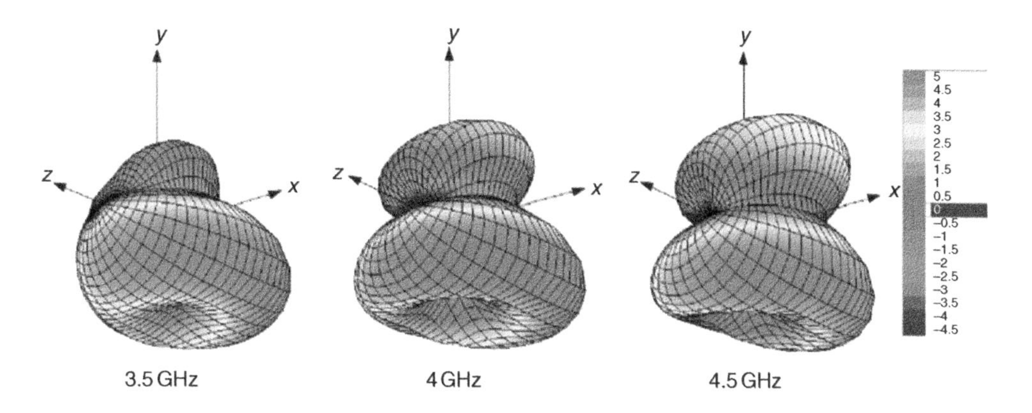

Figure 7.43 Simulated total field radiation patterns of the planar antenna at 3.5, 4, and 4.5 GHz in free space.

Furthermore, Figure 7.44 illustrates the current distributions on the antenna at frequencies of 3, 3.5, 4, and 4.5 GHz. At 3 and 3.5 GHz, most of the current is concentrated around the notch. This suggests that the notch has a significant effect on the antenna performance at lower operating frequencies. As a result, the impedance matching at 3 GHz is very sensitive to the notch dimensions. It should be noted that the current on the ground plane is weaker than at the radiator. This suppresses the effect of the RF cable on the antenna performance at the lower operating frequency. The bright shading denotes the stronger current and darker shading the weaker one. Similar to the phenomenon observed in the last section, the majority

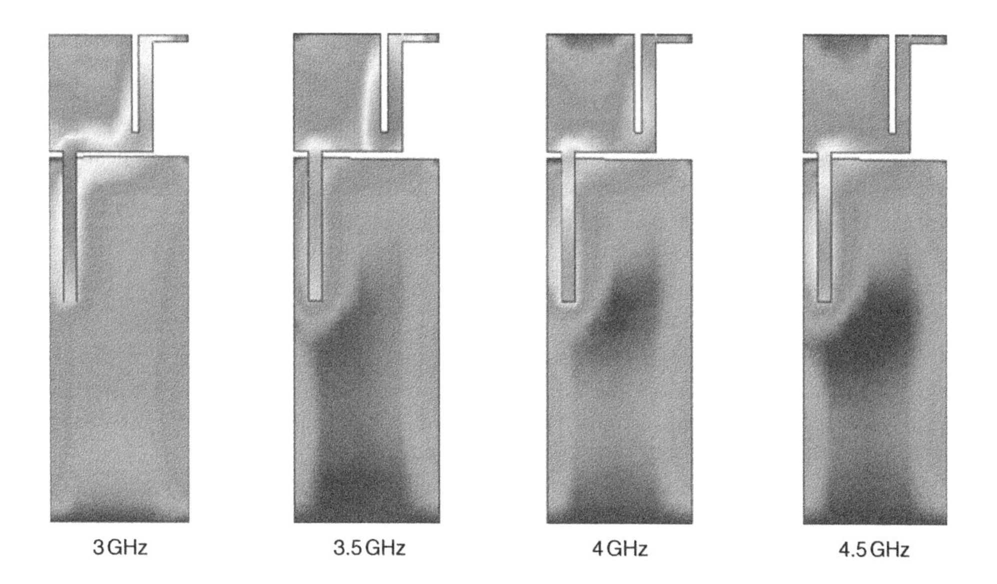

Figure 7.44 Simulated current distributions on the planar antenna in free space at 3, 3.5, 4, and 4.5 GHz.

of the current is concentrated around the notch and feeding strip while the current on the ground plane is quite weak. This suggests that the idea of reducing the ground-plane effect proposed in the previous section can be applied in WUSB UWB antenna designs.

The antenna is to be used in a WUSB dongle, which is usually placed beside a laptop. The laptop usually has a metallic and lossy cover, which affects the performance of antennas installed on or close to the cover. In order to analyze the effects of the laptop on the radiation performance, an IBM ThinkPad laptop has been selected for study. The keyboard panel of the laptop measures 250 mm × 300 mm × 25 mm. The screen measures 250 mm × 300 mm and is opened as if it is in use such that the angle between the screen and the keyboard panel is 100 °. The antenna is placed at four different positions on the laptop for analysis, as shown in Figure 7.45. The antenna at position P1 is placed 20 mm from the edge of the right-hand side of the back panel. The antennas at positions P2 and P4 are positioned at the centre of the back and right-hand side panel, respectively. The antenna at position P3 is placed 20 mm from the edge on the right-hand side panel. In this design, since the size of the ground plane of the antenna does not significantly affect the impedance and radiation performance, it can be electrically connected to the metal casing. Figure 7.45(b) shows the antennas under test, which were installed close to the laptop at positions P1 and P3.

Figure 7.46 plots the radiation patterns for the total field in the horizontal (x–y) plane at the four positions on the laptop. Table 7.6 shows the values of the average gain at the four positions from 3 GHz to 5 GHz. The simulated values are also given in parentheses. The simulation was performed using Zeland IE3D. From the results shown, it can be seen that the simulated and measured results are generally in good agreement.

By comparing the radiation patterns for P1 and P2, it can be observed that spanning $\phi = 0°–270°–180°$, the gain for the antenna at P2 is lower than that at P1 due to the severe blockage by the laptop. This results in a lower average gain for the antenna at P2 across the bandwidth. Similarly, comparing P3 and P4, the blockage is more severe for the antenna at P4, which results in lower average gain than that at P3. However, the average gain for the antenna at P1 is higher than that for the antenna at P3. This is most likely due to the presence of the screen which acts as a reflector when the antenna is placed at P1, whereas for the antenna at P3 the effect of the screen is minimal. Generally, the radiation patterns are stable across the broad impedance bandwidth of 3–5 GHz at each of the four positions. Therefore, from the results, it can be observed that the antenna has the best average gain performance when it is placed near the edge of the back panel.

7.4.2.3 Bent Antenna Design

The planar antenna is good for low-profile design. However, the antenna may be installed on the WUSB dongle to improve the radiation performance in horizontal planes. Here, an alternative design is provided, namely the bent small printed antenna for WUSB dongle applications.

Figure 7.47 shows the geometry of the bent antenna and the Cartesian coordinate system. The radiator is orthogonal to the ground plane etched on the underside of the PCB (RO4003, $\varepsilon_r = 3.38$ and 1.52 mm in thickness). The radiator consists of a rectangular section measuring 17 mm × 20 mm and a horizontal strip. A rectangular notch of $w_s \times l_s = 1\,\text{mm} \times 14\,\text{mm}$ is cut close to the horizontal strip of $w_{rs} \times l_{rs} = 3\,\text{mm} \times 3\,\text{mm}$ at a distance of $d = 3\,\text{mm}$. The radiator is fed by a microstrip line of 3.5 mm width located at $d_s = 2\,\text{mm}$ from the left-hand

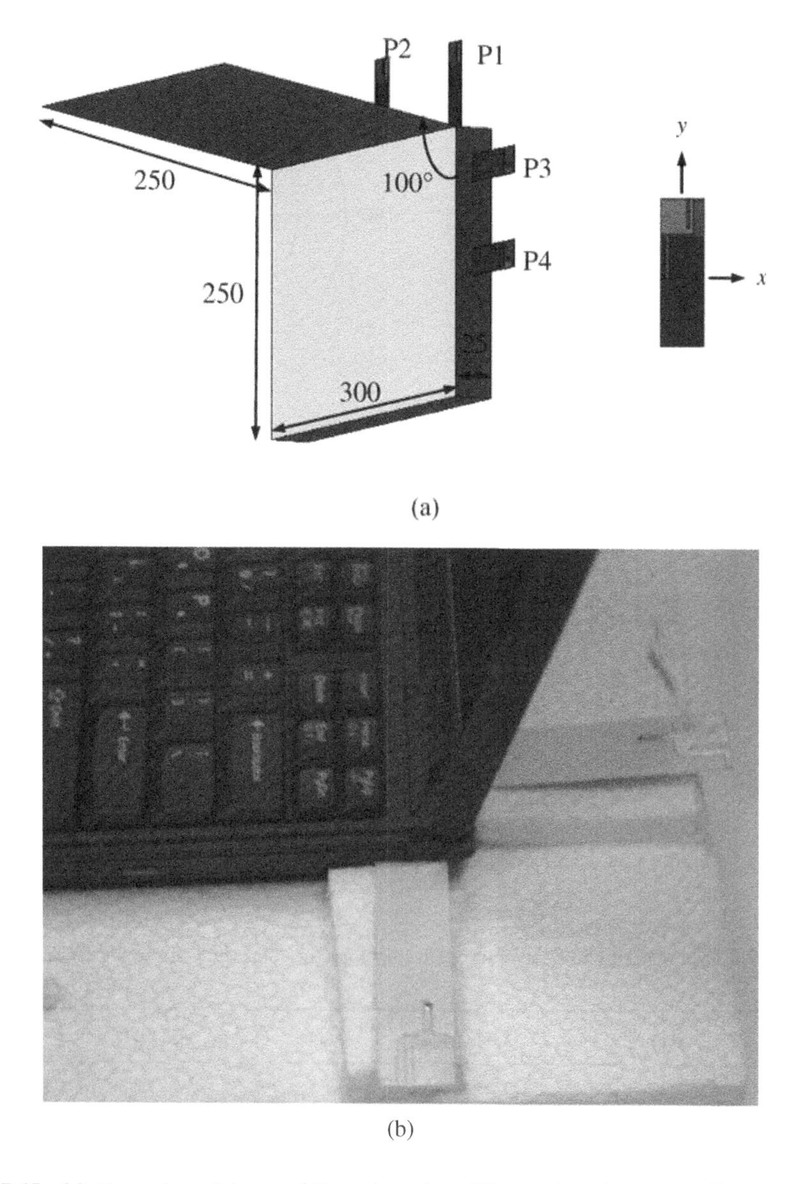

Figure 7.45 (a) Geometry of laptop (dimensions in millimetres) and antenna placement; (b) the antennas under test at p3.

side of the radiator with a feed gap $g = 1$ mm. The excitation is located at the edge of the microstrip line of length 19 mm. The ground plane has size $l_g \times w_g = 47\,\text{mm} \times 20\,\text{mm}$.

Figure 7.48 shows the impedance response obtained from the Agilent N5230A vector network analyzer. It can be seen that the antenna is well matched at 3.1–5 GHz for the return loss $|S_{11}| < -10\,\text{dB}$. It has better impedance response than that of the planar antenna shown in Figure 7.41.

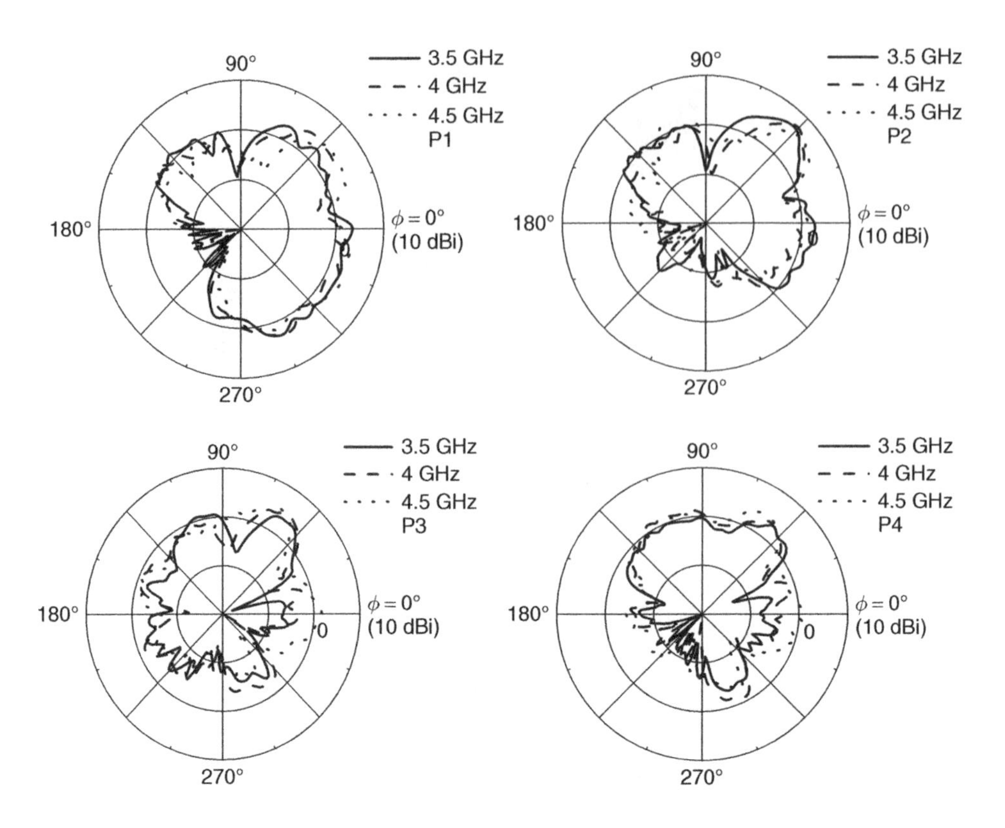

Figure 7.46 Measured total field radiation patterns of the planar antenna in the x–y plane for the different positions on the laptop.

Table 7.6 Measured and simulated average gain (dBi) for the total field of the planar antenna in the x–y plane.

Position	3 GHz	3.5 GHz	4 GHz	4.5 GHz	5 GHz
P1	−2.32	−2.48	−3.15	−3.17	−2.97
	(−0.74)	(−2.04)	(−1.38)	(−1.28)	(−1.71)
P2	−3.13	−4.27	−5.12	−5.08	−4.65
	(−0.88)	(−3.60)	(−2.60)	(−2.86)	(−3.93)
P3	−3.69	−5.32	−4.31	−4.06	−5.68
	(−2.72)	(−5.00)	(−4.02)	(−4.53)	(−5.50)
P4	−4.38	−6.18	−5.14	−4.18	−5.38
	(−4.10)	(−5.81)	(−6.66)	(−7.18)	(−6.52)

Figure 7.49 shows the measured total field radiation patterns of the antenna in the horizontal (x–y) plane at 3.5, 4, and 4.5 GHz. It can be noted that the pattern is more omnidirectional than that of the planar antenna discussed earlier. Table 7.7 tabulates the average gain of the total field for the bent antenna. It can be seen that the average gain varies around 0 dBi

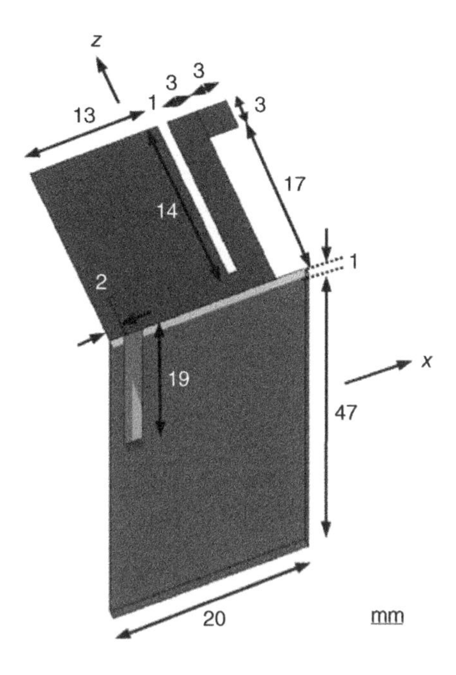

Figure 7.47 Geometry of the bent antenna.

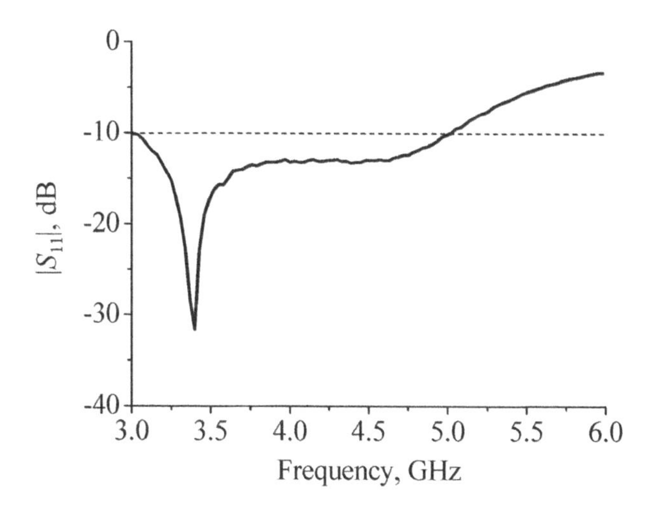

Figure 7.48 Measured return loss of the bent antenna.

to −3 dBi across the impedance bandwidth, which is much higher than that of the planar antenna as tabulated in Table 7.5, in particular at the higher frequencies, because the bent antenna has a strong radiation in the horizontal plane like a monopole vertically installed on a WUSB dongle.

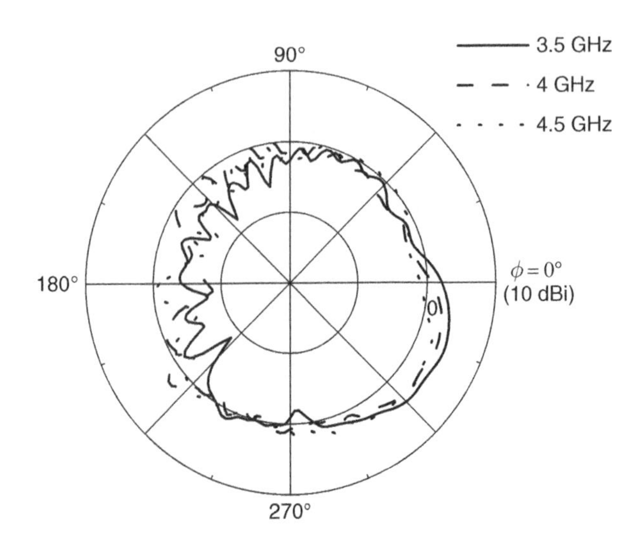

Figure 7.49 Measured radiation patterns for the total field of the bent antenna in the x–y plane in free space.

Table 7.7 Average gain for the total field in the x–y plane.

f, GHz	3	3.5	4	4.5	5
Average gain, dBi	−3.23	−1.35	−0.85	−0.94	−0.90

Figure 7.50 shows the simulated 3D radiation patterns for the total field of the bent antenna at 3.5, 4, and 4.5 GHz in free space. Compared with the 3D radiation patterns shown in Figure 7.43, those of the bent antenna exhibit more omnidirectional characteristics, which will be more practical for mobile applications.

Figure 7.51 shows that the current distributions on the bent antenna at frequencies of 3, 3.5, 4, and 4.5 GHz are similar to those of the planar antenna (Figure 7.44). At 3 GHz and 3.5 GHz, it can be observed that most of the current is concentrated around the notch and feeding strip. This implies that the ground-plane effect has been suppressed using such a design concept.

Figure 7.52 compares the effects of the laptop on the radiation performance of the bent antenna. The antenna was placed on the laptop at four different positions similar to those shown in Figure 7.45. Figure 7.52 plots the radiation patterns for the total field of the bent antenna in the horizontal (x–y) plane at the four positions on the laptop. Table 7.8 compares the values of the average gain at the four positions from 3 GHz to 5 GHz. The simulated values are also given in parentheses. The simulated and measured results are generally in good agreement. As in the case of the planar antenna, it can be found that within the span of $\phi = 0°–270°–180°$, the gain for the antenna at P2 is lower than that for the antenna at P1 due to the severe blockage by the laptop. This results in a lower average gain for the antenna at P2 across the bandwidth. Similarly, comparing P3 and P4, the blockage is more severe at P4, which results in a lower average gain. However, the average gain for the antenna at P1 is higher than that for the antenna at P3. Generally, the radiation patterns are stable across

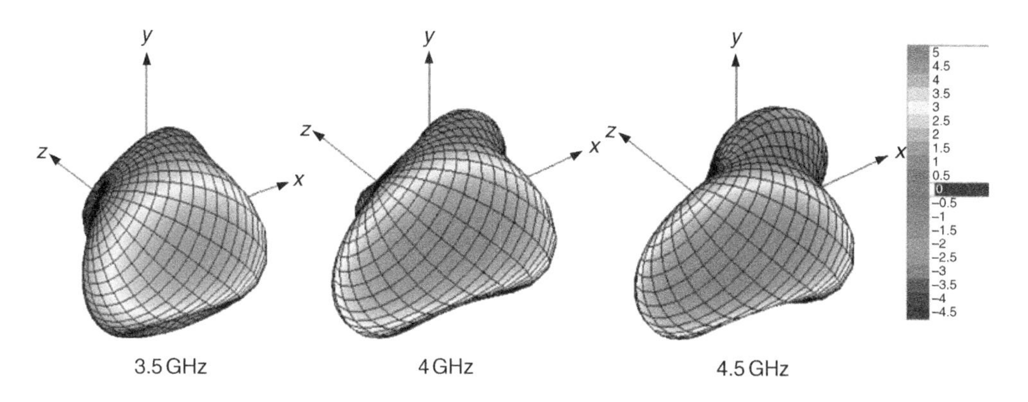

Figure 7.50 Simulated 3D radiation patterns for the total field of the bent antenna in free space.

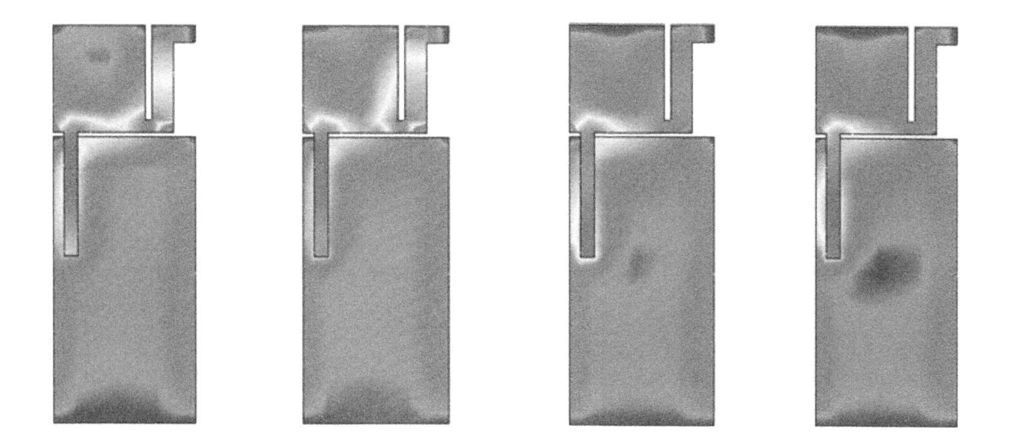

Figure 7.51 Simulated current distributions on the bent antenna.

Table 7.8 Measured and simulated average gain (dBi) for the total field of the bent antenna in the *x*–*y* plane.

Position	3 GHz	3.5 GHz	4 GHz	4.5 GHz	5 GHz
P1	−2.85 (−2.80)	−1.94 (−0.87)	−1.90 (−0.64)	−2.48 (−0.47)	−2.63 (−1.00)
P2	−4.97 (−3.49)	−5.40 (−3.19)	−3.91 (−2.08)	−5.12 (−2.99)	−5.79 (−3.65)
P3	−3.44 (−1.95)	−3.25 (−2.21)	−2.14 (−1.08)	−2.44 (−1.78)	−2.81 (−2.74)
P4	4.24 (−3.02)	−4.47 (−3.32)	−2.28 (−2.20)	−2.27 (−2.19)	−2.71 (−2.83)

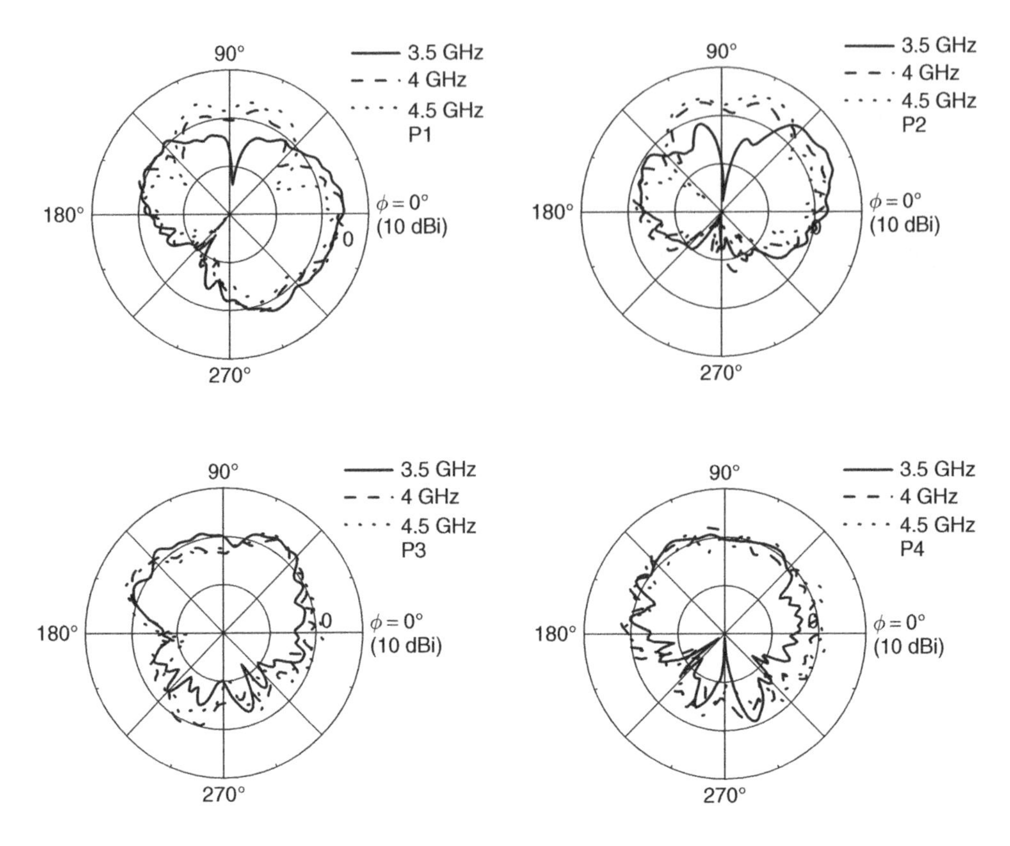

Figure 7.52 Measured total field radiation patterns of the bent antenna in the x–y plane for the different positions on the laptop.

the broad impedance bandwidth at each of the four positions. The results for the planar and bent antenna suggest that the bent antenna is able to give a better average gain performance and its radiation pattern is more robust to the presence of the screen and keyboard panel.

Compared with the planar antenna, the bent antenna is capable of providing almost the same average gain at 3 GHz and higher average gain at the higher frequencies. It can be concluded that the bent antenna is a better candidate than the planar antenna if it can be vertically installed.

7.5 Summary

UWB technology is emerging in short-range high-data-rate wireless connections, high-accuracy image radar, and localization systems. Due to the extremely broad bandwidth and carrier-free features, antenna design is facing many challenges. Conventional design considerations are insufficient to evaluate the design. Therefore, this chapter started with a discussion of the special design considerations for UWB antennas. In accordance with these considerations, antennas suitable for portable and mobile UWB applications were proposed. The latest developed antennas were presented, with illustrations and simulated or measured data. In addition, a new concept for designing small UWB printed antennas with reduced ground-plane effects was introduced and applied to wireless UWB designs as a practical engineering case study. In short, the most important issue in UWB antenna designs is to select the most suitable and not the 'best' solutions for the specific applications according to the system requirements.

References

[1] Federal Communications Commission, First Report and Order, February 14, 2002.
[2] Z.N. Chen, X.H. Wu, H.F. Li, N. Yang, and M.Y.W. Chia, Considerations for source pulses and antennas in UWB radio systems. *IEEE Transactions on Antennas and Propagation*, 52 (2004), 1739–1748.
[3] D. Lamensdorf and L. Susman, Baseband-pulse-antenna techniques. *IEEE Antennas and Propagation Magazine*, 36 (1994), 20–30.
[4] M. Kanda, Transients in a resistively loaded linear antenna compared with those in a conical antenna and a TEM horn. *IEEE Transactions on Antennas and Propagation*, 28 (1980), 132–136.
[5] L.T. Chang and W.D. Burnside, An ultrawide-bandwidth tapered resistive TEM horn antenna. *IEEE Transactions on Antennas and Propagation*, 48 (2000), 1848–1857.
[6] R.T. Lee and G.S. Smith, On the characteristic impedance of the TEM horn antenna. *IEEE Transactions on Antennas and Propagation*, 52 (2004), 315–318.
[7] P.E. Mayes, Frequency-independent antennas and broad-band derivatives thereof. *Proceedings of the IEEE*, 80 (1992), 103–112.
[8] T.W. Hertel and G.S. Smith, On the dispersive properties of the conical spiral antenna and its use for pulsed radiation. *IEEE Transactions on Antennas and Propagation*, 51 (2003), 1426–1433.
[9] J.D. Kraus, *Antennas* (2nd edition), pp. 340–358. New York: McGraw-Hill, 1988.
[10] C.W. Harrison, Jr. and C.S. Williams, Jr., Transients in wide-angle conical antennas. *IEEE Transactions on Antennas and Propagation*, 13 (1965), 236–246.
[11] S.S. Sandler and R.W.P. King, Compact conical antennas for wide-band coverage. *IEEE Transactions on Antennas and Propagation*, 42 (1994), 436–439.
[12] S.N Samaddar and E.L. Mokole, Biconical antennas with unequal cone angles. *IEEE Transactions on Antennas and Propagation*, 46 (1998), 181–193.
[13] T.T. Wu and R.W.P. King, The cylindrical antenna with nonreflecting resistive loading. *IEEE Transactions on Antennas and Propagation*, 13 (1965), 369–373.
[14] D.L. Senguta and Y.P. Liu, Analytical investigation of waveforms radiated by a resistively loaded linear antenna excited by a Gaussian pulse. *Radio Science*, 9 (1974), 621–630.
[15] J.G. Maloney and G.S. Smith, A study of transient radiation from the Wu-King resistive monopole – FDTD analysis and experimental measurements. *IEEE Transactions on Antennas and Propagation*, 41 (1993), 668–676.
[16] H. Meinke and F.W. Gundlach, Taschenbuch der Hochfrequenztechnik, pp. 531–535. Berlin: Springer-Verlag, 1968.
[17] G. Dubost and Zisler, *Antennas à Large Bande*, pp. 128–129. Paris, New York: Masson, 1976.
[18] S. Honda, M. Ito, H. Seki, and Y. Jinbo, A disk monopole antenna with 1:8 impedance bandwidth and omnidirectional radiation pattern. *Proceedings of the International Symposium of Antennas and Propagation*, Sapporo, Japan, pp. 1145–1148, 1992.
[19] M. Hammoud, P. Poey, and F. Colombel, Matching the input impedance of a broadband disc monopole. *Electronics Letters*, 29 (1993), 406–407.
[20] G.H. Brown and O.M. Woodward, Experimentally determined radiation characteristics of conical and triangular antennas. *RCA Review*, 13 (1952), 425–452.
[21] M.J. Ammann and Z.N. Chen, Wideband monopole antennas for multi-band wireless systems. *IEEE Antennas and Propagation Magazine*, 45 (2003), 146–150.
[22] Z.N. Chen and M.Y.W. Chia, *Broadband Planar Antennas: Design and Applications*. Chichester: John Wiley & Sons, Ltd, 2006.
[23] Z.N. Chen and M.Y.W. Chia, Impedance characteristics of trapezoidal planar monopole antenna. *Microwave and Optical Technology Letters*, 27 (2000), 120–122.

[24] J.A. Evans and M.J. Ammann, Planar trapezoidal and pentagonal monopoles with impedance bandwidths in excess of 10:1. *Proceedings of the IEEE Antennas and Propagation Society International Symposium*, Vol. 3, pp. 1558–1561, July 11–16, 1999.
[25] X.H. Wu, Z.N. Chen, and N. Yang, Optimization of planar diamond antenna for single/multi-band UWB wireless communications. *Microwave and Optical Technology Letters*, 42 (2004), 451–455.
[26] Z.N. Chen, Impedance characteristics of planar bow-tie-like monopole antennas. *Electronics Letters*, 36 (2000),1100–1101.
[27] M.J. Ammann, Square planar monopole antenna, *IEE National Conference on Antennas and Propagation*, pp. 37–40, 31 March 31 – April 1, 1999.
[28] M.J. Ammann, Impedance bandwidth of the square planar monopole. *Microwave and Optical Technology Letters*, 24 (2000), 185–187.
[29] M.J. Ammann and Z.N. Chen, An asymmetrical feed arrangement for improved impedance bandwidth of planar monopole antennas. *Microwave and Optical Technology Letters*, 40 (2004), 156–158.
[30] Z.N. Chen, M.J. Ammann, and M.Y.W. Chia, Broadband square annular planar monopoles. *Microwave and Optical Technology Letters*, 36 (2003), 449–454.
[31] Z.N. Chen, Experiments on input impedance of tilted planar monopole antenna. *Microwave and Optical Technology Letters*, 26 (2000), 202–204.
[32] K.G. Thomas, N. Lenin, and R. Sivaramakrishnan, Ultrawideband planar disc monopole. *IEEE Transactions on Antennas and Propagation*, 54 (2006), 1339–1341.
[33] S. Su, K. Wong, and C. Tang, Ultra-wideband square planar antenna for IEEE 802.16a operating in the 2–11 GHz band. *Microwave and Optical Technology Letters*, 42 (2004), 463–466.
[34] X.H. Wu and Z.N. Chen, Comparison of planar dipoles in UWB applications. *IEEE Transactions on Antennas and Propagation*, 53 (2005), 1973–1983.
[35] M.J. Ammann and Z.N. Chen, A wideband shorted planar monopole with bevel. *IEEE Transactions on Antennas and Propagation*, 51 (2003), 901–903.
[36] E. Lee, P.S. Hall, and P. Gardner, Compact wideband planar monopole antenna. *Electronics Letters*, 35 (1999), 2157–2158.
[37] X.H. Wu, A.A. Kishk, and Z.N. Chen, A linear antenna array for UWB applications. *Proceedings of the IEEE Antennas and Propagation Society International Symposium*, Vol. 1A, Washington, DC, July 3–8, 2005, pp. 594–597.
[38] E. Antonino-Daviu, M. Cabedo-Fabres, M. Ferrando-Bataller, and A. Valero-Nogueira, Wideband double-fed planar monopole antennas. *Electronics Letters*, 39 (2003), 1635–1636.
[39] A. Cai, T.S.P. See, and Z.N. Chen, Study of human head effects on UWB antenna. *IEEE International Workshop on Antenna Technology (iWAT)*, Singapore, pp. 310–313, March 7–9, 2005.
[40] N.P. Agrawall, G. Kumar, and K.P. Ray, Wide-band planar monopole antenna. *IEEE Transactions on Antennas and Propagation*, 46 (1998), 294–295.
[41] T. Yang and W.A. Davis, Planar half-disk antenna structures for ultra-wideband communications. *Proceedings of the IEEE Antennas and Propagation Society International Symposium*, Vol. 3, pp. 2508–2511, June 2004.
[42] C.Y. Huang and W.C. Hsia, Planar elliptical antenna for ultra-wideband communications. *Electronics Letters*, 41 (2005), 296–297.
[43] P.V. Anob, K.P. Ray, and G. Kumar, Wideband orthogonal square monopole antennas with semi-circular base.*Proceedings of the IEEE Antennas and Propagation Society International Symposium*, Vol. 3, pp. 294–297, July 2001.
[44] J.W. Lee, C.S. Cho, and J. Kim, A new vertical half disc-loaded ultra-wideband monopole antenna (VHDMA) with a horizontally top-loaded small disc. *Antennas and Wireless Propagation Letters*, 4 (2005), 198–201.
[45] H.S. Choi, J.K. Park, S.K. Kim, and J.Y. Park, A new ultra-wideband antenna for UWB applications. *Microwave and Optical Technology Letters*, 40 (2004), 399–401.
[46] S.Y. Suh, W.L. Stutzman, and W.A. Davis, A new ultrawideband printed monopole antenna: the planar inverted cone antenna (PICA). *IEEE Transactions on Antennas and Propagation*, 52 (2004), 1361–1364.
[47] Z.N. Chen, M.J. Ammann, M.Y.W. Chia, and T.S.P. See, Circular annular planar monopoles with EM coupling. *IEE Proceedings – Microwaves, Antennas and Propagation*, 150 (2003), 269–273.
[48] D. Valderas, J. Meléndez, and I. Sancho, Some design criteria for UWB planar monopole antennas: Application to a slotted rectangular monopole. *Microwave and Optical Technology Letters*, 46 (2005), 6–11.
[49] Z.N. Chen and M.Y.W Chia, Impedance characteristics of EMC triangular planar monopoles. *Electronics Letters*, 37 (2001), 1271–1272.

[50] Z.N. Chen, Broadband roll monopole. *IEEE Transactions on Antennas and Propagation*, 51 (2003), 3175–3177.

[51] M.J. Ammann, Improved pattern stability for monopole antennas with ultrawideband impedance characteristics. *Proceedings of the IEEE Antennas and Propagation Society International Symposium*, Vol. 1, pp. 818–821, June 2003.

[52] Z.N. Chen, M.Y.W. Chia, and M.J. Ammann, Optimization and comparison of broadband monopoles. *IEE Proceedings – Microwaves, Antennas and Propagation*, 150 (2003), 429-435.

[53] Z.N. Chen, A new bi-arm roll antenna for UWB applications. *IEEE Transactions on Antennas and Propagation*, 53 (2005), 672–677.

[54] C.Y. Huang and W.C. Hsia, Planar elliptical antenna for ultra-wideband communications. *Electronics Letters*, 41 (2005), 296–297.

[55] D.H. Kwon and Y. Kim, CPW-fed planar ultra-wideband antenna with hexagonal radiating elements. *Proceedings of the IEEE Antennas and Propagation Society International Symposium*, Vol. 3, pp. 2947–2950, June 2004.

[56] J. Liang, C.C. Chiau, X. Chen, and C.G. Parini, Printed circular disc monopole antenna for ultra-wideband applications. *Electronics Letters*, 40 (2004), 1246–1247.

[57] K. Chung, H. Park, and J. Choi, Wideband microstrip-fed monopole antenna with a narrow slit. *Microwave and Optical Technology Letters*, 47 (2005), 400–402.

[58] J. Liang, L. Guo, C.C. Chiau, and X. Chen, CPW-fed circular disc monopole antenna for UWB applications. *IEEE International Workshop on Antenna Technology: Small Antennas and Novel Metamaterials (iWAT)*, pp. 505–508, March 7–9, 2005.

[59] J. Yeo, Y. Lee and R. Mittra, Design of a wideband planar volcano-smoke slot antenna (PVSA) for wireless communications. *Proceedings of the IEEE Antennas and Propagation Society International Symposium*, Vol. 2, pp. 655–658, June 22–27, 2003.

[60] T.W. Hertel, Cable-current effects of miniature UWB antennas, *Proceedings of the IEEE Antennas and Propagation Society International Symposium*, Vol. 3A, pp. 524–527, July 3–8, 2005.

[61] K. Kiminami, A. Hirata, and T. Shiozawa, Double-sided printed bow-tie antenna for UWB communications. *Antennas and Wireless Propagation Letters*, 3 (2004), 152–153.

[62] Y. Zhang, Z.N. Chen, and M.Y.W. Chia, Characteristics of planar dipoles printed on finite-size PCBs in UWB radio systems. *Proceedings of the IEEE Antennas and Propagation Society International Symposium*, Vol. 3, pp. 2512–2515, June 2004.

[63] A. Koksal and F. Kauffman, Moment method analysis of linearly tapered slot antennas. *Proceedings of the IEEE Antennas and Propagation Society International Symposium*, Vol. 1, pp. 314–317, June 1991.

[64] H.Y. Wang, D. Mirshekar-Syahkal, and I.J. Dilworth, Numerical modeling of V-shaped linearly tapered slot antennas. *Proceedings of the IEEE Antennas and Propagation Society International Symposium*, Vol. 2, pp. 1118–1121, July 1997

[65] R.N. Simons and R.Q. Lee, Linearly tapered slot antenna radiation characteristics at millimeter-wave frequencies. *Proceedings of the IEEE Antennas and Propagation Society International Symposium*, Vol. 2, pp. 1168–1171, June 1998.

[66] J.B. Muldavin and G.M. Rebeiz, Millimeter-wave tapered-slot antennas on synthesized low permittivity substrates. *IEEE Transactions on Antennas and Propagation*, 47 (1999), 1276–1280.

[67] M.F. Catedra, J.A. Alcaraz, and J.C. Arredondo, Analysis of arrays of Vivaldi and LTSA antennas. *Proceedings of the IEEE Antennas and Propagation Society International Symposium*, Vol. 1, pp. 122–125, June 1989.

[68] E. Gazit, Improved design of the Vivaldi antenna. *IEE Proceedings – Microwaves, Antennas and Propagation*, 135 (1988), 89–92.

[69] J.D.S. Langley, P.S. Hall, and P. Newham, Multi-octave phased array for circuit integration using balanced antipodal Vivaldi antenna elements. *Proceedings of the IEEE Antennas and Propagation Society International Symposium*, Vol. 1, pp. 178–181, June 1995.

[70] S.G. Kim and K. Chang, Ultra wideband exponentially-tapered antipodal Vivaldi antennas. *Proceedings of the IEEE Antennas and Propagation Society International Symposium*, Vol. 3, pp. 2273–2276, June 2004.

[71] X.M. Qing and Z.N. Chen, Antipodal Vivaldi antenna for UWB applications. In *Proceedings of the European Electromagnetics Symposium*, UWB SP7, Magdeburg, Germany, July 12–16, 2004.

[72] T. Yang and W.A. Davis, Planar half-disk antenna structures for ultra-wideband communications. *Proceedings of the IEEE Antennas and Propagation Society International Symposium*, Vol. 3, pp. 2508–2511, June 2004.

[73] D.H. Kwon and Y. Kim, CPW-fed planar ultra-wideband antenna with hexagonal radiating elements, *Proceedings of the IEEE Antennas and Propagation Society International Symposium*, Vol. 3, pp. 2947–2950, June 2004

[74] J. Liang, C.C. Chiau, X. Chen, and C.G. Parini, Printed circular ring monopole antennas. *Microwave and Optical Technology Letters*, 45 (2005), 372–375.

[75] H.S. Choi, J.K. Park, S.K. Kim, and J.Y. Park, A new ultra-wideband antenna for UWB applications. *Microwave and Optical Technology Letters*, 40 (2004), 399–401.

[76] K. Chung, H. Park, and J. Choi, Wideband microstrip-fed monopole antenna with a narrow slit. *Microwave and Optical Technology Letters*, 47 (2005), 400–402.

[77] Z.N. Chen, N. Yang, Y.X. Guo, and M.Y.W. Chia, An investigation into measurement of handset antennas. IEEE Transactions on Instrumentation and Measurement, 54 (2005), 1100–1110.

Index

Antennas for Portable Devices Zhi Ning Chen

Printed and bound by CPI Group (UK) Ltd, Croydon, CR0 4YY

16/04/2025

14658476-0002